Student Study Guide

TO ACCOMPANY

The Chemical World

CONCEPTS AND APPLICATIONS

KOTZ / JOESTEN / WOOD / MOORE

Dean Nelson
University of Wisconsin-Eau Claire

SAUNDERS GOLDEN SUNBURST SERIES

SAUNDERS COLLEGE PUBLISHING
Harcourt Brace College Publishers
Fort Worth Philadelphia San Diego New York
Orlando Austin San Antonio Toronto
Montreal London Sydney Tokyo

Printed in the United States of America.

Dean Nelson: Student Study Guide to accompany THE CHEMICAL WORLD: CONCEPTS AND APPLICATIONS, by Kotz/Joesten/Wood/Moore

ISBN 0-03-097198-5

456 017 987654321

Preface

The authors of your textbook, The Chemical World: Concepts and Applications, have put a great deal of effort in presenting complex topics in an interesting, logical and readable manner. It is the intent of this study guide to direct and focus your study efforts to those areas which are fundamental. This study guide will aid you in mastering those areas most likely to be difficult to understand and conceptually tie together. Hopefully it will be able to stimulate your thoughts in patterns helpful to your comprehension of a very interesting and dynamic subject.

This study guide will help summarize each chapter, remind you of key vocabulary terms, provide you with practice to help you gain confidence as you prepare for evaluations and give you a means of assessing your learning progress.

The format of this study guide matches that of your textbook. There is a study guide chapter for each text chapter and references to the text will be by chapter and page. Each chapter in this study guide will follow the same pattern and will contain most of the following sections:

1. Study hints
2. True/False questions
3. Symbols presented in the chapter
4. Terminology terms
5. Concept Map(s) of the chapter topic
6. A concept test
7. Ties to previous topics
8. Worked example problems
9. Questions to test your understanding of analytical problems
10. A crossword puzzle of terms and concepts
11. Supplementary readings of current interest

Concept maps may be new to you but evidence has shown that they are effective as a learning tool. A brief introduction to their use may help you use them.

Concept maps can be thought of as visual outlines. The emphasis in this study guide is to have them act as an organizer of the material in each chapter. Concept maps start with a **key concept** which is broken down into several **general concepts**, each with a link to the key concept. This breakdown continues with a series of links down to specific concepts and examples. These maps also allow for visual cross linking of concepts which often facilitates seeing relationships between sections which were not obvious or expected.

The following example concept map is of an easy to understand situation, but will illustrate how these maps are to be used. The idea to be understood is burning. This requires that certain materials be present, these are then identified - fuel, air and starting temperature. The next lower level gives materials or properties that would work, such as wood, oxygen and a match. Also note that there are essential requirements, for example the heat from the match must be applied to the fuel. Maps can also identify that only specific parts of something are important, for example only the oxygen in the air is used.

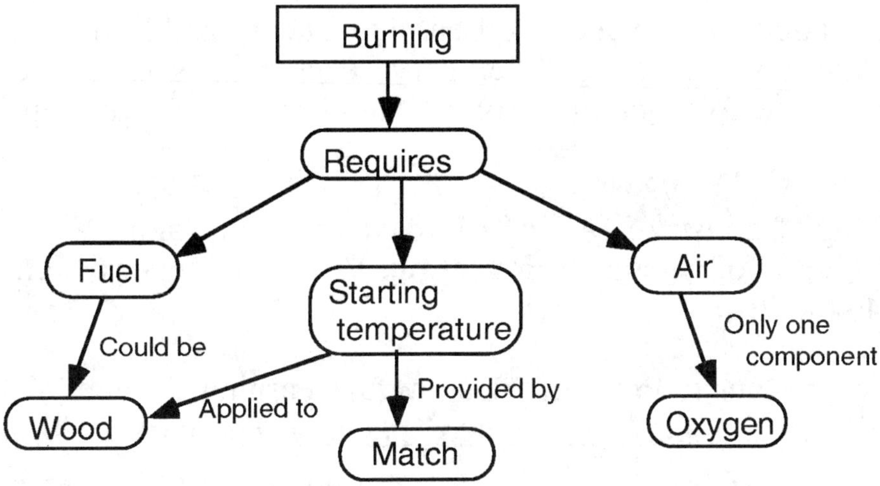

As you read the concept map for each chapter look for the concepts that relate to the key idea through some specific property. You will then be able to recognize how the various sections of the chapter relate and which concepts are a basis to understanding others. These maps should also aid you if you get lost in a topic by allowing you to find the current place on the map, and then, by reading up you can find where you failed to understand an idea.

Concept maps can be as sketchy or as detailed as one wishes. It often works well to have a very general map for a large section and then break each subsection down into its own map.

Developing concept maps can be an excellent learning tool. The maps presented in this study guide, in many instances, can be developed further. It may be worth your time to do this. The lines on the map represent links and these can be labeled if you wish. Don't hesitate to write on the maps and add your own notes.

Some useful background information on concept mapping can be found in the following references:

Applications of Advances in Learning Theory and Philosophy of Science to the Improvement of Chemistry Teaching, Joseph D. Novak, Journal of Chemical Education, Vol 61, Number 7, July 1984, pp 607-612.

Making Chemistry Learning More Meaningful, Jazlin V. Ebenezer, Journal of Chemical Education, Vol 69, Number 6, June 1992, pp 464-467.

Concept Maps to Promote Meaningful Learning, Joseph W. Cliburn,Jr. Journal of College Science Teaching, February 1990, pp 212-217.

Novak, Joseph D and D. Bob Gowin. Learning How to Learn. New York: Cambridge University Press, 1984.

The immediate following section, *To Student Progress* , provides you with some proven study techniques. Please take a few minutes to study it.

TO STUDENT PROGRESS

Chemistry is somewhat different from most other subject areas because chemistry itself is subdivided into many categories and because it is so closely related to many other disciplines. The subject itself is subdivided into sections. You will see areas of inorganic chemistry, analytical chemistry, organic chemistry, and several others. Additionally we have interdisciplinary areas such as geo-chemistry, bio-chemistry, chemical-physics, and theoretical chemistry. One of the complications encountered in the first course of chemistry is that you are exposed to nearly all of these subdivisions. As a consequence of this the question often becomes how much of this am I really responsible for knowing in some detail; what must I master, and what is presented for general background information?

The study of chemistry requires that you be able to draw on past learning and apply it to current and new situations. You will need to master certain basic relationships and this means more than just knowing what they are. There are certain skills that will enable you to make the best possible progress in your study of chemistry:

1) The most significant aid to your learning chemistry (or any other subject for that matter) is the development of excellent study skills. Many students have some background in some of the chemistry they encounter during the first term at the college level and all too often they use past success to carry them early in the course. Unfortunately it often is students with a working background that find themselves in trouble because they were coasting and not actively participating in the class. The ability to follow a class presentation does not necessarily prove you have the depth of knowledge your instructor will expect of you. A common phrase instructors hear is " But I followed your problem solving in class with no trouble."

2) During this first year of chemistry you will encounter a very large and new vocabulary. It is important that you learn more than just

the words. You need to understand what the definition means. Think of an application for the term beyond what was presented in the chapter.

3) Develop sound problem solving techniques. Good problem solvers often start by making a diagram of the problem so they can visualize it in total. Solving problems is a set of steps. Identify what they are and then apply them in sequence. Resist the temptation to skip steps as you work your way through to the solution. Make sure you know why you do each step in solving a problem. Remember, problems also exist in a format other than numerical.

4) Keep in mind you are in a chemistry class when you undertake to solve quantitative problems. Understand what the chemical aspects of the problem are and then apply the mathematics to it. This chemistry course uses mathematics as a tool to solve chemistry problems- look at the problems in terms of the chemistry, and then ask what mathematical application will supply a valid chemical answer. When done, always ask yourself if the answer seems reasonable.

5) Chemistry as a discipline is rapidly developing because the material is organized in a logical sequential order. It is this organization that you can make use of in your study of the subject.Your ability to comprehend the subject will be enhanced by developing an organized approach. Look for patterns in problems. You will find that relationships are repeated. Also remember that two or more problems that appear different, may simply be the same problem with a change in emphasis. Learn to develop linkages between topics and problem types. Chemistry is an organized and interrelated subject- look for this as you move from topic to topic. Some memorization is essential and will be expected of you, but this is only for commonly used factual material. Most of your effort should be directed to a fundamental understanding of the concept or process, with memorization at the bottom of the learning hierarchy.

6) Participate! Be an active learner. Chemistry will not be learned by osmosis (such as sleeping on your text). What isn't used is soon lost. You will not know you don't know until you try something. If possible join a study group and use each other to explain concepts to someone else. List the topics that may be on an exam. Keep a list of key concepts and use those previously studied to relate to current topics whenever possible. When an exam or assignment is returned to you, check out your errors - find out where your logic was incorrect or what principle you missed.

7) This list of study hints started with an extremely important key to making solid progress and this last one is equally important. Do Not Procrastinate ! This becomes so easy to do. We all seem to have more than enough to do and tend to apply ourselves to pressing issues while letting other tasks slide by until they become more pressing. We may think this saves time, but in actuality it costs us time because we need to review so much more to refresh ourselves about past material. Read ahead so you can anticipate what is coming in class. Then you will have some idea of what sections you really do not understand and you can therefore pay close attention when those explanations come up. Go to class with a brief pre-outline. Review your class thoughts and notes within a few hours after each class as the forgetting curve is very steep at the beginning. As you review your notes, tie them to your outline and the appropriate section of the textbook. Review often and work assigned problems on a daily basis. Chemistry is a course that constantly builds on the topics of the previous days. When you realize you do not understand something, ask questions. Do not wait, because what you don't understand today may be a key to comprehension of tomorrow's material. When you seek help for questions, go prepared. Know where you are stuck - it saves you and your instructor time if you can identify the part of the problem causing you trouble.

If you have a laboratory associated with this class, make every effort to relate what is happening in the laboratory to what is taking place in the lecture. At the same time be aware that there may not always be a direct correlation between the two. Some topics are better presented in the laboratory and may receive very little discussion in lecture.

Learning is an individual effort. Whatever you do to make it easier to learn will make the course more enjoyable. **Get involved right away and enjoy chemistry.**

Table of Contents

1	The Nature of Chemistry	1
2	The Nature of Matter	8
3	Chemical Elements	36
4	Chemical Compounds	57
5	Chemical Reactions	100
6	Stoichiometry	132
7	Thermodynamics	163
8	Kinetics and Equilibrium	196
9	Electron Configurations and Periodicity	234
10	Covalent Bonding	266
11	Molecular Geometry and Isomerism	295
12	Gases, Intermolecular Forces, and Liquids	322
13	Energy, Organic Chemicals, and Polymers	348
14	Solids	374
15	Solution Chemistry	397
16	Acids and Bases	428
17	Electrochemistry	463
18	Elements from the Land, Sea, and Air	491
19	Nuclear Chemistry	510
20	Chemistry of the Atmosphere	536
21	Chemistry of Life	563

Contents in Detail

1 The Nature of Chemistry 1
- Focus your study 1
- Symbols 3
- Terms 3
- Concepts test 3
- Answers to tests 4
- Suggested readings 6

2 The Nature of Matter 8
- Focus your study 8
- Symbols 19
- Terms 20
- Concept Maps 21
- True/False test 22
- Concepts test 23
- Analytical test 25
- Crossword puzzle 27
- Ties to previous chapters 28
- Answers to tests 29
- Suggested readings 35

3 Chemical Elements 36
- Focus your study 36
- Symbols 42
- Terms 43
- Concept Maps 44
- True/False test 46
- Concepts test 47
- Analytical test 48
- Crossword puzzle 49
- Ties to previous chapters 51
- Answers to tests 51
- Suggested readings 56

4 Chemical Compounds 57
- Focus your study 57
- Symbols 81
- Terms 81
- Concept Maps 83
- True/False test 85
- Concepts test 86
- Analytical test 89
- Crossword puzzle 90
- Ties to previous chapters 92
- Answers to tests 92
- Suggested readings 99

5 Chemical Reactions 100
- Focus your study 100
- Symbols 114
- Terms 115
- Concept Maps 116
- True/False test 118
- Concepts test 120
- Analytical test 121
- Crossword puzzle 122
- Ties to previous chapters 124
- Answers to tests 124
- Suggested readings 131

6 Stoichiometery 132
- Focus your study 132
- Symbols 145
- Terms 145
- Concept Maps 146
- True/False test 148
- Concepts test 149
- Analytical test 150
- Crossword puzzle 152
- Ties to previous chapters 154
- Answers to tests 154

7 Thermodynamics 163
- Focus your study 163
- Symbols 172
- Terms 173
- Concept Maps 175
- True/False test 177
- Concepts test 179
- Analytical test 182

	Crossword puzzle	184
	Ties to previous chapters	186
	Answers to tests	186
	Suggested readings	195
8	Kinetics and Equilibrium	196
	Focus your study	196
	Symbols	210
	Terms	210
	Concept Maps	212
	True/False test	214
	Concepts test	216
	Analytical test	218
	Crossword puzzle	221
	Ties to previous chapters	223
	Answers to tests	223
	Suggested readings	233
9	Electron Configurations and Periodicity	234
	Focus your study	234
	Symbols	248
	Terms	248
	Concept Maps	250
	True/False test	252
	Concepts test	254
	Analytical test	255
	Crossword puzzle	256
	Ties to previous chapters	258
	Answers to tests	258
	Suggested readings	265
10	Covalent Bonding	266
	Focus your study	266
	Symbols	276
	Terms	276
	Concept Maps	278
	True/False test	281
	Concepts test	282
	Analytical test	284
	Crossword puzzle	285
	Ties to previous chapters	287
	Answers to tests	287
	Suggested readings	294
11	Molecular Geometry and Isomerism	295
	Focus your study	295
	Symbols	303
	Terms	303
	Concept Maps	305
	True/False test	307
	Concepts test	309
	Analytical test	310
	Crossword puzzle	311
	Ties to previous chapters	313
	Answers to tests	313
	Suggested readings	320
12	Gases, Intermolecular Forces and Liquids	322
	Focus your study	322
	Symbols	331
	Terms	332
	Concept Maps	334
	True/False test	335
	Concepts test	336
	Analytical test	337
	Crossword puzzle	338
	Ties to previous chapters	340
	Answers to tests	340
13	Energy, Organic Chemicals and Polymers	348
	Focus your study	348
	Symbols	356
	Terms	357
	Concept Maps	358
	True/False test	361
	Concepts test	363
	Analytical test	364
	Crossword puzzle	365
	Ties to previous chapters	367
	Answers to tests	367
	Suggested readings	372

14 Solids	374
Focus your study	374
Symbols	380
Terms	381
Concept Maps	382
True/False test	384
Concepts test	385
Analytical test	386
Crossword puzzle	387
Ties to previous chapters	388
Answers to tests	389
Suggested readings	396
15 Solution Chemistry	397
Focus your study	397
Symbols	407
Terms	408
Concept Maps	409
True/False test	411
Concepts test	413
Analytical test	414
Crossword puzzle	416
Ties to previous chapters	418
Answers to tests	418
Suggested readings	427
16 Acids and Bases	428
Focus your study	428
Symbols	442
Terms	442
Concept Maps	444
True/False test	448
Concepts test	450
Analytical test	451
Crossword puzzle	452
Ties to previous chapters	453
Answers to tests	454
Suggested readings	462
17 Electrochemistry	463
Focus your study	463
Symbols	473
Terms	473
Concept Maps	475
True/False test	478
Concepts test	479
Analytical test	480
Crossword puzzle	481
Ties to previous chapters	483
Answers to tests	483
Suggested readings	490
18 Elements from the Land, Sea, and Air	491
Focus your study	491
Symbols	496
Terms	497
Concept Maps	498
True/False test	499
Concepts test	500
Analytical test	501
Crossword puzzle	502
Ties to previous chapters	504
Answers to tests	504
Suggested readings	509
19 Nuclear Chemistry	510
Focus your study	510
Symbols	517
Terms	518
Concept Maps	520
True/False test	522
Concepts test	523
Analytical test	524
Crossword puzzle	526
Ties to previous chapters	528
Answers to tests	528
Suggested readings	535
20 Chemistry of the Atmosphere	536
Focus your study	536
Symbols	544
Terms	544
Concept Maps	546
True/False test	547
Concepts test	548

Analytical test	550
Crossword puzzle	551
Ties to previous chapters	553
Answers to tests	553
Suggested readings	561
21 Chemistry of Life	563
Focus your study	563
Symbols	573
Terms	574
Concept Maps	576
True/False test	580
Concepts test	582
Analytical test	584
Crossword puzzle	586
Ties to previous chapters	588
Answers to tests	588
Suggested readings	597

Chapter 1 Nature of Chemistry

This chapter highlights some of the basic aspects of chemistry - the approach to problems, the types of problems which chemists can address and the interaction with other disciplines. The basic ideas of evaluating the benefits and risks of the products of chemistry are outlined.

Focus Your Study Efforts

Although short, this chapter provides a framework for the course in chemistry. As you progress through this course in chemistry use the overview presented here as a reference to establish the significance of topics discussed later.

What is chemistry?

Chemistry is what chemists do. We do not just make materials- which is certainly important- but we also find uses for these materials, ways to modify them and certainly to understand them in a chemical context.

Another major effort made by chemists is the analysis of materials for content and structure. This is especially true of natural products which may be useful but are in very limited supply. Before synthetic alternatives can be produced a very complex problem of analysis of the natural material must be made. This analysis gives us clues about how to do a synthesis and what parts of a compound are "chemically important".

Always remember chemistry is a branch of science. It has strong relationships with many disciplines and this means answers to some questions may involve other branches of science such as biology, physics and geology. Consequently you will often see references to mixed disciplines such as biochemistry, chemical physics and geochemistry. Solutions to a problem may involve scientists from many disciplines. Chemists work with, and expend great effort in collecting, data. The data collected must be complete and accurate. It is sometimes the small detail that provides the clue to solving a problem - remember this as you do any laboratory work. A crucial requirement of our data is that it be reproducible and withstand the test of our peers.

Data analysis is a key to understanding.

Data are divided into one of two classifications; qualitative or quantitative. Both are important and provide valuable information. Ultimately we would like to obtain numerical values, or quantitative information. The fact that a chemical explodes (qualitative data) is significant and knowing that may save your life. The amount of energy released by that explosion (quantitative) becomes very important to the uses and handling of the material.

Chemical science (in fact all problem solving) involves a series of defined steps:

1. State an important solvable problem.
2. Narrow the problem to a realistic size.
3. Research current information about the problem.
4. Hypothesize an explanation for the problem.
5. Design and perform valid experiments to test the hypothesis.
6. Summarize your data and identify patterns and exceptions.
7. Formulate a summarizing theory.
8. Subject your work to review by your peers.

Another point this chapter addresses refers to the openendedness of chemical data. Information is continuously being updated and this may have significant impact on how we evaluate a chemical material. Time allows us to add to our information database and refine our understanding of our knowledge. Long term results often vary from short term effects. We need to be receptive to new data and interpretations and also be willing to explore long term studies of important issues. This may mean we have to change our minds if data justifies it.

Evaluation of advantages and disadvantages of our chemical (as well as other science) work has led to a new study area known as Risk/Benefit Analysis. This study has been subdivided into three areas:

1. assessment
2. communication
3. management

Symbols found in Chapter 1

Page	Symbol	
10	CFC's	chlorofluorocarbons

Terms found in Chapter 1

The following terms are listed for your convenience. They should not become a list for you to memorize. You should have a working definition for each. If you have doubt about any particular term check its definition in the glossary and/or review it in the chapter.

Synthesis	Precursor	Quantitative
Qualitative	Hypothesis	Theory
Law	Toxicology	Epidemiology

Testing Your Understanding of Concepts in Chapter 1.

1. Why are transformations of molecules so important to chemists?

2. Based on comments about Taxol in this chapter explain why uncontrolled clear cutting of the Amazon rain forests is of so much concern to many people.

3. How do synthesis and semisynthesis vary in relationship to problem solving in chemistry?

4. How do Laws differ from Theories?

5. How do scientific laws differ from civil laws?

6. A chemist collected the following pieces of data. Identify which are qualitative and which are quantitative.

   A. The material had a yellow-green color.
   B. The density of the material is 3.68 g/mL.
   C. The material is a solid at room temperature.
   D. At 0°C, 29.3 grams will dissolve in 100 mL. of water.
   E. The material is only slightly soluble in alcohol.
   F. The material exists as geometric crystals.

7. What disciplines are involved in a "Risk Management" study?

8. Respond to the comment: Laws should be passed that require new chemicals to have zero risk to society before use is allowed.

Answers to Understanding of Concepts Questions:

1. Chemistry is a study of changes of materials and the interactions of a material with other substances. We are interested in what happens to a given material as it interacts with others. A material which does nothing does not much interest a chemist other than to determine why.

2. Taxol is a natural plant product which just recently was found to have a very important use. Many useful materials come from plants. As we destroy habitats and the plant life in them we may also be destroying the source of many useful chemicals without knowing it.

3. This is a matter of where the chemist starts. If the starting point is from a related material the process is classified as a semisynthesis. This is preferred because it usually makes the problem easier to solve.

4. Laws have withstood the test of time and do not change. They summarize a fact or set of facts. Theories are created to explain related information and undergo changes as new facts become known.

5. Scientific laws are a statement of the way materials do behave. They are not something that can be adjusted or violated. Civil laws on the other hand are statements of how things should be done. Civil laws are sometimes violated - have you ever exceeded the speed limit?

6. If we remember that quantitative data involves values or numbers the distinction becomes easy. A,C,E and F do not have numeric information so they are <u>qualitative.</u> B and D both are numeric and thus <u>quantitative.</u>

7. Many diverse disciplines are important in considering risk management. One or more sciences, such as chemistry. Sociology, philosophy, economics and political science are other studies that are almost always involved.

8. The idea of "zero risk" would, at the least, be extremely expensive. It would be nearly impossible to prove since it may take years before long term information is available. In many

cases the ability to do valid tests is not available. This would make it nearly impossible to bring new materials to market.

Some Supplementary Readings Relating to This Chapter:

1. Taxol To Be Made from Renewable Materials by Organic Semisynthesis. Chemical and Engineering News , October 12, 1992, page 30.

2. Surprise! A Fungus Factory For Taxol? Science Vol 260 , 9 April 1993, page 154.

3. Wresting Anticancer Secrets From Garlic and Soy Sauce. Science Vol 257, 4 September 1992, page 1349.

4. A Biopesticidal Tree Begins to Blossom, Science Vol 255, 28 February 1992, page 1070.

5. Risk Within Reason, Science Vol 248, 4 May 1990, page 559.

6. Incorporation of New Science into Risk Assessment, Science , 14 December 1990, page 1497.

7. Pressure Mounts for Change in Risk Assessment of Problem Chemicals. Chemical and Engineering News , September 17, 1990, page 17.

8. Risk Assessment of Pesticides. Chemical and Engineering News , January 7, 1991, page 27.

9. Dioxin Toxicity: New Studies Prompt Debate, Regulatory Action, Chemical and Engineering News , August 12, 1991, page 7.

10. The Ozone Backlash, Science Vol 260, 11 June 1993, page 1580.

11. Science and Society, Science Vol 260, 9 April 1993, page 143.

12. Science's Real Role In Policy-Making, Chemical and Engineering News , May 31, 1993, page 9.

13. Oxygenated fuel cost may outweigh effectiveness, Chemical and Engineering News , April 12, 1993, page 28.

14. Dispute Over Exxon Valdez Cleanup Date Gets Messy, Science Vol 260, 7 May 1993, page 749,.

Chapter 2 The Nature of Matter

The basic ideas developed in this chapter involve properties of the materials we constantly work with. How we can identify, separate, measure and classify these substances are key factors you will encounter.

Focus Your Study Efforts

Terminology is a significant part of general chemistry. It is important that you develop a vocabulary for the major terms as they come up. This means a real understanding, not just the ability to restate a definition. Many terms are introduced in this chapter.

This chapter also introduces many prefixes and conversion factors. You will use some of these often while others come up only on occasion. Make an effort to learn the common ones.

Measurement in chemistry

Objectives:

1. You should recognize the relationships between the units in common use, English, and the units of science, metric.
2. You should be able to make reasonable estimates for values in one system given values in the other system.

Most of the measurement done in science use the metric system which is really very straightforward. Do not let this different system intimidate you - you will soon learn to appreciate it. Conversion between the English and metric systems is what frustrates us the most and this is where numerous conversion factors must be used. See text Tables 2.3 & 2.4 for details.

There are four common conversions between measurement systems:

1 ounce	= 28.3 grams (g)
1 fluid ounce	= 29.6 milliliters (mL)
1 inch	= 25.4 millimeters (mm)
1° C	= 1.8° F

Approximate values are often all that is needed when comparing values between the metric and English system. Two conversions will usually get you reasonable comparisons.

For a celsius degree change, the fahrenheit change is about 2 degrees.The number 28, gives a reasonable approximation for the other three. For each English unit change, the metric change is about 28 units.

For example, if you need to approximate how many mm in 20 inches, just multiply by 28, or it is about 560 mm. Likewise 250 mL is about 8.9 fluid ounces.

Become familiar with only a few conversions between the systems of measurement. Don't try to remember the number of mm in an inch, foot, yard etc. If you know a base value the others can easily be figured out.

Use of properties in identification.

Objectives:

1. You should be able to identify properties that distinguish materials from each other and help identify various materials.

Identification of materials is done through the use of chemical and/or physical properties. If a property of the substance changes, a chemical reaction occurs and we are looking at a chemical change. Examples would be: a material which spontaneously ignites when exposed to air, or a material which when added to vinegar produces a gas. In contrast, physical properties do not result in changing the substance. The fact that a material is a gas at room temperature, or that a substance sinks when added to water are physical properties which help us identify the material.

Sometimes it is difficult to determine if a property or change is chemical or physical. As a result of this chemists sometimes use the term characteristic property, which refers to any property that aids in identifying or classifying without trying to decide if it is chemical or physical. A characteristic property is simply any property that can not be the same for all materials.

The term solid state can refer to two kinds of solids. We recognize both a candle and a nail as solids but we also recognize extraordinary differences between them. The iron in the nail is a crystalline material meaning it has definite shape and distinctive properties, such as melting point. The candle, an amorphous solid, can easily deform and if we tried to melt it, it would gradually soften over a wide temperature range.

Size of scale in chemistry

Objectives:

1. You should recognize that chemical reactions occur on a scale very different from the scale we normally work with.

As you study chemistry remember that the realm people relate to involves the macroscopic scale but the real interactions of the chemicals occurs at the submicroscopic level.

Since chemists, along with everyone else, live in a macroscopic world but work with materials that change at the submicroscopic level, we develop models that enable us to understand the world we work with. These models are developed assuming perfect (ideal) behavior. We then may make an effort to express these models with mathematical equations. You need to learn what the chemistry behind each mathematical relationship is.

Some useful properties.

Objectives:

1. You should be able to use density to relate a materials mass and volume.
2. You should be able to distinguish between mass and density.
3. You should be able to convert between the Kelvin and Celsius, and Fahrenheit temperature scales.

Density is a ratio of two measurable quantities, mass and volume. Since it is a mathematical relationship the equation can be rearranged to solve for any one of the three quantities if the other two are known. Density, like many other values, can be expressed in many different formats so it is extremely important that you include units on any density values.

Be careful not to confuse density and mass (weight). Frequently students use the term mass when they really mean density.

When a sample of the alloy, plumbus solder, weighing 75.6 g was placed in a cylinder containing 65.0 mL of water the volume in the cylinder measured 73.0 mL. What is the density of the plumbus solder?

To solve for density we must know the mass and volume of the object. Since the mass was given we only need the volume. From basic physics we recall that a material will displace a volume equal to its own volume. Therefore the volume of the alloy is something we can calculate.

$$\text{Vol. of alloy} = \text{Vol. of water + alloy} - \text{Vol. of water}$$
$$= 73.0\ \text{mL} - 65.0\ \text{mL} = 8.0\ \text{mL}$$

With these two pieces of data density can be determined.

$$d = \frac{\text{Mass}}{\text{Volume}} = \frac{75.6\ \text{grams}}{8.0\ \text{mL.}} = 9.4\ \text{g/mL.}$$

Keep in mind that some units can be interchanged, for example mL and cm^3. If you are working with water a mL (cm^3) is equivalent to a gram since water has a density of 1.0 g/mL.

A piece of uranium has a mass of 12.83 grams. The density of uranium is 19.07 g/cm^3. What volume does it have?

This is a density problem so we start with the formula d= m/v.

We also must rearrange the formula to solve for our unknown which is volume. v = m/d.

A check of the numbers show we have a mass of 12.83 grams which is less than the density shows a cubic centimeter would occupy so we expect an answer to be less than 1 cm^3.

$$v = \frac{12.83\ \text{grams}}{19.07\ \text{grams cm}^{-3}} = 0.67\ \text{cm}^3$$

This is also 0.67 mL since 1 cm^3 = 1 mL.

The density relationship is convenient when measuring liquids. If we know the density of the liquid we can obtain a certain mass by using a volume measurement. The solvent carbon tetrachloride, CCl_4, has a density of 1.60 g/mL. If we wanted 88.0 grams of this solvent we could measure out 55.0 mL as shown below:

$$d = \frac{\text{Mass}}{\text{Volume}} \qquad \text{Volume} = \frac{\text{Mass}}{d} = \frac{88.0\ \text{g}}{1.60\text{g/mL}} = 55.0\ \text{mL}$$

The energy imparted to matter will determine the physical state it has. A basic premise of science (known as the Kinetic Molecular Theory) is that all particles of matter are in constant random motion. As the motion increases, particles get farther apart, interactions among them weaken and the physical state of the material changes.

When energy is added or removed from a substance a state change may occur. When the particle motion of a solid becomes great enough, the structure will break down and it becomes liquid at a specific temperature known as the melting point. This same temperature is also the freezing point if energy is removed rather than added. Likewise the boiling point and condensation temperature are the same except the state change is between liquid and gaseous states.

Of all the temperature scales, the two most important to you in this chemistry course will be the Celsius and Kelvin. The Celsius scale is the one you will use to make all of your measurements. On this scale the coldest temperature possible, absolute zero, is -273.15°C. Since negative temperature values do not work in some mathematical arguments the Kelvin scale is required. The Kelvin scale is zero K at absolute zero. Any Kelvin temperature will be 273.15 degrees higher than the same Celsius value.

The ability to relate the Celsius temperature to the Fahrenheit temperature just because of our extensive use of the latter measurement.

Table salt melts at 801°C. Convert this temperature first to Kelvins and then to °F.

Since a °C and Kelvin are the same size we just add or subtract the difference as needed. In this case we know that we need to add 273.15°C to the melting point of 801°C.

$$801°C + 273.15°C = 1074\ K$$

To convert to °F we can apply the standard conversion formula.

$$°F = \frac{9\ °F}{5\ °C}\ (°C)\ + 32°F \qquad \text{note } 9/5 = 1.8$$

$$= \frac{9\ °F}{5}\ (801)\ + 32\ °F \ = 1442\ °F + 32°F = \ 1474\ °F$$

Could you have estimated the last value without doing the problem?

The materials of chemistry.

Objectives:

1. You should be able to recognize what a chemist means by a pure material.
2. You should be able to distinguish between heterogeneous and homogeneous mixtures.
3. You should be able to elements, compounds and ionic materials and write notations for each type.

When chemists talk about a pure substance they mean either an element (submicroscopic part is the atom or in special cases such as H_2 or S_8 it may be a molecule) or a compound (submicroscopic part is the molecule). Pure substances, at the microscopic level, can not be subdivided any further and still retain the original properties. Pure substances have definite properties and a fixed composition.It is important to remember that even a very complex material can be a pure substance.

If a material is not pure it is a mixture. Mixtures can be separated by physical processes, such as solubility differences; like putting salt and sand in water. Other common methods of separation include distillation, chromatography and selective freezing.

If a mixture is stirred enough it becomes more and more uniform. Since complete uniformity is difficult to obtain, mixtures are nearly always heterogeneous.

Making a mixture by dissolving one material in another, such as salt in water, with stirring, makes for a very uniform system which is defined as homogeneous. These mixtures are known as solutions.

Sometimes it is difficult to determine if what appears to be a solution is really homogeneous. Finely divided particles may be present and may, over time settle, if so we have a suspension. If these particles are extremely small the system may be a colloidal dispersion.

When asked to evaluate if a material is a mixture or a pure solution, make use of your everyday knowledge. Ink or a tree, when given some thought, can readily be identified as mixtures.

The fundamental substances in chemistry are the elements. These, (currently 109 are known) are combined by chemical reaction to make all the other pure substances. The elements have some repeating properties that are highlighted in the periodic table.

Be careful when writing symbols for the elements. Some have two letters in the symbol and others only one. If there are two, only the first letter is capitalized.

Chemical compounds result when elements undergo chemical reactions. Starting materials are called reactants and the resulting materials are referred to as products. Arrows, $\rightarrow$,are used to separate them in chemical equations. You also may see notes over the arrows which give more information. These notes may be for example, 100°C, meaning heat to that temperature, or Δ meaning just heat.

When elements form molecules, the material is represented with a formula. When writing formulas be sure each symbol used has only one capital letter.

You will see some materials represented with charge symbols attached to them. These represent ions and result from unequal charge distribution when the product was formed. An example would be Na^+Cl^-.

The importance of units in problem solving.

Objectives:

1. You should be able to recognize common conversion factors.
2. You should be able to apply conversion factors to problem solving.

Chemistry is a quantitative science so measurement is a cornerstone to our work. As chemists, we employ the SI modified version of the metric system. You need to learn (memorize) some of these units and conversions. If you see units appearing frequently those are the ones you need to learn. There are many, so be selective.

Dimensional analysis is a workable method to use for problem solving. It makes use of conversion factors and units to get from one set of values to another. A major advantage is that the conversion factors can be used to guide you in the step by step sequencing of the solution. The conversion factors are always fractions which are equal to 1. This method can be divided into three basic steps:

1. Determine what you want to solve for and write down the units followed by an = sign.
2. Write down a known factor with its units.
3. Multiply by conversion factors that change from the current to the sought units.

> A students height and weight, were measured to be 6.50 feet and 255 pounds. What are the students height and weight in units of cm and grams?
>
> This is actually two problems so we solve them separately. Both are conversion problems which lend themselves to dimensional analysis procedures.
>
> Solving dimensional analysis problems involves three basic steps:
>
> 1. What do we want to know including units?
> 2. What do we already know, including units?
> 3. What conversions factors get us from one unit set to another?
>
> Height
>
> Known - 6.50 feet
> Want to know - number of cm.
> Conversions - feet to inches and inches to cm.
>
> $$\text{cm} = 6.50 \text{ ft} \times \frac{12 \text{ in}}{1 \text{ ft}} \times \frac{2.54 \text{ cm}}{1.00 \text{ in}} = 198 \text{ cm}$$
>
> Weight
>
> Known - 255 pounds
> Want to know - number of grams
> Conversions - pounds to ounces and ounces to grams

$$g = 255 \text{ lbs} \times \frac{16 \text{ oz}}{1 \text{ lb}} \times \frac{28.37 \text{ g}}{1.00 \text{ oz}} = 116{,}000 \text{ g}$$

The most frequently used units for general chemistry are:

Length:	meter	centimeter	millimeter
Volume:	liter	milliliter	
Mass:	kilogram	gram	milligram

Composition of material is often expressed in units of parts per. This varies depending upon what is being expressed. Solutions used in medical use will be different from those used in environmental issues. If the units are parts per hundred we have an expression of percent. Other parts per unit you will use in this course are parts per million and parts per billion.

A caustic solution used in cleaning was found to have a density of 1.324 g/cm^3. Tests found it to be 45.3% active ingredient. If 1.500 liters of this solution was used what mass of active ingredient was present?

This problem is more complex because it has several parts which need be considered. This problem gives us several known factors which must be evaluated. One key point is that the base solution has 45.30 grams in 100.0 grams of solution and this fact comes from the percent information. It is important to keep the word solution in the labels.

Known - % active ingredient, solution density, solution volume
Want to know - grams of active ingredient
Conversions - this time they will be determined as we need them.

$$g = \frac{45.30 \text{ g active ingredient}}{100.0 \text{ g soln}} \times \frac{1.324 \text{ g soln}}{1 \text{ cm}^3 \text{ of soln}} \times 1.500 \text{ L soln} \times$$

$$\frac{1000 \text{ cm}^3 \text{ soln}}{1 \text{ L soln}} = 899.6 \text{ g}$$

Notice that only one additional conversion factor was needed.

Symbols found in Chapter 2

Page	Symbol	
16	nm	nanometer
37	mm	millimeter
37	m	meter
18	R	gas constant
21	mL	milliliter
38	L	liter
37	kg	kilogram
38	g	gram
38	mg	milligram
18	P	pressure
18	T	temperature in gas formula
23	K	kelvin temperature unit
22	°C	celsius temperature unit
22	°F	fahrenheit temperature unit
18	V	volume
21	g/cm^3	grams per cubic centimeter
37	s	seconds of time
18	n	number of moles of material
37	mol	mole
37	SI	System International units of measure

22	BP	boiling point
22	FP	freezing point
21	cm^3	cubic centimeters

Terms in Chapter 2

The following terms are listed for your convenience. They should not become a list for you to memorize. You should have a working definition for each. If you have doubt about any particular term check its definition in the glossary and/or review it in the chapter. Some of these terms are included in the crossword puzzle later in the chapter.

Physical property	Macroscopic scale	Submicroscopic scale
Kinetic-Molecular Theory		Crystalline solid
Amorphous solid	Ideal behavior	Non-ideal behavior
Density	Temperature	Boiling Point
Physical change	Chemical property	Reactants
Absolute zero	Suspension	Colloidal
Heterogeneous	Homogeneous	Pure substance
Element	Compound	Mixture
Atom	Ion	Molecule
Chemical change	Products	Ore
Electrolysis	Nanotechnology	
International System of Units		Dimensional Analysis
Percent		

Concept Map for The Nature of Matter

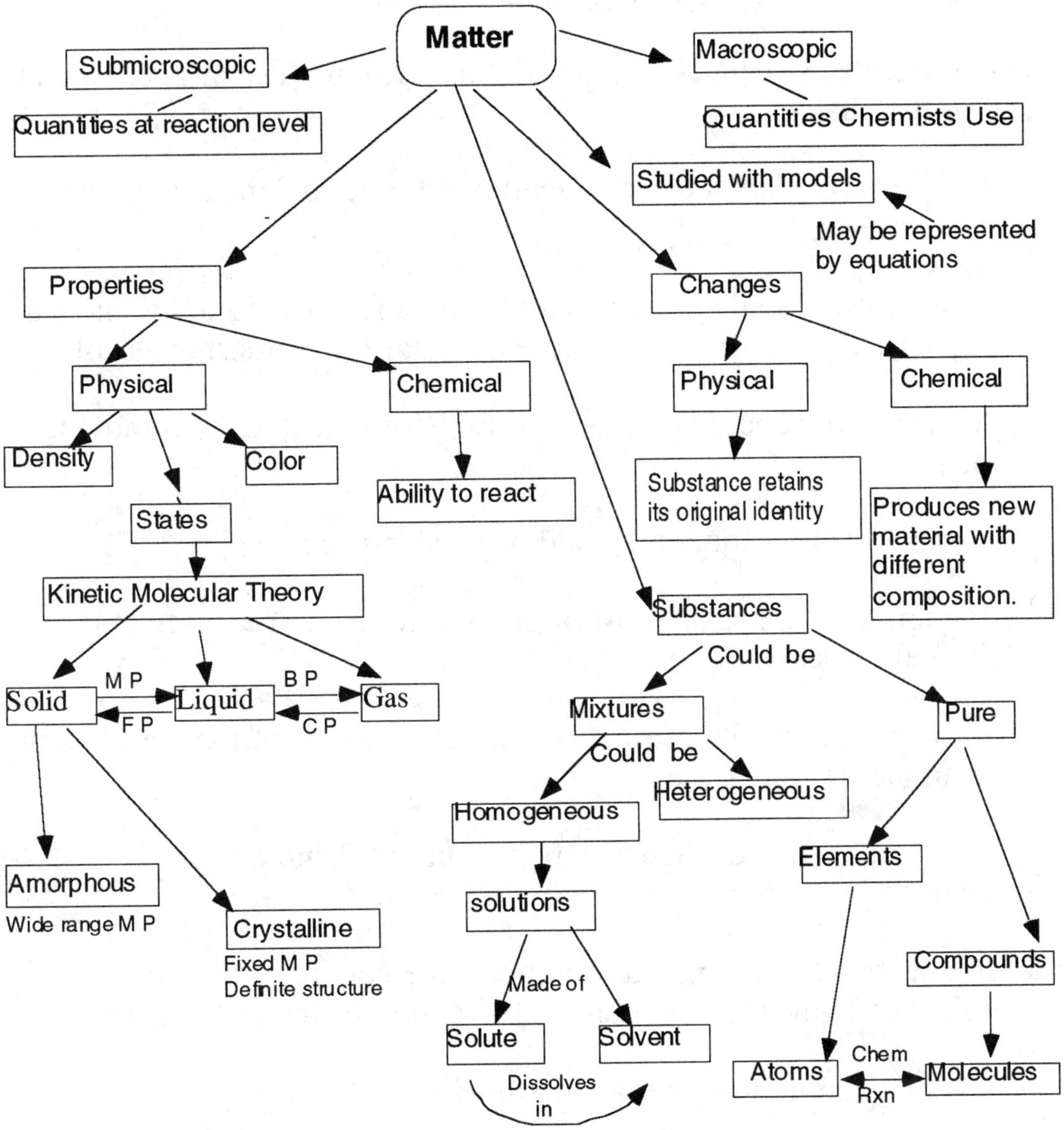

Please see the Preface for a discussion about the use concept maps.

Nature of Matter True/False Statements

Answers are at end of the chapter

Indicate if each statement is true or false as written. The answers are at the end of the chapter.

1. The physical state a piece of matter is found in depends upon its temperature.

2. When viewed under a microscope a material was found to have a regular shape. This indicates the material is an amorphous solid.

3. A gas sample could be converted to the liquid state by removing energy.

4. A mathematical equation could be a scientific model.

5. The temperature scale most often used to make measurements is the Kelvin scale.

6. Temperature is a measure of the kinetic energy of the particles in a sample of matter.

7. When a physical change occurs new pure substances are produced.

8. If objects have a temperature difference heat may flow from the object of higher temperature to the one with lower temperature.

9. Solutions are homogeneous mixtures.

10. The particles in a suspension are smaller than those in a colloidal dispersion.

11. A pure substance cannot be separated by physical methods.

12. BR is a valid symbol for an element.

13. Formulas represent compounds and symbols represent elements.

14. The fact that sodium metal reacts with water gives evidence of one of sodium's chemical properties.

15. When sodium reacts with water we have a physical change.

16. An ore is a pure substance.

17. Electrolysis causes a chemical change in the material subjected to the electric current.

18. There are one million, 10^6, mg in a kg.

19. A 5.0% salt solution has 5.0 grams of salt in a total of 95 grams of solution.

20. The density of a material can be used to help identify that material.

Testing Your Understanding of Concepts in Chapter 2.
Answers are at the end of the chapter.

1. Which of the following are physical properties?
   - A. The material is colorless.
   - B. The material is a liquid at room temperature.
   - C. The material melts at 15.0°C.
   - D. The material reacts with an acid solution.
   - E. The material dissolves in water.

2. A state of matter we do not come in contact with is a ________.

3. The freezing of water can be explained by what theory?

4. A chemist collected data on the density, boiling point and solubility in grams per 100 mL of solvent. These data would be classified as what type?

5. If the temperature of a substance was lowered by 12°C, how much lower would it be on the Kelvin scale? ________ The Fahrenheit scale _______?

6. Models we devise to explain a phenomenon often do not explain all possible situations because some materials exhibit _____________ behavior and models assume they are ideal.

7. The element copper, whose symbol is _______, has a density of 8.96 g/cm^3. The material Mg, which is ______________, has a density of 1.74 g/cm^3. If you had 500 grams of each which would require the larger container?

8. What is the significance of absolute zero,which is _____kelvins.

9. If you put 10.0 grams of fine sand in one container of 100 mL of water and 10.0 grams of sugar in another container with 100 mL of water which would form a homogeneous solution? Why?

10. A chemist had a material which contained charged particles. This material is then said to contain particles called ______.

11. A chemist received a material which could not be purified further. What kind of substance did the chemist receive?

12. The materials represented by the periodic table are _________.

13. Two chemicals are put together and a brand new substance is produced. What kind of change occurred? The starting materials are known as the __________.

14. What is a major advantage of the metric system over the English system?

15. The relationship 3 ft. = 1 yard is known as a ______ ________. When rearranged as a fraction it has a value of ______.

16. If you had a beaker containing 5.0 liters would it have more or fewer quarts than liters? __________ When you change systems of units such as mass (grams to ounces) do you change the quantity of material present? _____

17. Would it take you longer to run a kilometer or a mile at the same pace?

18. You need to obtain 50.0 grams of the liquid acetone. You do not have a balance available. Describe what piece of information you would need and how you could get the proper amount of acetone.

19. You were given a piece of pure metal. It was known to be either silver, whose symbol is ______, or Al which is _____________. Without destroying it what property could you measure to suggest what it is?

20. A common chemical has the formula, Na_3PO_4 . What atoms are present in this molecule? What is the ratio of atoms in any molecule of this compound?

Testing You on Some Analytical Concepts.

Answers are at the end of the chapter

1. A common antifreeze solution has a density of 1.11 g/mL at room temperature. If you had 250.0 mL, what mass would you have?

2. Given the following density values, if you need 50.0 grams of the two metals which would require the larger container, and how much larger would the container have to be? Metal A has a density of 7.19 g/cm^3 and metal B has a density of 11.4 g/cm^3.

3. Battery plates are 94.0% lead and 6.0% antimony. If you had 1.0 kg of the alloy what mass of antimony would be present?

4. A display on a building was flashing the temperature as 42.0°C. Would you suspect the display was working correctly or not? Explain your answer by converting the temperature to °F.

5. You have 0.52 gallons of a liquid. What volume in cm^3 is that equivalent to?

6. When 1000 kernels of popcorn weighing 125.0 grams were popped and then reweighed, the popped mass was 106.0 grams. How many pounds did the popcorn lose?

7. There is an amateur radio band with a wavelength of 80 meters. Is this longer or shorter than a football field which is 100 yards? By how much?

8. It takes 20 nails to build a box. The nails cost $1.17 per pound. If a nail weighs 0.08 pounds what is the cost of nails per box built?

Crossword - Nature of Matter

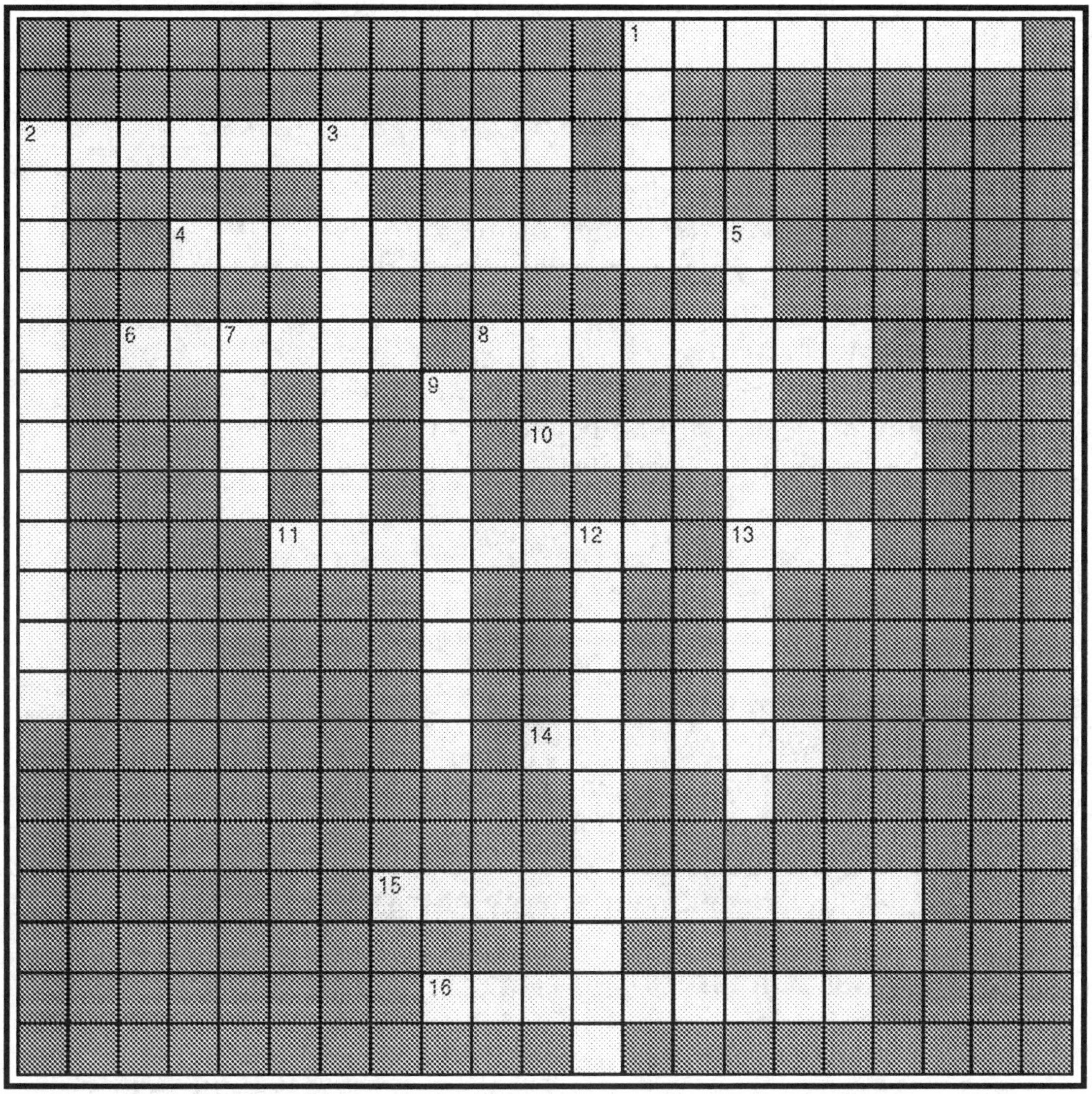

Crossword - Nature of Matter Clues

ACROSS

1 Smallest part of a non-ionic compound
2 True solutions are this
4 Data which is expressed with numbers
6 Sometimes called a forth state of matter
8 System which does not behave as expected
10 A starting material in a chemical reaction
11 Allotrope of graphite
13 Material which contains a valuable mineral
14 Temperature scale with no negative values
15 Solid with a regular geometric shape
16 Solid which does not have a distinct melting point

DOWN

1 Means 1/1000
2 A non-uniform mixture
3 Abbreviated with nm
5 Reaction caused by using an electric current
7 Smallest particle used in chemical reactions
9 Result of a chemical reaction
12 Measure of average kinetic energy of particles in a material

Ties to Previous Chapters

This chapter illustrates some of the points mentioned in chapter 1. The use of measurement to obtain information, new technologies such as nanotechnology and the methods of science to address specific problems have been illustrated.

Answers to True/False Statements:

1. True. Temperature measures average energy present and energy determines the motion of molecules or physical state.

2. False. Crystalline materials have the definite shape.

3. True. Same as 1 above.

4. True. Many scientific models are mathematical equations.

5. False. The measurements are usually made in the celsius scale.

6. True. Same as 1 above.

7. False. Chemical changes result in new products with different composition.

8. True. Energy moves from a high state to a lower state.

9. True. To be a true solution the system must be homogeneous.

10. False. Colloidal dispersions are larger. The particles in a suspension would settle out much sooner because they are larger.

11. True. Physical methods of separation work on mixtures.

12. False. Symbols can have only one capital letter.

13. True.

14. True. This is a chemical reaction.

15. False. A reaction means a chemical change. New materials are formed.

16. False. An ore is a mixture that contains a material of economic importance plus other items such as sand.

17. True. This is a method of causing chemical changes to occur.

18. True. From milli to the unit is 1000 and from the unit to kilo is another 1000.

19. False. Percent means parts per 100 total. You would need 95 grams of the solvent with the 5 grams of salt.

20. True This is a physical property which is not the same for all items.

Answers to Understanding of Concepts Questions:

1. Physical properties do not involve changes that create new materials. Therefore A,B,C and E are physical. Since D has a reaction occurring it represents a chemical property.

2. Plasma. This is now being called the fourth state of matter. It involves very large amounts of energy.

3. The Kinetic Molecular Theory. The state matter is in is related to the kinetic energy of the particles in the matter.

4. Quantitative. All of these have numerical values.

5. 12 since these two degree sizes are identical. It would be 21.6°F lower since each °C is equivalent to 1.8°F.

6. Nonideal. This means they do not always behave in the same way.

7. Cu; Magnesium. The magnesium would need the larger container. Density is a measure of particle packing. The greater the density the more material in the same volume.

8. 0 Kelvin. This is the coldest possible. There is no motion of any particles.

9. The sugar could produce a homogeneous solution since it can dissolve. The sand does not dissolve. The sugar would have to be stirred to assure it is homogeneous.

10. Ions.

11. This would be a pure substance. It could be either an element or a compound.

12. Elements.

13. Chemical change. Reactants are the starting materials.

14. The metric system has consistent divisions. It is based on 10. The metric system also has distinct names for each type of measurement in contrast to the English where an ounce could be mass or volume.

15. Conversion Factor.These are equal to one.

16. More quarts. The quart is a bit smaller than the liter. The quantity of material is the same. The only difference is the size of the units of measurement.

17. A mile. The mile is about 1.6 kilometers.

18. If you determine the density then you could get the 50.0 gram by converting to volume and use a graduated container to measure the liquid.

19. Ag; Aluminum. Density would be a good measurement to make. It is non-destructive and easy to determine. The Al is less dense so would have the larger volume for the same mass.

20. Atoms of sodium, phosphorous and oxygen are present. The atom ratio is 3 sodium and 4 oxygens for each phosphorous.

Answers to Testing the Analytical Concepts

1. You would have 277.5 grams. d = Mass/Volume by rearrangement; Mass = d x Volume = 1.11 g/mL x 250 mL.

2. Volume = Mass/ Density. Metal A would have a volume of 6.95 mL and Metal B has a volume of 4.39 mL. Metal A would need a container which is 2.56 mL larger.

3. Percent refers to parts per 100.

$$\text{grams of Sb} = 1.0\text{ kg} \times \frac{1000\text{ g}}{\text{kg}} \times \frac{6\text{g Sb}}{100\text{g plate}} = 60 \text{ grams of Sb}$$

4. $$°F = \frac{9°F}{5°C}(°C) + 32°F = \frac{9}{5}\cdot°F \times 42 + 32°F = 107.6°F$$

This would be possible depending upon where you were.

5. $$cm^3 = 0.52\text{ gal} \times \frac{4\text{ qt}}{1\text{ gal}} \times \frac{941\text{ mL}}{1\text{ qt}} \times \frac{1\text{ cm}^3}{1\text{ mL}} = 1957\text{ cm}^3$$

6. Mass loss = 125.0 grams - 106.0 grams = 19.0 grams

$$\text{lb} = 19.0\text{ g} \times \frac{1.0\text{ lb}}{454\text{ g}} = 0.042\text{ lb}$$

7. $$\text{yds} = 80\text{ meters} \times \frac{100\text{ cm}}{1\text{ meter}} \times \frac{1.0\text{ in}}{2.54\text{ in}} \times \frac{1\text{ yd}}{36\text{ in}} = 87.5\text{ yds}$$

The football field is longer: 100.0 yds - 87.5 yds = 12.5 yds.

8. $$\$ = 20\text{ nails} \times \frac{0.08\text{ lb}}{1\text{ nail}} \times \frac{\$\,1.17}{1\text{ lb}} = \$\,1.87$$

Answer to Chapter 2 Crossword

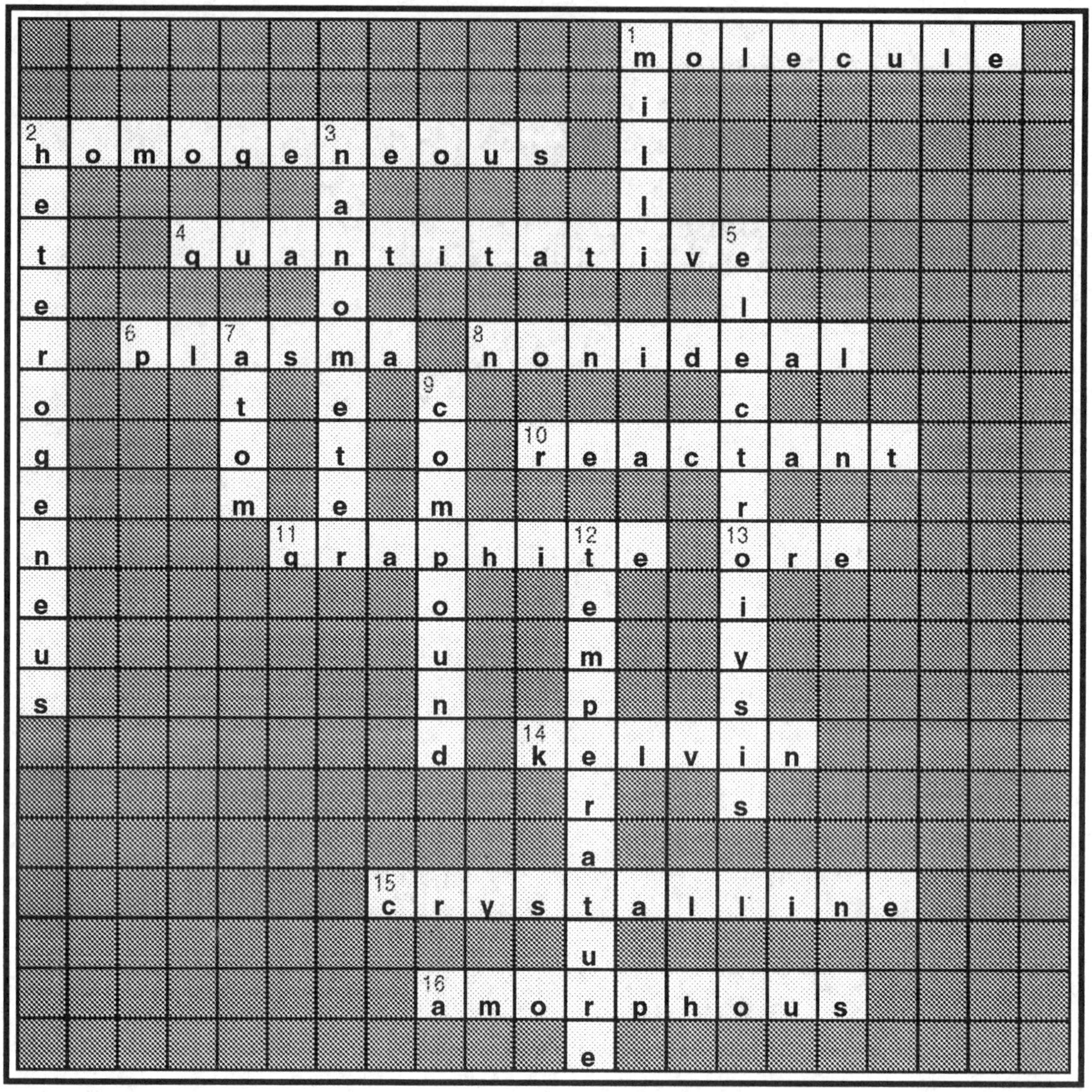

Some Supplementary Readings Relating to This Chapter:

1. A Metallurgical Tale of Irony, Science Vol 215, 8 January 1982, Page 153.

2. It Isn't Easy Being King, Science Vol 203, 16 February 1979, Page 637.

3. Mystery Matters - Saints Blood, Chem Matters, February 1993, Page 12.

4. The Smell of Danger, Chem Matters, October 1988, Page 9.

5. Buckyballs, Chem Matters, December 1992, Page 7.

Chapter 3 Chemical Elements

This chapter presents some information about various chemical elements. More importantly, it develops the basic ideas about the structure of the atom from which elements are made. The idea of chemical periodicity and the periodic table are also developed. This chapter introduces one of the most important organizational tools available to the chemist - the periodic table.

Focus Your Study Efforts

Elements and atoms.

Objectives:

1. You should be able to relate the idea of macroscopic amounts to their microscopic counterparts.
2. You should be able to explain the natural laws that apply to the study of elements.

Elements often have important uses of their own, but more frequently their greatest use is in the production of other compounds. This is a major reason that the study of the elements centers on the reactions they undergo.

Uses and handling of the elements depends upon both their chemical and physical properties.

As chemists we are forced to work with macroscopic amounts of materials. However, we really cannot explain a material's fundamental properties until we get to the atomic level. Since we cannot physically see this level, we develop models and theories to aid our understanding of atomic structure.

This chapter identifies several laws which help explain the elements, how they react, and their properties. Included are:

1. the law of conservation of matter.
2. the law of definite proportions of elements in a given compound.
3. the law of multiple proportions with respect to ratios of like atoms in different compounds.
4. the law of chemical periodicity.

Subatomic parts of atoms.

Objectives:

1. You should be able to discuss some of the early ways and tools used to study atomic structure.
2. You should be able to summarize the subatomic parts of interest to the chemist and the importance of these parts.
3. You should be able to summarize the types of radioactivity.
4. You should be able to explain the idea of isotopes and why they exist.

Evidence for subatomic parts of atoms appear in several ways:

1. Electroscope - evidence for opposite charges in matter.
2. Radioactivity - three common types; two of which are particles, alpha and beta; and one which is radiation, gamma.
3. Discharge tubes - cathode rays are negative in charge, have mass, and are electron beams such as in TV picture tubes.
4. Canal Rays - beams of massive positively charged particles.

Discharge tubes have surfaces where the particle beam starts from and moves to. These surfaces are known as electrodes. The negative electrode, known as the cathode, is the source of the negative (electron) beam. The positive electrode is the anode.

An atoms charge was measured at $-8.0108865 \times 10^{-19}$ C. How else could the charge be expressed?

A charge of $-1.60217733 \times 10^{-19}$ C is the fundamental negative charge. This charge is represented as -1.

$$\text{net charge} = \frac{8.0108865 \times 10^{-19}\text{ C}}{1.60217733 \times 10^{-19}\text{ C}} = 5 \text{ charge units}$$

Therefore -5, since the original charge was negative.

The chemist has an interest in two fundamental regions of the atom. The nucleus, is very small and contains the protons and neutrons. The regions of space about the nucleus, which provides most of the volume, and where the electrons are found.

The number of protons determines the atoms atomic number, symbolized by Z, and the atoms chemical identity. The number of protons always equal the number by electrons, so atoms have an electric neutrality. The electrons are fundamental to the chemical reaction.

The nucleus also contains neutrons which add mass. The atomic mass is equal to the number of protons plus the neutrons. The atomic mass number is symbolized by A. Neutrons are thought to help stabilize the nucleus.

Radioactivity is the result of an unstable nucleus adjusting its ratio of protons to neutrons. Evidence has shown that there are optimum values for this ratio. There are three common types of nuclear decay: alpha (α) , beta (β) and gamma (γ).

The period of time from about 1870 to 1930 was very productive in developing information on atomic structure. Several of the major people involved were: Henri Becquerel, Robert Millikan, J.J. Thomson,

Ernest Rutherford, Hans Geiger, H. Moseley, Marie Curie, Neils Bohr and James Chadwick.

If the number of protons in an atomic nucleus changes, that atom's identity changes.

Atomic weight and atomic mass are used interchangably between textbooks and teachers. Some prefer one term and some prefer the other. Be sure you understand that both terms are usually assumed to mean the same.

The number of neutrons in a nucleus can vary by several units. If nuclei vary in numbers of neutrons, the atoms will have different atomic masses. These variations of mass in the same element are what produces isotopes.

> A specific isotope of Au has a mass of 201 amu. How many protons, neutrons and electrons are present in this isotope? What element does this represent?
>
> Au identifies the element as gold.
>
> All gold isotopes must have 79 protons. Since the material is neutral there must be 79 electrons.
>
> Mass number = # protons + # neutrons
> # neutrons = Mass number - # protons
> = 201 - 79 = 122 neutrons

Most elements have different isotopes. A common example is carbon. There is carbon with 6 protons and 6 neutrons, C-12, and carbon with 8 neutrons, C-14. The latter isotope is radioactive and is used in the technique known as carbon dating. There is also a C-13 isotope.

There is another variation in elements which is very different. Atoms can be joined in different ways and this may lead to very different physical structures. An example of this in carbon is the graphite form and the diamond form. Molecules with these variations in structure are known as allotropes.

The atomic mass of an element usually is a fractional value because several natural isotopes may exist, and the atomic mass represents a weighted average of all isotopes found in nature.

> Silicon has three common isotopes. The amu mass values are 27.98577, 28.98566 and 29.98325 and the relative abundances of these isotopes are 92.18%, 4.71% and 3.12% respectively. What is the average atomic mass of Si?
>
> The average mass of the occurring isotopes depends upon which isotopes are present, the mass of the individual isotopes and their abundance.
>
> At. Wt. = Si-28 mass + Si-29 mass + Si-30 mass
>
> $$= (27.98577 \times 0.9218) + (28.98566 \times 0.0471) + (29.98325 \times 0.0312)$$
> $$= 25.79728 + 1.365225 + 0.93548$$
> $$= 28.09798 \text{ amu}$$

The periodic table is used to organize atom information.

> **Objectives:**
> 1. You should be able to explain the classification of elements based on the organization in the periodic table.
> 2. You should be able to explain the general patterns in the periodic table.

One of the most useful organizational tools in chemistry is the Periodic Table. The original table was based on atomic weights of the elements. The present version is based on the atomic number (number of protons) in the nucleus.

Elements are subdivided into three main groups:

- metals - about 80% of the elements are metals.
- nonmetals - along with hydrogen the elements on the right side of the table. Of these 22, 6 have little or no chemistry.
- metalloids - sometimes called semi-metals. These elements, under varying chemical environments, show properties expected of either metals or nonmetals.

Groups, often called families, are vertical columns in the periodic table. Elements within the same group have similar chemistry.

Groups have been given names that are used to identify them by relative properties. Common group names are alkali metals, alkaline earth metals, halogens, and noble gases.

Periods are the horizontal rows. The elements in periods undergo periodic changes from metallic to nonmetallic to nonreactive. Notice that some periods are much longer than others. They vary from only 2 elements in period 1, eight elements in others, 18 in others and 32 elements in the longest periodπ. More will be said about the length of the periods in later chapters.

Properties common to metals include a surface luster, solid state at room conditions, malleability, and ductility. The metals, however, show little tendency to react with each other.

Nonmetals show fewer overall similarities than the metals do. However, reactions of nonmetals with nonmetals is very common.

The periodic table is commonly divided into major divisions known as the main group elements, usually identified with a letter A after the group number, and the transition elements, including the lanthanides and actinides. The transition elements are identified with a letter B after the group number. The 3A group would be a main group set of elements and the 8B group would be a set of transition elements.

Every actinide element is radioactive in all isotopic forms. Many of these elements no longer exist in nature and are around only because they have been artificially made.

One element, carbon, is responsible for the production of millions of compounds. This element is the backbone for the area of chemistry known as organic chemistry.

Symbols found in Chapter 3

Page	Symbol	
60	α	alpha particle
60	β	beta particle
60	γ	gamma radiation
61	CRT	cathode ray tube
62	C	coulomb
68	Z	atomic number of an element
68	pm	picometer
69	A	atomic mass of element
69	${}_{-1}^{0}e$	beta particle
69	${}_{1}^{1}p$	proton
69	${}_{0}^{1}n$	neutron
71	D	deuterium isotope
71	T	tritium isotope
72	D_2O	heavy water

Terms in Chapter 3

The following terms are listed for your convenience. They should not become a list for you to memorize. You should have a working definition for each. If you have doubt about any particular term check its definition in the glossary and/or review it in the chapter. Some of these terms are included in the crossword puzzle later in the chapter.

Law of Conservation of Matter		electroscope
Law of Definite Proportions		radioactivity
Law of Multiple Proportions		alpha
beta	gamma	cathode ray
electron	cathode	anode
coulomb	canal ray	proton
neutron	nucleus	phosphor
atomic number	atomic mass unit	mass spectrometer
isotope	deuterium	metal
nonmetal	metalloid	wavelength
groups (families)	periods	transition elements
malleable	ductile	semiconductor
alkali metal	alkaline earth metal	actinides
lanthanides	chalcogens	halogens
noble gases	amalgam	

Concept Maps for Chemical Elements

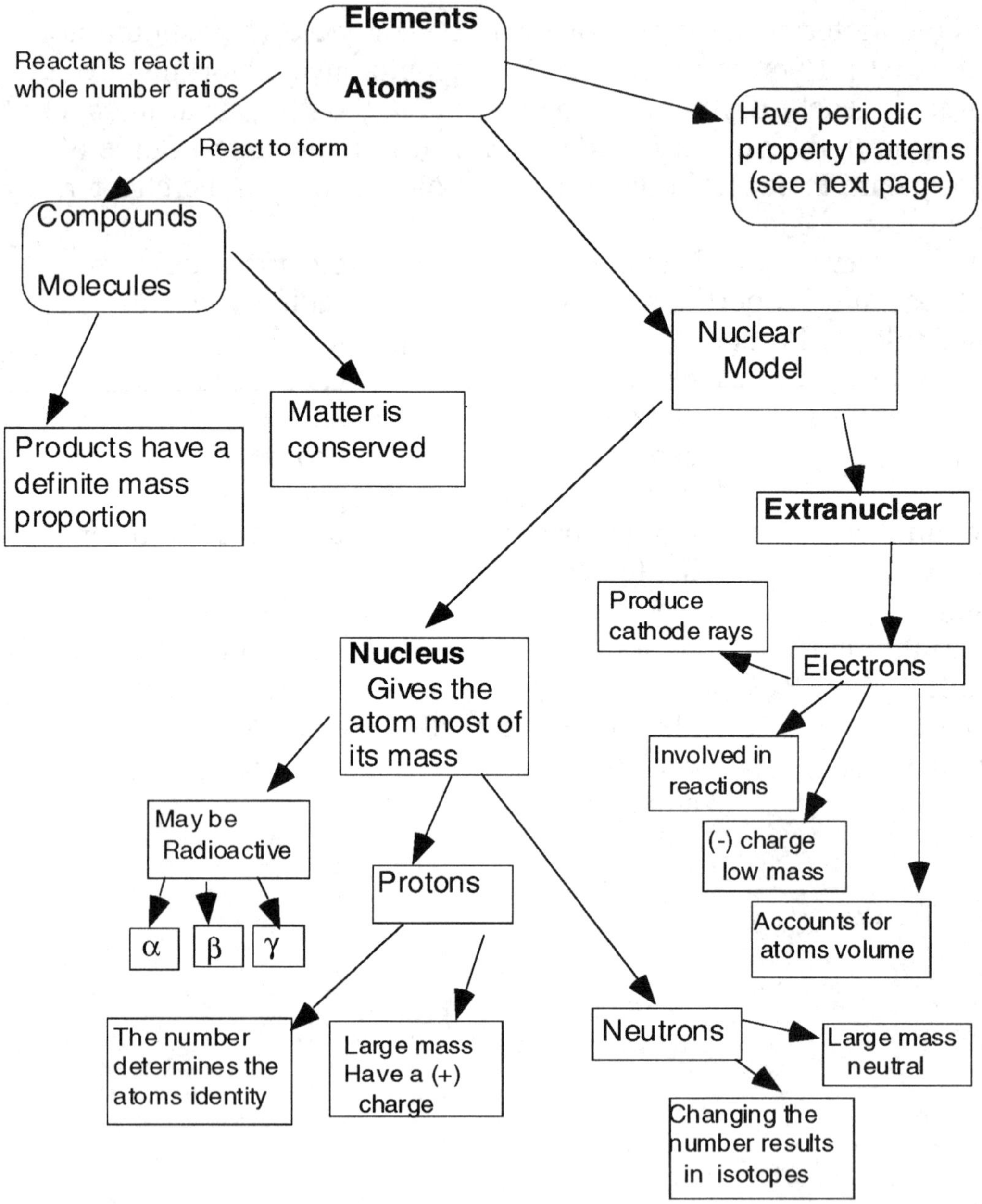

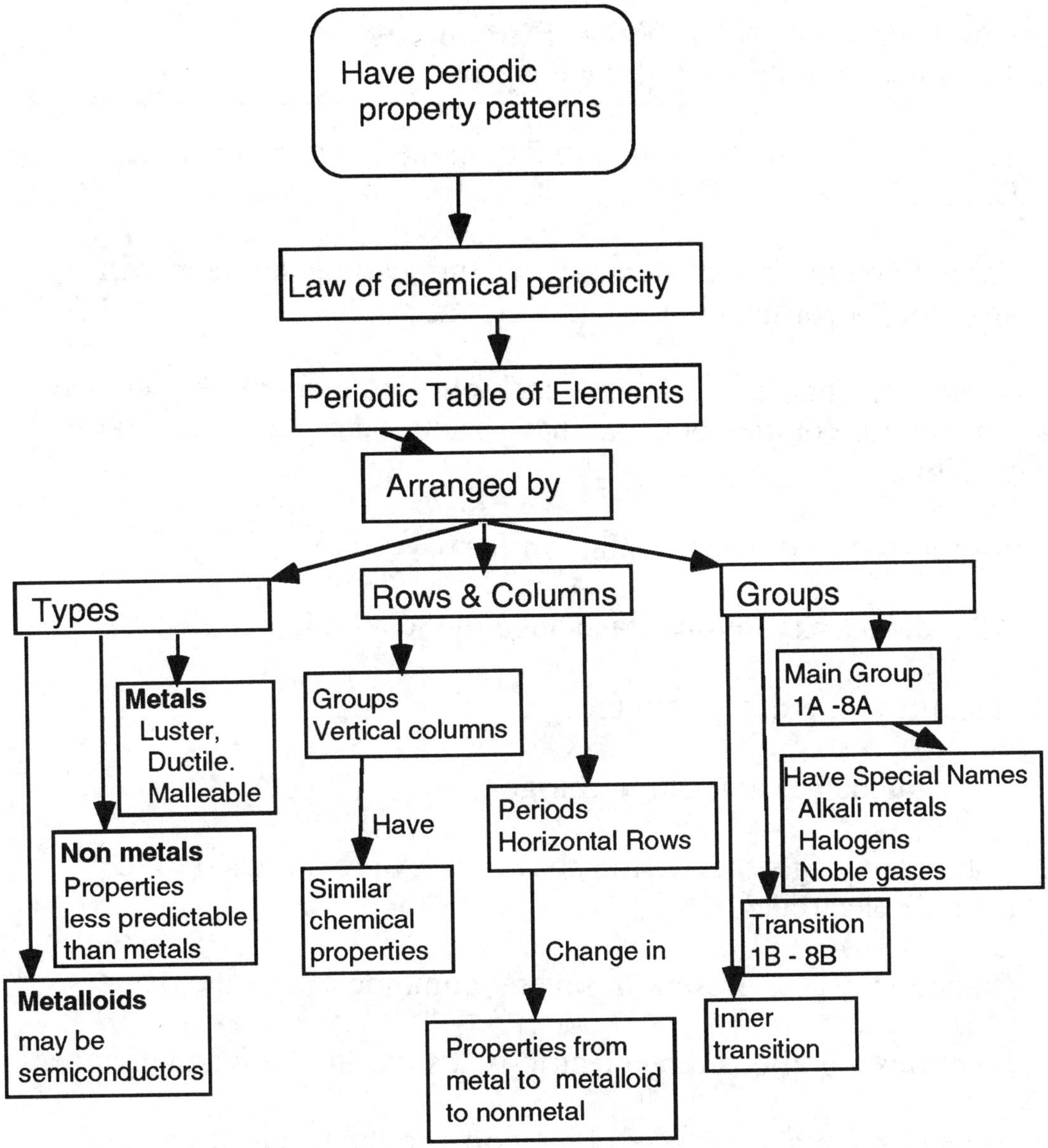
Have periodic property patterns
Law of chemical periodicity
Periodic Table of Elements
Arranged by
Types
Rows & Columns
Groups
Metals
Luster,
Ductile.
Malleable
Non metals
Properties less predictable than metals
Metalloids
may be semiconductors
Groups
Vertical columns
Have
Similar chemical properties
Periods
Horizontal Rows
Change in
Properties from metal to metalloid to nonmetal
Main Group
1A -8A
Have Special Names
Alkali metals
Halogens
Noble gases
Transition
1B - 8B
Inner transition

Chemical Elements True/False Statements

Answers are at the end of the chapter.

1. The basis for current atomic theory came from the ideas of John Dalton.

2. For a given kind of matter, say calcium carbonate, there will always be a constant mass composition.

3. Some elements can react with each other to form more than one compound. For elements that have this property any mass ratio is possible.

4. All radioactivity is particulate in form.

5. All radioactivity involves a change in the atomic nucleus.

6. Canal rays are due to protons.

7. The cathode has a negative charge.

8. Cathode rays get their name because they are attracted to the cathode electrode.

9. Cathode rays are present in several common electronic devices.

10. Electrons are affected by electric fields but not by magnetic ones.

11. Most of the mass and size of an atom are the result of the nucleus.

12. The fundamental basis for the periodic table is the atomic mass value of the various atoms.

13. If you know the atomic number of an element you know which element it is.

14. A mass spectrometer can separate isotopes.

15. All metalloids are metallic elements.

16. Strontium is known as an alkali metal.

17. If CO_2 is a correct formula, then CS_2 is also reasonable.

Testing Your Understanding of Concepts in Chapter 3.
Answers are at the end of the chapter.

1. All elements in group II have what general formula with the halogens?

2. What accounts for the fact that few atomic weights are whole numbers?

3. What do the elements Al, Sb and As have in common?

4. What is the difference between hydrogen and deuterium?

5. The claim is made that some element has five natural isotopes. What is one way this could be verified?

6. What is the significance of the name for element number 101?

7. What was the significance of the work done by H.G.J. Moseley?

8. Summarize the patterns present in the periodic table.

9. Identify four general characteristics of metals.

10. The 6A elements have general formulas for their oxides which are XO_3 and XO_2, such as SO_3 and SO_2. What law do these formula relationships exhibit?

11. Give an example element for each of the following:

Not a main group element.	________
An alkaline earth element.	________
Forms an XO_2 type compound	________
Used as a semiconductor	________
Used in color TV picture tubes	________
In the same group as the most abundant element in the earths atmosphere.	________
One of many nonreactive gases	________

12. When a material is said to have a charge of -1 what amount of charge is this?

Testing You on Some Analytical Concepts.
Answers are at the end of the chapter.

1. How are carbon-12 and carbon-14 alike and yet different?

2. What is the number of neutrons in an atom of iron with a mass of 60?

3. The element chlorine has natural isotopes with mass values of 34.969 and 36.966. The relative abundances are 75.4% and 24.6%. What is the atomic mass?

4. Sulfur has an average atomic weight of 32.064. It has natural isotopes of mass 31.972, 32.971, 33.968 and 35.967. What general statements can be made about the relative abundances?

5. The average atomic weight of nitrogen is 14.007. There are two isotopes of mass 14.0031 and 15.0001. What are the relative abundances of the isotopes?

Crossword - Chemical Elements

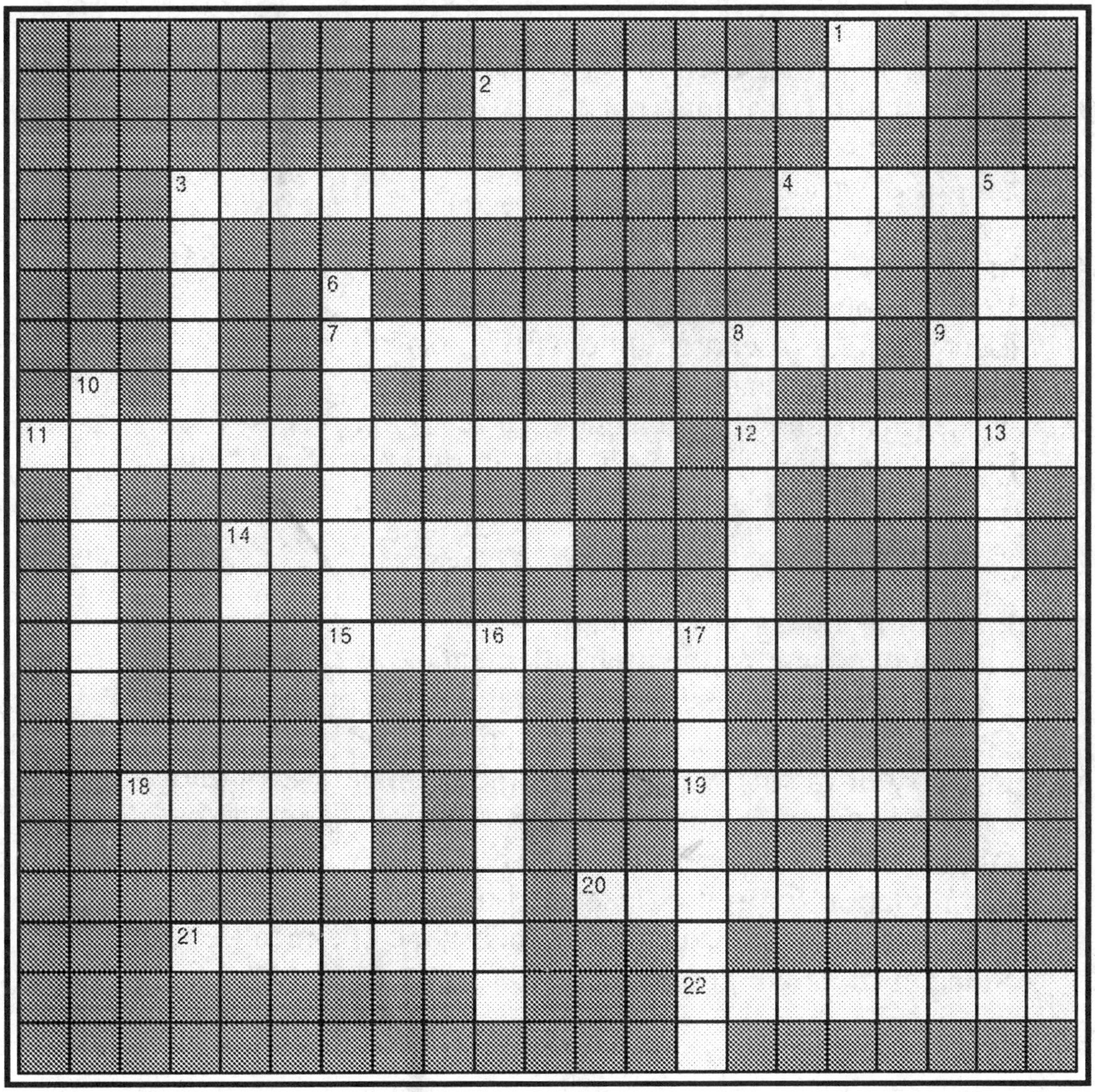

Crossword - Chemical Elements Clues

ACROSS

2 Isotope of hydrogen
3 Mercury is part of these alloys
4 Positive nuclear decay particle
7 Some of these are used in TV picture tubes
9 Abbreviation for atomic mass unit
11 This is the result of having an unstable nucleus
12 Unit of electrical charge
14 Horizontal rows in the periodic table
15 Instrument that can be used to detect isotopes of elements
18 Vertical columns in the periodic table
19 Electrode in a discharge tube with a positive charge
20 Family of very reactive nonmetal elements
21 Nuclear particle which has no charge
22 Forms of elements that very only in number of neutrons

DOWN

1 Part of the atom which contains nearly all the mass
3 Family of very reactive metal elements
5 Smallest part of an element that retains the properties
6 Device that can be used to detect charge
8 Property of metals which means they can be drawn into wire
10 Electrode where beams of negative particles come from
13 Ability to be deformed into thin sheets
14 Abbreviation for unit of size used to measure nuclear distances
16 Result of elements which have undergone chemical reaction
17 Some of these are good semiconductors

Ties to Previous Chapters

Section 2.3 of chapter 2 was an introduction to elements and atoms which has been expanded upon in this chapter.

Units of measurement introduced in section 2.7 are used in this chapter with the assumption you are familiar with them. If units are causing you problems review this section of chapter 2.

Answers to True/False Statements:

1. True. The work of John Dalton was the start of modern discussion of the atom and its structure. Many others had contributed some, but his work was the significant revival.

2. True. This is the Law of Definite Proportions.

3. False. The first statement is correct, however the second is not. The ratio is small whole numbers as stated in the Law of Multiple Proportions.

4. False. Gamma is a form of radiation, not a particle.

5. True. Unless the radiation is only gamma then there is no change, but either alpha or beta can cause a change.

6. True.

7. True. Cathode rays are a beam of electrons.

8. False. They come from the cathode and move toward the (+) anode.

9. True. Some examples are TV picture tubes and non-LCD screens.

10. False. Magnetic fields will easily deflect an electron beam.

11. False. The first part of the statement is true but the volume of the atom is due to the space occupied by electrons.

12. False. The periodic table is based on the atomic number or number of protons in the atom.

13. True. The atomic number is the number of protons and it is the number of protons which determines an atoms identity.

14. True. The mass spectrometer works on the basis of mass and since isotopes vary in mass they will be separated.

15. False. Some, such as As, are nonmetallic.

16. False. Sr is an alkaline earth metal.

17. True. Since oxygen and sulfur are in the same group they will react in a similar manner.

Answers to Understanding of Concepts Questions:

1. MX_2. M represents a metal and X a halogen.

2. The atomic weights are weighted averages of the isotopes found in nature. When the relative amounts are considered the fractional values are common.

3. They are all metalloids.

4. These are isotopes of hydrogen. The deuterium has a neutron and hydrogen does not so deuterium is heavier.

5. The mass spectrometer would be one instrument that could give the verification.

6. This element is named in honor of the person that is given credit for the periodic table.

7. Moseley provided evidence that the periodic table was based on the number of protons in the atom rather than the atomic mass.

8. Elements in vertical columns have similar chemical properties. As you progress from left to right in a row the elements change from very metallic to elements which can show both metallic and nonmetallic properties, to nonmetals, and finally to nonreactive elements.

9. Metals have a luster, they are almost always solids at room conditions, they conduct electric current, they are ductile and malleable.

10. This is the Law of Multiple Proportions.

11. Fe or any of the transition or inner transition elements.
Mg or any element in group 2A
Carbon or group 4A elements
Ge or some other metalloids
Some of the lanthanide elements
Phosphorus or a group 5A element
Ar or any noble gas element

12. The charge is a coulomb. $1.60217733 \times 10^{-19}$ C

Answers to Testing the Analytical Concepts

1. Both elements are carbon so they have 6 protons and 6 electrons. They vary in mass. At Wt. = # protons + # neutrons. Carbon-12 has 6 neutrons and Carbon-14 has 8.

2. At. Wt. = # protons + # neutron
   Iron must have 26 protons
   # neutrons = At. Wt. - # protons = 60 - 26 = 34

3. At.Wt. = 34.969(.754) + 36.966(.246) = 26.365 + 9.094
   = 35.458 amu

4. There is not enough data to calculate values for this. However a general inspection shows that the average mass is very close to 32 so the isotope of mass 31.972 must predominate to a large extent.

5. 14.007 = (14.0031)(x) + (15.0001)(1-x)
   x = 99.61%

   It is 99.61% of the 14.007 isotope and 0.39% of the 15.0001 isotope.

Answer to Chapter 2 Crossword

																1 n				
									2 d	e	u	t	e	r	i	u	m			
																c				
			3 a	m	a	l	g	a	m						4 a	l	p	h	5 a	
			l													e			t	
			k			6 e										u			o	
			a			7 l	a	n	t	h	a	n	i	8 d	e	s		9 a	m	u
	10 c		l			e								u						
11 r	a	d	i	o	a	c	t	i	v	i	t	y		12 c	o	u	l	o	13 m	b
	t					t								t					a	
	h			14 p	e	r	i	o	d	s				i					l	
	o			m		o								l					l	
	d					15 s	p	e	16 c	t	r	o	17 m	e	t	e	r		e	
	e					c			o				e						a	
						o			m				t						b	
		18 g	r	o	u	p	s		p				19 a	n	o	d	e		l	
						e			o				l						e	
									u		20 h	a	l	o	g	e	n	s		
			21 n	e	u	t	r	o	n				o							
									d				22 i	s	o	t	o	p	e	s
													d							

Some Supplementary Readings Relating to This Chapter:

1. Naming the Heavy Elements, Chemical and Engineering News, September 14,1992, page 4.

2. Three-Dimensional Periodic Table of Element Proposed, Chemical and Engineering News, January 1, 1990, page 18.

3. Beyond Uranium, Chemical and Engineering News, November 19, 1990, page18.

Chapter 4 Chemical Compounds

This chapter gets us to the heart of chemistry - the results of chemical reactions and the many products that the elements can produce. Some of the special characteristics of various elements are presented and the ways these characteristics can affect the chemistry are highlighted. Naming of materials resulting from reactions is presented. This will be very important to future chapters. The use of the chemical mole as the connection between the macro and submicroscale is developed. Again this chapter presents a large vocabulary and you will need to be very familiar with most of it.

Focus Your Study Efforts

This chapter will expect you have an understanding of topics presented in chapters 2 and 3. You may find it advantageous to refer back to these two chapters if you seem to be unsure of some ideas.

There are several new concepts introduced in this chapter and they follow a sequential order. Be sure you understand each concept as it is discussed as it will be used later in the chapter.

This chapter develops five major ideas: properties of elements that are molecules; types of molecular compounds; naming of chemical compounds and the rules which apply; the concept of the chemical mole which ties the world of the chemist to the world of the particle, and finally; using the mole concept to determine chemical formulas.

There will be several places that ideas having very similar terminology will come up. It will be important for you to know the distinctions between these terms.

Representing elements.

Objectives:

1. You should be able to identify the elements that have more than one atom in the element symbol.
2. You should be able to identify allotropic forms of an element.

Most elements can be represented as a single atom. There are however, seven common elements which will always be represented as a pair when they **exist unreacted**. These elements are H_2, O_2, N_2, F_2, Cl_2, Br_2, and I_2. A key point to remember is that they are diatomic when by themselves. When they are reacted with another element to form a compound; NaCl or $SnCl_4$ for example, the subscript can vary. A few other elements are represented in other than the single atom format most of the time. The common ones are P_4 and S_8.

Several elements exist in a form involving different atom arrangements in the structure. Carbon is an excellent example since we all are familiar with the graphite and diamond form of this element and know the extensive differences they have. These are known as <u>allotropic forms</u>. Remember that allotropic forms are due to atom arrangement in making a large structure. Do not confuse allotropes with isotopes.

A recent and very exciting, new allotropic form of carbon is making headlines in chemistry. These are large numbers of carbon atoms, for example C_{60}, that form cage type systems. These are known as the fullerenes, or more commonly, "buckyballs". Much research is being done with these systems because the potential for many new and useful products is promising. The number of carbon atoms that may be involved and the possible shapes vary to a large extent.

Compounds - the result of chemical reactions.

Objectives:

1. You should be able to distinguish between molecular and ionic substances.
2. You should be able to identify the various ways organic substances can be written.

Compounds are divided into two major groupings: **molecular** compounds which are made from molecules but no charged particles, and the **ionic** substances, which are made of positively and negatively charged particles that just balance out to zero net charge.

Molecular compounds are represented by a molecular formula. Most organic compounds, such as the carbohydrates, fit into this class of materials. Also compounds made only from nonmetal elements will most likely be in this grouping.

The field of organic chemistry is very large. Currently over 10 million organic compounds are registered with Chemical Abstracts. The most common elements in these molecules are carbon, hydrogen and oxygen. These molecules can, as you might imagine, become very complex. As a result of this, their formulas can be written in several ways.

The molecular formula of the organic compounds just identifies the elements present and the numbers of each. Only some parts of the organic molecule are chemically active and these are known as functional groups. Another way the organic molecular formula can be written is in the condensed form which shows the functional groups present. A third way, and the most informative, is the structural formula and this shows how the various atoms are attached to each other.

Chemical bonds - types and ways to represent them.

Objectives:

1. You should be able to distinguish between chemicals that have molecular and ionic bonds.
2. You should be able to identify the various kinds of ions.
3. You should be able to use the periodic table to predict bond properties for some elements.

Chemists make extensive use of lines between atoms in molecular formulas. These lines are used to represent forces that bind the particles together. The chemical bond in molecules is the result of sharing of valence electrons between the reacting elements. Each line represents a chemical bond so if there are two lines between atoms it means the forces are about twice as great.

Chemical reactions result in the formation of new materials with new properties and the original materials are no longer present. This change in properties is one of the ways we verify that a chemical reaction has occurred.

Some elements, rather than sharing electrons, have an electron gain and loss between the particles. The metallic elements tend to lose valence electrons and the nonmetal elements gain them. Generally, the further the elements are from each other in the periodic table, the greater the tendency to form ions by electron gain and loss rather than have electron sharing.

The ionic compounds are made from cations (positive ions resulting from electron loss by metals) and anions (negative ions which result when electrons are gained by nonmetals).

Only some of the electrons present in an atom can participate in chemical bonding. Electrons exist in regions of space around the nucleus but only those in the outermost region participate in bond formation. The noble gas elements, those found in column 8A, are very unreactive. This nonreactive behavior is related to the number of valence electrons they have. One of the driving forces in chemical reactions is for the various atoms to attain the electron structure of the noble gases. When possible they obtain this structure by electron gain or loss.

Based on the periodic table what would be the expected charges on ions of:

(A) Sulfur (B) Phosphorus (C) Indium (D) Cesium
(E) Iodine (F) Barium

(A) Sulfur is in group 6A so it will gain 2 electrons -2
(B) Phosphorus is in group 5A so it will gain 3 electrons -3
(C) Indium is in group 3A and will lose 3 electrons +3
(D) Cesium is in group 1A and will lose 1 electron +1
(E) Iodine is a 7A element and will gain 1 electron -1
(F) Barium is a 2A element and will lose 2 electrons +2

The periodic table can be very useful in predicting how many electrons a given element may gain or lose. The metals will lose electrons until they have the number equal to that of the nearest noble gas. For example, calcium has 20 electrons and if it lost two it would have 18 which equals the number of electrons argon has. On the other hand, bromine which has 35 electrons, is only one electron short of the number that krypton has, so it readily gains an electron.

Predicting the number of electrons the transition elements will use in bond formation is much less precise, so, for these elements we find it best to just memorize the charges the ions can have.

Polyatomic ions - groups of atoms with a charge.

Objectives:

1. You should be able to identify polyatomic ions.
2. You should be able to identify oxoanions.
3. You should be able to name polyatomic anions containing hydrogen.
4. You should be able to name the anions using the proper prefixes.

Some ions involve more than one atom. These ions are polyatomic. A polyatomic ion is a group of atoms which has a charge and the group reacts as a unit. There is only one common positive polyatomic ion, it is the ammonium ion, NH_4^+.

There are about 30 common polyatomic ions (text page 117) and it is important that you make an effort to memorize their names and charges. These ions will show up many times during this chemistry course.

Polyatomic ions have some keys to help you recognize them. Notice that those with an H at the beginning are known as hydrogen _____, such as hydrogen carbonate, if two H atoms are present they are dihydrogen. Another key to look for is oxygen. If two oxoanions exist the one with the greater amount of oxygen will have an -ate ending and that with lesser oxygens has an -ite ending.

You should learn the prefixes and suffixes that are common to oxoanions. These include, -ate, -ite, hydrogen-, per- and hypo- . Anytime there is more than one oxyanion of a nonmetal the ate and ite endings are used. The hydrogen prefix is common with sulfur, phosphorous and carbon in oxyanions such as HSO_4^{-1}. The oxohalogen ions, such as BrO_3^{-1} is where prefixes such as per- and hypo- appear depending upon the amount of oxygen involved.

Writing chemical formulas.

Objectives:

1. You should be able to name ionic compounds and binary compounds.
2. You should be able to write formulas for selected binary compounds.
3. You should be able to explain why ionic materials dissolve better in polar solvents than in non-polar solvents.
4. You should be able to use the rules to properly name selected transition metal compounds.
5. You should be able to name and use numerical prefixes.
6. You should be able to identify and name simple hydrate compounds.

All ionic compounds must be electrically neutral. The positive and negative charges must offset. This does not mean that the number of positive and negative particles must be equal however. If a material such as aluminum which will lose three electrons to form Al^{+3} reacts with something like iodine which gains one electron to form I^{-1} then a neutral molecule will have 3 iodines for each aluminum or AlI_3.

When writing formulas for binary compounds (only two **different kinds** of elements present) of a metal and nonmetal a simple crossover technique is useful. Write the cation symbol first and place its charge above it then write the anion symbol and place its charge above it. The cation charge becomes the anion subscript (use no sign) and the anion charge becomes the cation subscript. For example barium with nitrogen would be:

$$\overset{+2}{Ba} \quad \overset{-3}{N} \qquad Ba_3N_2$$

From this we can deduce that three bariums each with a +2 charge gives a total + charge of 6 and this is offset by the two nitrogens each with a -3 charge for a total - charge of 6 so we have a neutral molecule.

Write correct formulas for the binary compounds made from the following combination of elements. (Assume the products are ionic)

(A) Cesium and Iodine (B) Barium and Phosphorus
(C) Indium and Sulfur (D) Indium and Phosphorus

There are several ways to complete this problem but the following method is simple and complete.

The first thing which must be done is write correct symbols for the materials. Follow this by putting the charge above the symbol and then use the charge (with no sign) as the subscript for the other symbol.

+1 -1
Cs I CsI When subscripts are 1 we do not need to write them. Also we do not write iodine as diatomic because it has reacted and the correct number of atoms is determined by the material it reacts with.

+2 -3
Ba P Ba_3P_2 This is now neutral since for Ba, 3 x +2 =+6 and for P, 2 x -3 is -6.

+3 -2
In S In_2S_3

+3 -3
In P InP

Write the correct names for the following:

(A) Al_2S_3 (B) KI (C) C_7H_{16}
(D) XeF_6 (E) $Mg(HCO_3)_2$ (F) $CoCl_3$
(G) SO_3 (H) $Ca_3(PO_4)_2$ (I) IF_7
(J) NH_4ClO_4 (K) $Cu(OH)_2$ (L) P_2O_5

These represent examples of many possible types of compounds. First determine the general type and then apply the appropriate rule(s) for naming.

(A) A binary compound of a metal and nonmetal. Name the metal and drop the ending of the nonmetal and add the -ide suffix.
Aluminum sulfide
(B) Same rules as (A)
Potassium iodide
(C) This is a hydrocarbon since it contains only hydrogen and carbon. Name the compound on the basis of the number of carbon atoms present and add the -ane ending. Seven carbons is hept.
Heptane
(D) A binary compound of two nonmetals. Name the first element and drop the ending of the second nonmetal and add the -ide suffix. You need to identify the number of atoms present.
Xenon hexafluoride
(E) This is not a binary compound. It has an oxyanion. You will need to memorize the names for several oxyanions. Use the name of the metal and then add the oxyanion name.
Magnesium hydrogen carbonate
(F) This is another binary metal-nonmetal compound, however the metal is a transition metal so we need to identify the charge it has. Since Cl as an anion gains one electron it will have a -1 charge for a total of -3 negative charge. This means we have cobalt (III). Use same rule as in (A).
Cobalt(III) chloride

(G) See (D) for rules

Sulfur trioxide

(H) See (E) for rules

Calcium phosphate

(I) See (D) for rules

Iodine pentafluoride

(J) This is made of two polyatomic ions. The second, ClO_4, needs extra consideration because the halogens with oxygen can form several combinations. NH_4^{+1} is the only positive polyatomic ion.

Ammonium perchlorate

(K) See (F) and (E) for rules

Copper(II) hydroxide

(L) See (D) for rules

Diphosphorus pentoxide

Ionic compounds are not molecules. They are aggregates of cations and anions held together by their opposite charges. These aggregates exist in three dimensions and this results in a positive ion being surrounded by a series of negative ions and negative ions surrounded by a series of positive ions.

Several factors are important in the formation of the ionic structure and the strength of the forces present. The ion sizes, charge on the ion, and the distance between the ions are significant. The effects of these interactions are shown in the equation for Coulomb's law.

$$\text{force} = \frac{k\,(n_+e)(n_-e)}{d^2}$$ (also see text page 107)

The ionic structures formed are known as crystal lattices. We represent the formulas for ionic materials with what is known as the formula unit. This is the simplest atom ratio that has the properties of the material in the lattice.

Ionic crystals have several properties which make it possible to identify them. They are hard, brittle, and have a regular geometric shape. They also have distinct cleavage planes. They will have a sharp melting point and when molten will conduct an electric current if electrodes are placed in the melt. When molten, the ions undergo a process called dissociation.

Ionic crystals often are water soluble. When they dissolve, they break up into the charged ions which are surrounded by water molecules. This solution will also conduct an electric current. The ionic material is known as an electrolyte.

Water makes a good solvent for ionic materials because of its shape, which is not linear, but V shaped.

The negative end of the molecules of water are attracted to the cations and the positive end to the anions of the crystal.

Atoms and the ions that can be formed from them are very different particles. An iron atom, and an ion of iron, such as Fe^{+2}, do not resemble each other either chemically or physically.

The challenge of learning to name compounds can be minimized if you learn to recognize the type of material which needs to be named. Different rules apply with different situations. We can have main group metal with nonmetal, transition group metal with nonmetal, metal with polyatomic anions, nonmetal with nonmetal and the group of organic compounds whose binary-type compounds are known as hydrocarbons.

When metals are involved in a compound they are named first and their name is used. The nonmetal part has a change in the name, with the normal ending of the name deleted and the suffix -ide added. For example $MgBr_2$ is magnesium bromide and Al_2S_3 is aluminum sulfide. When the ammonium ion , NH_4^{+1}, is present in place of a metal it is also named first,as an example, NH_4F would be named ammonium fluoride.

When a transition metal ion is part of the compound it becomes necessary to identify a specific charge on the metal ion. The transition metals usually have two and often more possible charges. In the case of iron two common charges are possible as seen in the compounds $FeCl_2$ and $FeCl_3$. The charge on the metal ion is indicated in the name as a roman numeral within a set of parenthesis. Since compounds must be neutral the charge on the metal ion can be deduced from the compound. In the iron examples, if the charge of chloride is (-1) then the first iron must be (+2) and the name is Iron(II) chloride. The second compound is Iron(III) chloride. The number in () is the metal ion charge, not the number of anions present. The material CoS is cobalt(II) sulfide.

When a metal reacts with a polyatomic anion, such as the oxoanion, the same rules apply as outlined above. The specific anion name is added to the name of the metal ion present. KCN is potassium cyanide and $Mg(BrO_2)_2$ is magnesium bromite. With a transition metal metal compound, such as $NiCO_3$, the name is nickel(II) carbonate.

When two nonmetals react, the compound name is made by using the actual name of the first atom and changing the name of the second by adding the -ide suffix. The number of each type of atom is placed in the name as a prefix except that the mono prefix is normally ignored. The material CS_2 is carbon disulfide and N_2O_5 is dinitrogen pentoxide.

It would be a good idea to learn some of the common numerical prefixes such as those for 2,3,4,5,6,7,8 and 10 which are di-, tri-, tetra-, penta-, hexa-, hepta-, octa- and deca-.

Probably the largest group of binary compounds is the hydrocarbons. This is a group of organic compounds made only from carbon and hydrogen. There are several classes of hydrocarbons which will be introduced in a following chapter, but one very important class which sets the pattern for all others is the alkanes. The first four alkanes have what chemists call common names since they were identified prior to the introduction of systematic naming. After the first four, the names are derived from the number of carbon atoms present in the molecule, so C_9H_{20} is nonane since the prefix for 9 is non-. The last part of the name comes from the series name, alkane.

Another type of compound you will often come in contact with is the group of hydrates.These are molecules which have varying amounts of water present as part of their structure. The amount of water can vary from one-half to ten or more molecules in the structure. The compounds are named as two parts. The normal name is given to the first part and then the hydrate part of the name is attached. The material $CaCl_2 \cdot 2H_2O$ is named calcium chloride dihydrate. The · between the two parts indicates that the two are attached and the water is an integral part of the system. Do not confuse the · with a multiplication sign. Hydrates vary in their stability toward heat but when heated the water part can be driven off to leave the anhydrous material.

The amount of water in a hydrate can be determined by weighing the material before and after heating and the amount of water can be calculated from the mass loss.

Some molecules will spontaneously lose some of their hydrated waters when exposed to a dry atmosphere. Other molecules ,when placed in a moist atmosphere, will spontaneously pick up water to form a hydrate.

It is important that you do not let yourself be overwhelmed by the diversity in chemical naming. You will not have to use all of the naming at once. Learn to recognize the particular type of material and then apply the appropriate naming technique.

The chemical mole.

Objectives:

1. You should be able to explain the mole concept.
2. You should be able to convert moles to mass and mass to moles.
3. You should be able to use chemical formulas to determine the molar mass of the substance.

The mole is the chemist's unit of quantity. The number of particles in a mole is not exact. When looking at the most exact value 6.0221367 x 10^{23} notice that there are still 16 uncertain places in the value. The mole number, known as Avogadro's number, is still far from precise.

The important thing to remember is that a mole is a fixed number of particles - the number at this point in time is not important. What is significant is that when a chemist identifies the number of moles of material involved, everyone knows that a specific number of particles is represented. This is important because what reacts, in chemical reactions, are particles.

The number of particles in a mole is related to the mass unit via a specific isotope of carbon. A mole of any material has as many particles as are present in 12 grams (exactly) of carbon-12.

The atomic mass values for all the elements are based on carbon-12 as the standard. When the atomic weight amount of any element is weighed out the mass will give Avogadro's number of particles or one mole. This means that 6.94 grams of lithium and 200.59 grams of mercury will have the same number of atoms.

It is imperative for you to be able to convert from gram amounts to mole amounts and the reverse. This mass-to-mole conversion is one of the basic steps in chemistry.

During 1991, 22.65 billion pounds of chlorine was produced in the United States. How many moles is this?

We need to remember that chlorine as an element exists in the diatomic form, Cl_2.

To determine the number of moles we can use the equation:

$$\text{moles} = \frac{\text{Wt. material involved}}{\text{Mole weight}}$$

$$\text{Moles produced} = \frac{22.65 \times 10^{9} \text{ lbs} \times 454 \text{ g lb}^{1-}}{71.0 \text{ g} \cdot \text{mol}^{1-}} = 1.45 \times 10^{11} \text{ mol}$$

As you look at atomic mass values you can see they vary in the number of digits available. How many should you use in problem solving? Look at your problem data and then use an atomic mass value which has one more significant digit than your best data value.

Given the compound:

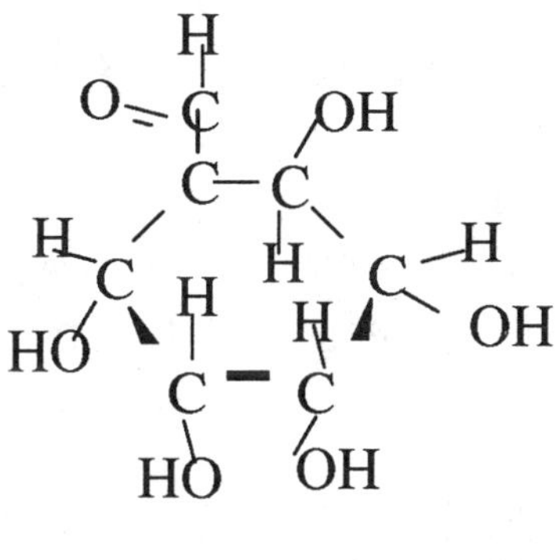

(A) How many of each type of atom is present?
(B) What is the condensed formula?
(C) What is the molar mass?

(A) C= 7 H= 12 O= 6
(B) $C_7H_{12}O_6$
(C) C = 7 x 12.0 = 84.0
H = 12 x 1.0 = 12.0
O = 6 x 16.0 = 96.0
Mol mass = 192.0

Mole to mass and the reverse conversions is a classic use of conversion factors in dimensional analysis.

Which of the following would require the larger container. How much larger would the container have to be. 100 moles of ethanol or 100 moles of water?

Data we need include formulas, mole weights and the density values.

<u>Ethanol</u> C_2H_5OH Density = 0.86 g/cm^3
Molar mass C = 2 x 12.0 = 24.0
H = 6 x 1.0 = 6.0
O = 1 x 16.0 = 16.0
total = 46.0 g/mol

<u>Water</u> H_2O Density = 1.00 g/cm^3
Molar mass = 18.0

100 moles of each would be:
Ethanol = 100 moles x 46.0 g/mol = 4600 g
Water = 100 moles x 18.0 g/mol = 1800 g

Volumes of each:
Ethanol = 4600g/ 0.86 g/cm^3 = 5349 cm^3
Water = 1800g/ 1.00 g/cm^3 = 1800 cm^3

The ethanol needs a larger volume by:

$5349\ cm^3 - 1800\ cm^3 = 3549\ cm^3$

If for some reason you wanted to know the number of atoms present in a sample all you need to do is multiply the number of moles of material available by Avogadro's number.

If you had 0.100 moles of water, how many grams and how many molecules would you have?

This question points out two important aspects of the mole which chemists make use of consistently. Moles represent particles which can be measured "counted" as grams and a mole is a very large number of particles so taking a small part of it is not a problem.

grams of water = 0.100 mole x 18.0 g/mol = 1.80 g water

molecules of water = 0.100 mol x 6.022×10^{23} molecules/mol

= 6.022×10^{22} molecules

When working with molecules, we need to recognize that they form as the result of a chemical reaction which bonds atoms together. As a result, the mass of a molecule, molar mass or molecular weight, is the sum of the atomic weights of all atoms present in the molecule. The molecular mass of zinc nitrate, $Zn(NO_3)_2$, will be the sum of the masses of 1 zinc, 2 nitrogen and 6 oxygen atoms.

The molar mass for a molecule such as calcium chloride dihydrate, $CaCl_2 \cdot 2H_2O$, would be equal to the masses of 1 calcium, 2 chlorines, 4 hydrogens and 2 oxygens. Remember the · in the formula between the parts of the molecule does not mean multiply. It means that 2 water molecules are attached to each calcium chloride. The molar mass is the sum of the two parts.

Convert each of the following to either grams or moles.

(A) 63.0 g Al (B) 3.1 mol of bromine
(C) 0.5 mol hydrogen gas (D) 100 mg of calcium metal
(E) 5×10^{-2} mol of $CaCO_3$ (F) 30.0 g of chromium(III) oxide
(G) 14.7 g of calcium chloride dihydrate

All of these involve the mole to gram relationship.

$$\text{moles} = \frac{\text{grams involved}}{\text{mol weight of material involved}}$$

To solve each you need to determine the molar mass.

(A) $\text{moles} = \frac{63.0 \text{ g Al}}{27.0 \text{ g Al/mol}} = 2.33 \text{ mol Al}$

(B) grams Br_2 = 3.1 mol Br_2 x 159.8 g Br_2/ mol= 495.4 g

(C) Remember hydrogen is diatomic.
grams H_2 = 0.5 mol x 2.0 g/mol = 1.0 g

(D) $\text{moles of Ca} = 100 \text{ mg Ca} \times \frac{1\text{g Ca}}{1000 \text{ mg Ca}} \times \frac{1 \text{ mol Ca}}{40.0 \text{ g Ca}}$

= 0.0025 mol

(E) gram of Ca = 0.05 mol $CaCO_3$ x 100 g $CaCO_3$ /mol
= 5.0 grams of Ca

(F) Formula for chromium(III) oxide is Cr_2O_3 and the mol mass is 152 g/mol.

$$\text{moles } Cr_2O_3 = \frac{30.0 \text{ g } Cr_2O_3}{152 \text{ g } Cr_2O_3/\text{mol}} = 0.197 \text{ mol}$$

(G) The molar mass of $CaCl_2 \cdot 2H_2O$ = 147.0 g / mol

$$\text{moles of } CaCl_2 \cdot 2H_2O = \frac{14.7 \text{ grams}}{147.0 \text{ g/mol}} = 0.1 \text{ mole}$$

When dealing with ionic materials we can only calculate the mass of the relative numbers of atoms present in the simplest formula. Instead of molecular weight we have formula weight. You will however, find that people get sloppy and often interchange formula weight, molecular weight and molar mass.

Remember when you convert from mass values to mole values you are converting from a unit the chemist can work with, grams, to a unit that is involved in the chemical reaction, particles. These particles can be atoms, ions, molecules or particles such as electrons.

Quantities related to formulas.

Objectives:

1. You should be able to use a chemical formula to calculate the percent composition of a material.
2. You should be able to use percent composition data to determine the simplest formula for a compound.

There are two other common calculations which are very useful to the chemist. One is determining the percent composition of a material from its chemical formula and the other is to determine the simplest formula from percent composition data.

The percent composition refers to the mass of a given element compared to the molar mass of the compound times 100%. Using calcium carbonate as an example the percent of calcium present is 40%.

The atomic mass of calcium is 40 g and the molar mass of the compound is 100 g.

Calculate the percentage composition (weight percent)of table sugar, $C_{12}H_{22}O_{11}$.

$$\text{Wt.\%} = \frac{\text{Total Wt. of each element}}{\text{Mol Wt. of the compound}} \times 100\%$$

Mol Wt. $C_{12}H_{22}O_{11} = 12 \times 12.0 + 22 \times 1.0 + 11 \times 16.0$

$= 144 + 22 + 176 = 342g$

(note that you now have the weight of each element and mol wt)

$$\%\ \text{Carbon} = \frac{144\ g}{342\ g/mol} \times 100\% = 42.1\%$$

$$\%\ \text{Hydrogen} = \frac{22\ g}{342\ g/mol} \times 100\% = 6.4\%$$

$$\%\ \text{Oxygen} = \frac{176\ g}{342\ g/mol} \times 100\% = 51.5\%$$

What is the percent of water in calcium chloride dihydrate?

To solve we need the formula, the compound mole weight and the mole weight of water.

Formula is $CaCl_2 \cdot 2H_2O$

Mole weight is 40.0 + 71.0 + 36.0 = 147.0

Mole mass for water is 36.0 (we are interested in the total water)

$$\%\ H_2O = \frac{\text{Wt } H_2O}{\text{Mole wt cmpd}} \times 100\% = \frac{36.0}{147.0} \times 100\% = 24.5\%$$

If you had 93.0 grams of lead(IV) oxide how many grams of lead could be reclaimed?

This is just a percent composition problem.

$$\begin{matrix} & +4 & -2 \\ \text{Lead(IV) oxide} & Pb & O \end{matrix} \quad \text{thus } PbO_2 \text{ (use simplest ratio)}$$

Mole weight = 1 x 207.2 + 2 x 16.0 = 239.2

$$\% \text{ Pb} = \frac{207.2 \text{ g Pb}}{239.2 \text{ g } PbO_2} \times 100\% = 86.62\%$$

Thus 100 grams of the compound would have 86.62 g of lead.

$$\text{g Pb in sample} = 93.0 \text{ g } PbO_2 \times \frac{86.62 \text{ g Pb}}{100 \text{ g } PbO_2} = 80.52 \text{ g Pb}$$

The reverse of the percent composition problem is determining the simplest formula (empirical formula) from the percent composition data. We need to convert the percent data to grams and then to relative moles of each element. A convenient way to think of this problem is assume you have a 100 gram sample of material. The percent composition for calcium carbonate is 40.0% calcium, 12.0% carbon and 48.0% oxygen. Since percent represents part per 100, we can think of the calcium carbonate as containing 40.0 grams of calcium,12.0 grams of carbon and 48.0 grams of oxygen. Referring back to chapter 3, we remember that moles are determined by dividing the mass involved by the mass of a mole. In this case the moles of each element would be:

calcium = 40.0 g / 40.0 g mol^{-1} = 1 mol
carbon = 12.0 g / 12.0 g mol^{-1} = 1 mol
oxygen = 48.0 g / 16.0 g mol^{-1} = 3 mol

If we use these mole values as the subscripts in the formula the simplest, or empirical formula, becomes $CaCO_3$.

A common chemical used in baking has the following percentage composition: Na = 27.5% , H = 1.2% , O = 57.0% , C = 14.3%. What is the empirical formula for the compound?

For convenience we can assume a 100.0 gram sample of the compound which then gives the relative weights for each element. These can be converted to relative mole amounts.

Na = 27.5 g / 23.1 g mol^{-1} = 1.19 mol
H = 1.2 g / 1.0 g mol^{-1} = 1.20 mol
O = 57.0 g / 16.0 g mol^{-1} = 3.56 mol
C = 14.3 g / 12.0 g mol^{-1} = 1.19 mol

These relative mole amounts can be normalized to the lowest denominator by dividing each value by the smallest value.

Na = 1.19 / 1.19 = 1.0
H = 1.20 / 1.19 = 1.01
O = 3.56 / 1.19 = 2.99
C = 1.19 / 1.19 = 1.0

Since these values are all nearly whole numbers, certainly within experimental error, they represent the subscripts for each element. The empirical formula then is

$(NaHCO_3)_x$

When 0.500 grams of an alkaloid compound was analyzed the following elements and the given masses of each element made up the compound sample: nitrogen = 0.0868 g, carbon = 0.3700 g and hydrogen = 0.0433 grams. What is the empirical formula of the alkaloid?

The data given allows us to calculate the moles of each element present.

Moles C = 0.3700 g / 12.0 g mol^{-1} = 0.0308 mol

Moles H = 0.0433 g/ 1.0 g mol^{-1} = 0.0433 mol
Moles N = 0.0868 g / 14.0 g mol^{-1} = 0.0062 mol

Normalization of the moles gives the subscript ratios:

C = 0.038 / 0.0062 = 4.97
H = 0.0433 / 0.0062 = 7.00
N = 0.0062 / 0.0062 = 1.00

The empirical formula is $(C_5H_7N)_x$

When 0.357 g of nitrogen reacted with oxygen the product weighed 0.970 grams. Determine the empirical formula of the nitrogen oxide.

The data do not give us explicitly the mass of oxygen present in the compound but this can be derived from the data available.

Compound mass - nitrogen mass = oxygen mass
0.970 g - 0.357 g = 0.613 g

The moles of nitrogen and oxygen can now be determined.

moles N = 0.357g / 14.0 g mol^{-1} = 0.0255 mol
moles O = 0.613g / 16.0 g mol^{-1} = 0.0383 mol

Normalization of the ratio gives:

Nitrogen = 0.0255 / 0.0255 = 1.00
Oxygen = 0.0383 / 0.0255 = 1.50
$(N_1O_{1.5})_x$

Since this subscript ratio does not contain whole numbers it needs to be converted.This can be accomplished by multiplying both values by 2.

$(N_2O_3)_x$

The empirical formula may or may not be the molecular formula for the material. The determination of the molecular formula requires that the molecular mass of the compound be known. If the empirical formula mas and molecular mass are the same the empirical formula is the molecular formula. If they are not the same there will be a whole number ratio relating them. This number becomes a multiplier for the subscripts of the empirical formula.

Analysis of a material to control moths found it has a molar mass of 147.0g. Further analysis found the material was 48.99 % C, 2.72% hydrogen and 48.30% chlorine. What is the molecular formula?

Again for convenience we will assume we have a 100 gram sample so there are 48.99 g of carbon, 2.72 g of hydrogen and 48.30 g of chlorine.

			normalized value
C	$48.99g / 12.0\ g\ mol^{-1}$	= 4.08 mol	4.08 / 1.36 = 3
H	$2.72g / 1.0\ g\ mol^{-1}$	= 2.72 mol	2.72 / 1.36 = 2
Cl	$48.30g / 35.5\ g\ mol^{-1}$	= 1.36 mol	1.36 / 1.36 = 1

The empirical formula is $(C_3H_2Cl)_x$

The empirical weight is 73.5 g

$$\text{The value of x} = \frac{\text{Mol weight}}{\text{Empirical Wt.}} = \frac{147.0}{73.5} = 2.0$$

This means the molecular formula is $C_6H_4Cl_2$

Symbols found in Chapter 4

Page	Symbol	
99	C_{60}	Represents one of many "buckyball" structures
103	Li^+	A cation. This will commonly be written as M^{n+}
103	F^-	An anion. This will commonly be written a X^{n-}
107	n_-	Charge in Coulomb's law for negative ion
107	n_+	Charge in Coulomb's law for positive ion
107	k	A constant in Coulomb's law formula. Be aware that many constants will be represented by k.
121	mol	Abbreviation for the chemical mole
122	*M*	(Italicized) Symbol for the chemical mole amount
	M	Molarity, refers to a solution concentration.
134	$NiCl_2 \cdot 6H_2O$	Type of formula used for hydrates

Terms in Chapter 4

The following terms are listed for your convenience. They should not become a list for you to memorize. You should have a working definition for each. If you have doubt about any particular term check its definition in the glossary and/or review it in the chapter. Some of these terms are included in the crossword puzzle later in the chapter.

Diatomic molecule	Allotrope	Fullerene (buckyball)
Carbohydrates	Molecular formula	Anhydrous
Condensed formula	Functional group	Structural formula
Covalent bond	Ion	Ionic compound
Cation	Anion	Monatomic ion
Valence electron	Polyatomic ion	Electrostatic force
Coulomb's law	Crystal lattice	Formula unit
Electrode	Electrolytes	Oxyanion
Binary compound	Hydrocarbon	Alkane
"Greenhouse effect"	Mole	Avogadro's number

Molar mass	Molecular weight	Formula weight
Percent composition	Empirical formula	Solvent
Precipitate	Filtration	Superoxide
Hydrated compound		

Concept Maps for Chemical Compounds

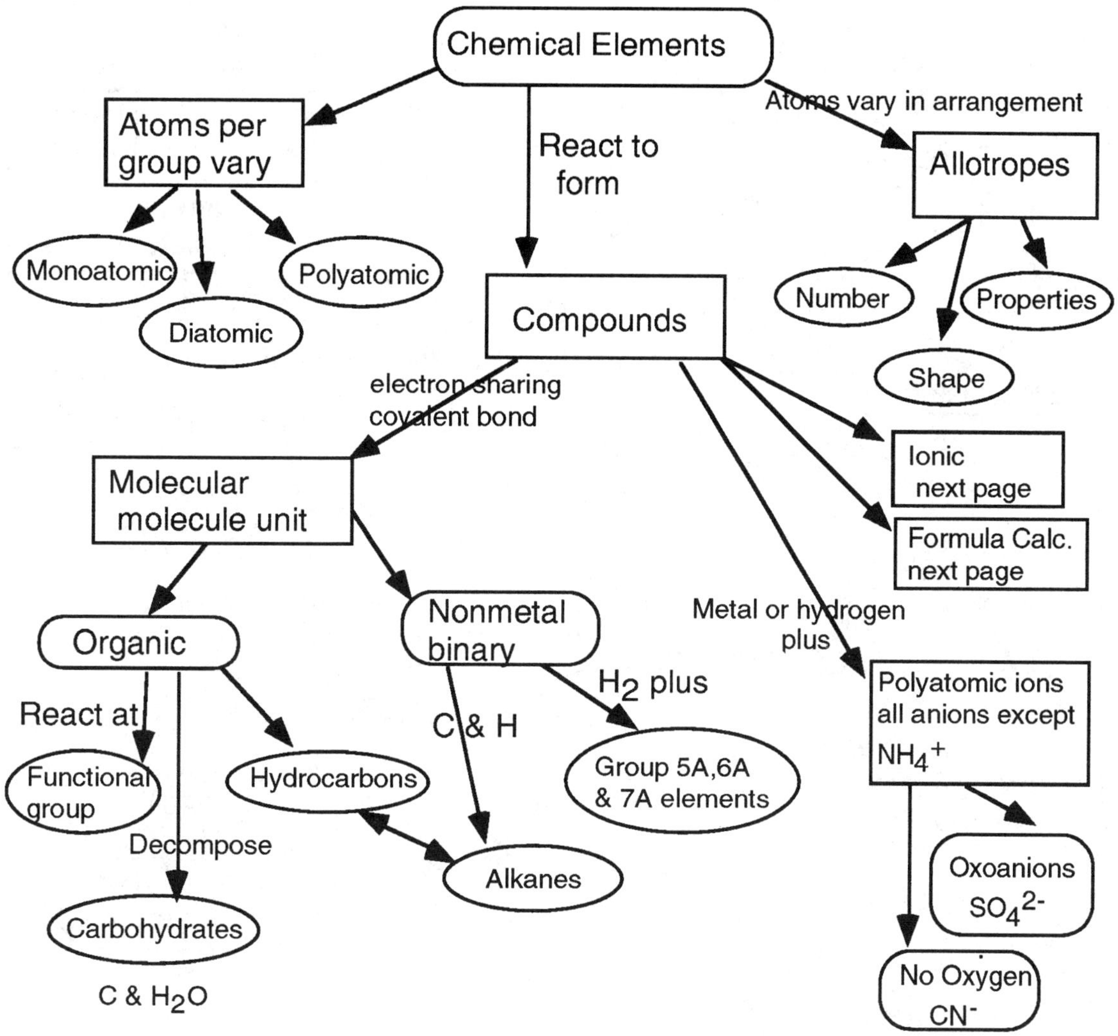

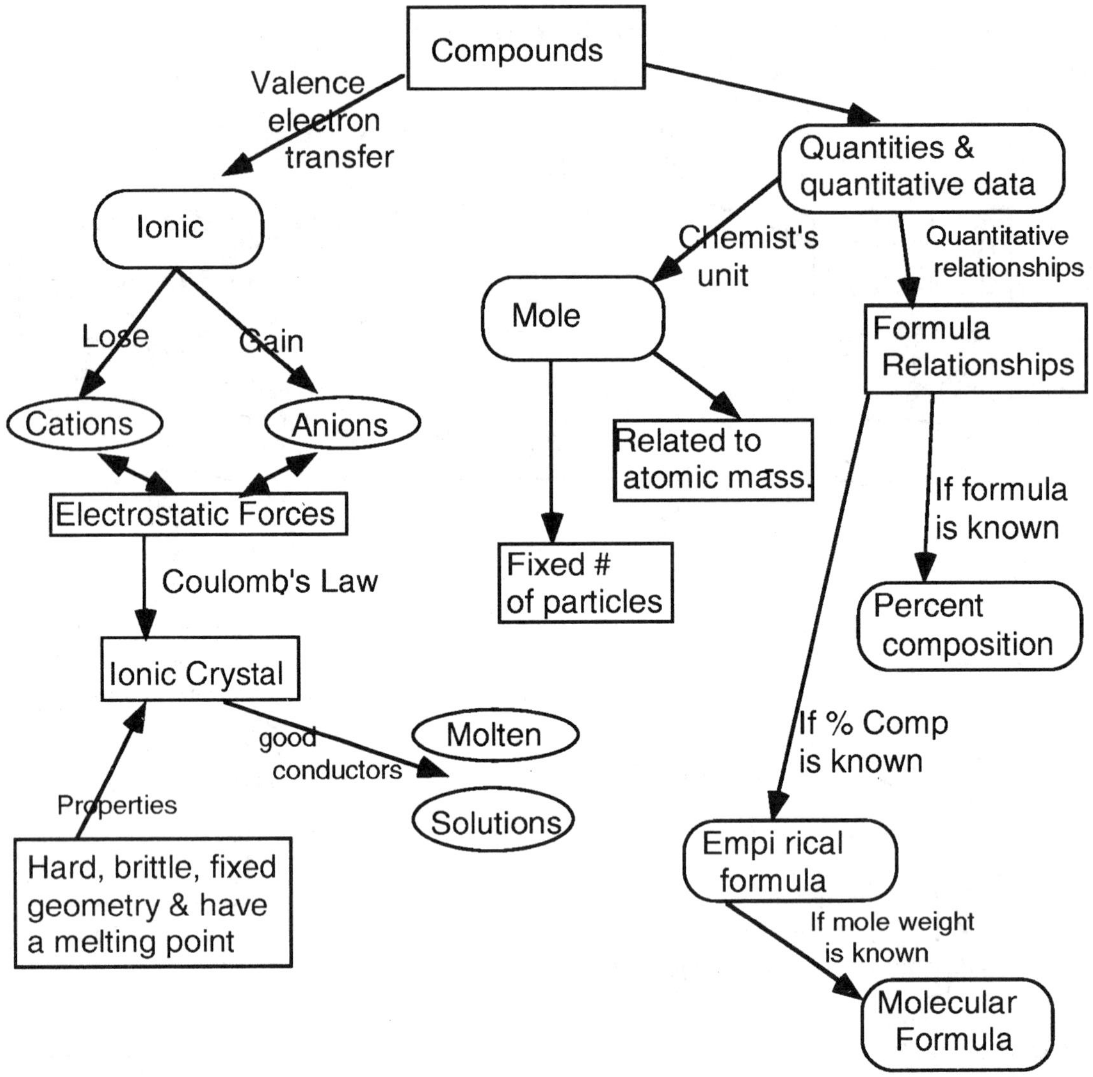
Compounds
Valence electron transfer
Ionic
Lose
Gain
Cations
Anions
Electrostatic Forces
Coulomb's Law
Ionic Crystal
good conductors
Molten
Solutions
Properties
Hard, brittle, fixed geometry & have a melting point
Quantities & quantitative data
Chemist's unit
Mole
Related to atomic mass
Fixed # of particles
Quantitative relationships
Formula Relationships
If formula is known
Percent composition
If % Comp is known
Empi rical formula
If mole weight is known
Molecular Formula

Chemical Compounds True/False Statements

Answers are at the end of the chapter.

1. Grey tin and white tin have two different crystal structures. These two forms of tin represent isotopes of the element.

2. The fullerenes, such as C_{60}, are capable of building cages around some other atoms.

3. When a certain organic compound was heated it decomposed and the only visible materials formed were carbon and some moisture. The material probably was a hydrocarbon.

4. The functional group is often the reactive part of an organic molecule. An example of a functional group is -OH which represents the alcohol.

5. Formulas for organic compounds can be represented in several ways. Of these the least informative to the chemist is the molecular formula.

6. When elements react to form compounds they often retain many of the properties they had as elements.

7. When a reaction occurs and atoms transfer valence electrons an ionic bond is formed.

8. Only valence electrons are involved in forming covalent bonds.

9. The location of an element in the periodic table may be useful in predicting the charge its ion will have.

10. All electrons in an atom are commonly available to form chemical bonds and are called valence electrons.

11. Forces in ionic bonds can be explained with the mathematical relationships represented by Coulomb's law.

12. Large ions tend to form stronger bonds than small ions of the same charge.

13. When metalloids react they may or may not form ions.

14. Both MgO and SBa would be considered acceptable formulas.

15. The names given to electrodes are cation and anion.

16. Metals and ionic solutions conduct via the same process.

17. CN^{-1} would not be considered an oxyanion but would be considered a polyatomic ion.

18. The alkanes are a class of the hydrocarbons.

19. It is possible for the empirical and formula weight of a material to have the same value.

20. Some anhydrous compounds can be converted to hydrates by adding water to them.

Testing Your Understanding of Concepts in Chapter 4.
Answers are at the end of the chapter.

1. The only common positive polyatomic ion is __________.

2. The element iodine exists as I_2 yet the formulas KI, BaI_2, and InI_3 are all correct. Why doesn't the iodine have a subscript of two in all cases?

3. There are several allotropes of sulfur and also several isotopes of this element. How do isotopes and allotropes differ?

4. What are some potential uses for the newest allotrope of carbon?

5. Water is a molecule with covalent chemical bonds. Draw a formula for water which shows this bonding.

6. Write a molecular, condensed and structural formula for the molecule methanol (methyl alcohol).

7. An aqueous solution of barium chloride conducts an electric current but a water solution of the sugar glucose does not. What is the major reason for the differences in solution behavior?

8. Name the following ions:

   (A) Mg^{+2} (B) S^{-2} (C) HS^{-1} (D) NO_3^{-1} (E) SO_3^{-2}

   (F) CH_3COO^{-1} (G) ClO_3^{-1} (H) IO_3^{-1} (I) $H_2PO_4^{-1}$

9. Which of the ions shown in question 8 are:

   (A) Not oxyanions
   (B) Cations
   (C) Not polyatomic ions

10. In what way and why are valence electrons important to chemical systems?

11. In a few words discuss the significant aspects of Coulomb's law.

12. Which of the following are incorrect? Why?

(A) KCl (B) HeAr (C) SMn (D) $AlBr_4$ (E) N_2

13. Why is the term "formula unit" used for ionic materials?

14. What are several significant properties of ionic compounds?

15. Sodium chloride dissolves well in water but not at all in hydrocarbons such as octane. What makes water such a good solvent compared to octane?

16. The element sodium exists by itself but the ion of sodium does not. Explain.

17. Some hydrocarbons are gases, others are liquids, and many are solids. What appears to be the main factor in determining the physical state a hydrocarbon material will have?

18. If one person has one mole of sugar, $C_6H_{12}O_6$, and another person has five moles of sugar, how many more grams of sugar does the second person have?

19. Using the balances available today would it be possible for you to weigh out 6.022×10^{19} molecules of $C_6H_{12}O_6$?

20. How many grams of gold would you need to have the same number of atoms that are in 12.01 grams of carbon ?

21. What is the molar mass of nickel(II) phosphate?

22. How could a precipitate, ppt., be separated from a solution it was made in?

23. How could you determine the number of waters of hydration in a given molecule?

Testing You on Some Analytical Concepts.
Answers are at the end of the chapter.

1. Determine the charge on the item at the right for each of the following materials:

$NiSO_4$	Ni =
$K_2Cr_2O_7$	Cr_2O_7 =
$Fe_3(PO_4)_2$	Fe =

2. A 5.0 pound bag of table sugar, $C_{12}H_{22}O_{11}$, contains how many moles of the chemical?

3. At 40°C the element Cs is a liquid with a density of 1.90 g/cm^3. If a graduated cylinder had 45.0 mL of the metal, how many moles were in the graduate?

4. What is the percent of silver in silver nitrate?

5. A particular noble gas compound was found to contain 77.55% Xe and 22.44% fluorine. What is the empirical formula for the compound?

6. When 1.50 grams of iron was reacted with a hydrocarbon 5.00 grams of product was produced. Analysis found that the product was 30.03% iron and 64.58% carbon. Further analysis found the molar mass to be 185.8 grams. What is the molecular formula for the compound?

Crossword - Chemical Elements

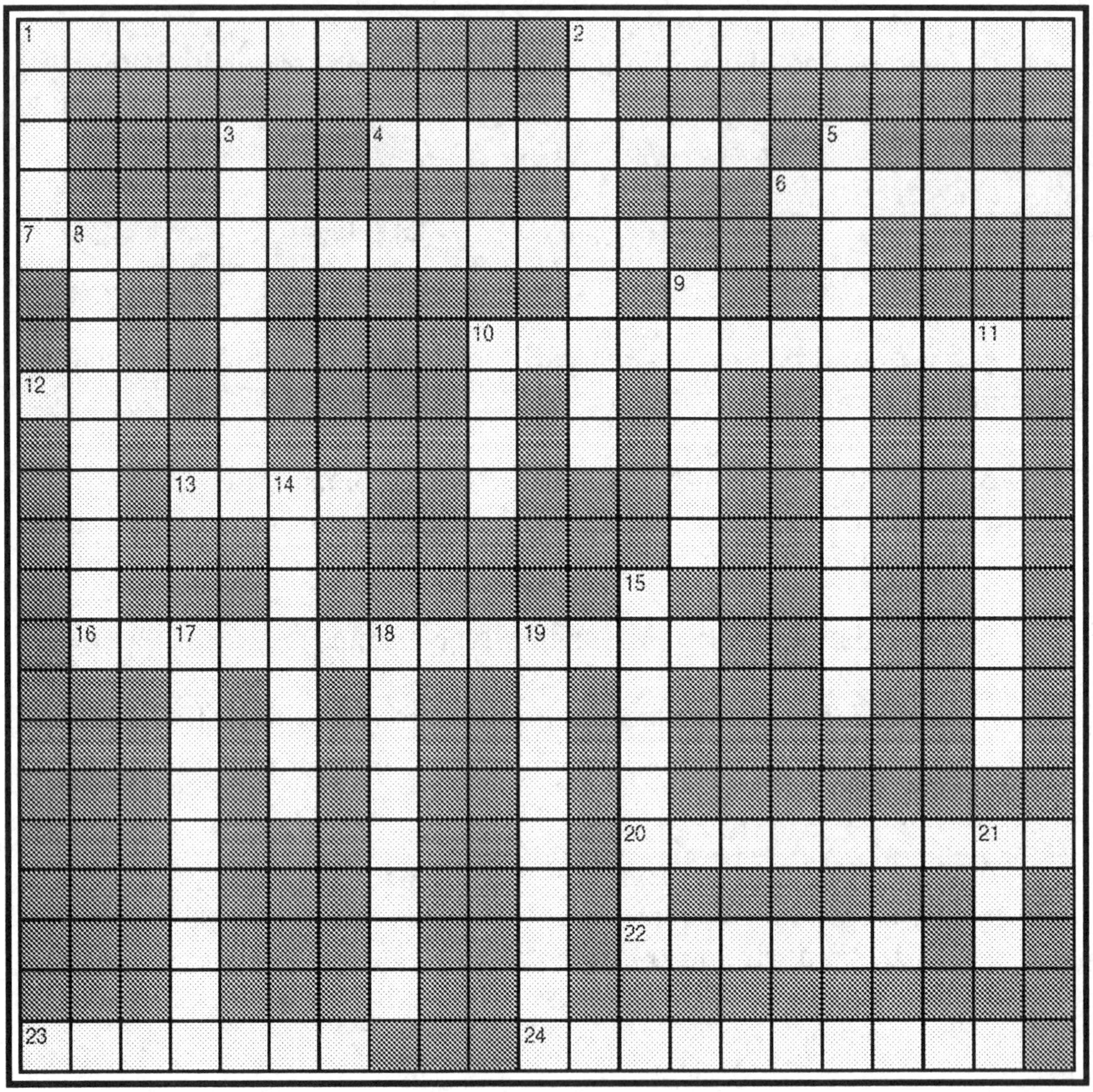

Crossword - Chemical Elements Clues

ACROSS

1 Refers to atoms that vary in number of neutrons in the nucleus.
2 Group that is the reactive part of an organic molecule.
4 Type of bonding that does not involve ions.
6 A class of hydrocarbon compounds.
7 A class of organic compounds that include sugars.
10 A solid material formed when some reactions occur.
12 Abreviation for the term mole.
13 A fixed number of particles and the chemists unit of quantity.
16 Charges that hold crystals together and vary with distance and charge on the particles.
20 Atoms that exist as single particles.
22 An ion that has a positive charge.
23 Electrons that are involved in forming the chemical bond.
24 Formula that shows the atom arrangement in the molecule.

DOWN

1 Type of material that contains charged particles.
2 Allotrope of carbon with 60 or more atoms
3 Number that represents the particles in a mole.
5 Solutions that will conduct an electric current.
8 Result of the arrangement of atoms to produce different structures.
9 Compound made of just two different kinds of atoms.
10 Prefix for an ion that has many atoms present.
11 Represents the simplest formula for a material.
14 Ionic crystals have this.
15 Hydrogen, chlorine and other atoms exist this way.
17 Surface where a reaction can take place when electric current is involved.
18 An ion that has oxygen in it and has a negative charge.

19 Compound without water or has had water of hydration removed.
21 A charged particle.

Ties to Previous Chapters

Some of the ideas which are well developed in this chapter were given an introduction in chapter 2. Sections 2.3, 2.4 and 2.5 are parts to review regarding the atom and their reactions to form molecules.

Chapter 3 was the background for the basic structure of the atom and general properties of elements. Section 3.2, 3.3, 3.4 and 3.5 are basic to the parts of the atom. Section 3.6 will provide some review of the periodic properties of atoms.

Answers to True/False Statements:

1. False. These represent allotropes. Isotopes vary in mass due to varying numbers of neutrons.

2. True. These structures have a rather large hole inside them.

3. False. Hydrocarbons contain C and H only. Moisture could not be a product of decomposition.

4. True. Both parts of this statement are correct.

5. True. This only shows what atoms and the relative amounts of each. With organic compounds it is very useful to know how the atoms are connected. This would show functional groups. The structural formula is the most informative.

6. False. When elements react they lose their original chemical properties and gain new ones.

7. True. Ionic bonding involves electron transfer and this results in the formation of ions.

8. True. The valence electrons are the only ones with low enough energy to interact.

9. True. The position in the periodic table is related to the number of valence electrons, which are the ones gained or lost.

10. False. Only those electrons in the incomplete energy level are valence electrons.

11. True. This law relates size, charge and distance.

12. False. Large ions increase the distance, this is an inverse relationship as shown by Coulomb's law.

13. True. Metalloids, however are hard to predict.

14. False. The first is correct but the second is not. The metal is always written first in the formula.

15. False. Names for the **electrodes** are anode and cathode.

16. False. Metals conduct because of electron movement. Ionic solutions conduct because of ion movement.

17. True. Since there is no oxygen present it is not an oxo ion but there is more than one kind of ion present so it is polyatomic.

18. True. They have a general formula of C_nH_{2n+2} .

19. True. Often the empirical and molecular formula are the same.

20. True. Many anhydrous materials will pick up moisture and become hydrates.

Answers to Understanding of Concepts Questions:

1. Ammonium, NH_4^{+1}

2. The subscript in compounds is determined by the number of particles which react with the other material. In this case, the various ions have different charges. It is always diatomic when it exists as the element.

3. Allotropes are the result of the arrangement of the atoms in the collective structure. Isotopes are the result of atoms having varying numbers of neutrons for the same number of protons.

4. The fullerenes may be useful for trapping particles that can be used in cancer treatment or timed release systems. They also may be able to act as lubricants or ball bearings, among others uses.

5. 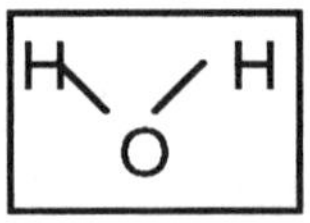

6. Molecular CH_4O
Condensed CH_3OH
Structural

```
      H
      |
  H - C - O - H
      |
      H
```

7. Barium chloride is made from elements that form ions, and these ions dissociate in water, so they are free to move. The glucose does not form any ions so it is molecular at all times.

8. (A) Magnesium (B) Sulfide (C) Hydrogen sulfide
(D) Nitrate (E) Sulfite (F) Acetate (G) Chlorate

(H) Iodate (I) Dihydrogen phosphate

9. Those which are not oxoanions are: A, B and C.
Those which are cations are: A.
Those which are not polyatomic are: A and B.

10. The electrons that react to form chemical bonds are the valence electrons. These are the electrons atoms have that are of low enough energy to be able to react under chemical conditions.

11. This law relates the size of the ions, the ion charges and the distance between the ions to the forces holding the particles together. The relationship shows that as distance increases the force decreases and when the ion charge increases the force increases. Since the distance factor is squared the distance becomes a more significant factor in the force of interaction.

12. B is incorrect because these are noble gases that do not react.
C is incorrect because the metal should be written first.
D is incorrect since Al can not have a +4 charge.

13. Ionic materials have no start and end. They form a crystal lattice that goes on indefinitely in three dimensions.

14. Ionic compounds have several distinctive properties including: definite melting point, rather high melting point, a regular geometric structure, they are hard, they are usually brittle and cleavage planes are present.

15. The water molecule has two ends with some charge. These charge areas allow the water molecules to interact with the ionic crystal and pull it apart. The hydrocarbons do not have this property.

16. The ion has undergone a chemical change that gives it a charge. This charge is present only because the electron lost was gained

by another particle. These particles then interact together and are dependent upon each other for electrical neutrality.

17. The physical state a hydrocarbon exists in depends upon the number of carbon atoms it contains.

18. One mole of the sugar is 180 grams so the second person would have 720 more grams of sugar than the first.

19. This could easily be done. That number of molecules is 0.0001 moles. Since one mole of the compound is 180 grams the mass would be 0.0180 grams. Present balances will easily weigh to the nearest 0.0001 gram.

20. 12.01 g of carbon is one mole. A mole of gold would be needed, and that is 197 grams of gold.

21. 366.1 grams per mole. Ni = 176.1 g, P = 62.0 g and O= 128.0 g

22. There may be several ways, but an easy method would be to filter the system. The ppt. will be caught in the filter paper and the rest of the solution will pass through.

23. One method would be to weigh a sample of the hydrate, then heat it until the water was driven off and reweigh. The mass difference would allow the moles of water to be calculated and the mass remaining would allow you to calculate the moles of anhydrous material still present.

Answers to Testing the Analytical Concepts

1. Ni is a transition element, so it may have more than one charge. The sulfate however is always -2. This means the Ni needs be +2.

Potassium is a 1A metal so it will have a +1 charge. The total charge for K is +2 that means the Cr_2O_7 needs be -2.
Iron could have different charges but phosphate is -3. The total negative is charge is -6, so three Fe must be +6 and each Fe is +2.

2. The molar mass of sugar is 342. Therefore 5.0 pounds is 2270 grams. The moles of sugar present is 6.6.

3. Molar mass for Cs is 132.9 grams.
Mass of Cs = 45.0 mL/1.90 g mL^{-1} = 23.7 grams.
Moles of Cs = 23.7 g/132.9 g mol^{-1} = 0.178 mol.

4. $AgNO_3$ Molar mass = 107.9 + 14.0 + 48.0 = 169.9 grams.
%Ag = [107.9 / 169.9] x 100% = 63.5 % Ag

5. Xe_xF_y A 100 gram sample would have 77.55 g Xe and 22.44g F.
moles Xe = 77.55 / 131.3 g/mol = 0.59 mol
moles F = 22.44 / 19.0 g/mol = 1.18 mol

This is a 1 to 2 ratio so empirical formula is $(XeF_2)_x$

6. Since a hydrocarbon is involved there must also be some hydrogen present. By difference the % hydrogen is 5.39%. The 5.00 gram sample will have the following amounts:

grams Fe = 5.00 x 0.3003 = 1.50(this was also given)
grams C = 5.00 x 0.6458 = 3.23
grams H = 5.00 x 0.0539 = 0.270

Moles Fe = 1.50 / 55.8 = 0.0269
Moles C = 3.23 / 12.0 = 0.269
Moles H = 0.270 / 1.0 = 0.270

Empirical formula = $(FeC_{10}H_{10})_x$

Empirical weight = 185.8

The molecular formula is also $FeC_{10}H_{10}$.

Answer to Chapter 4 Crossword

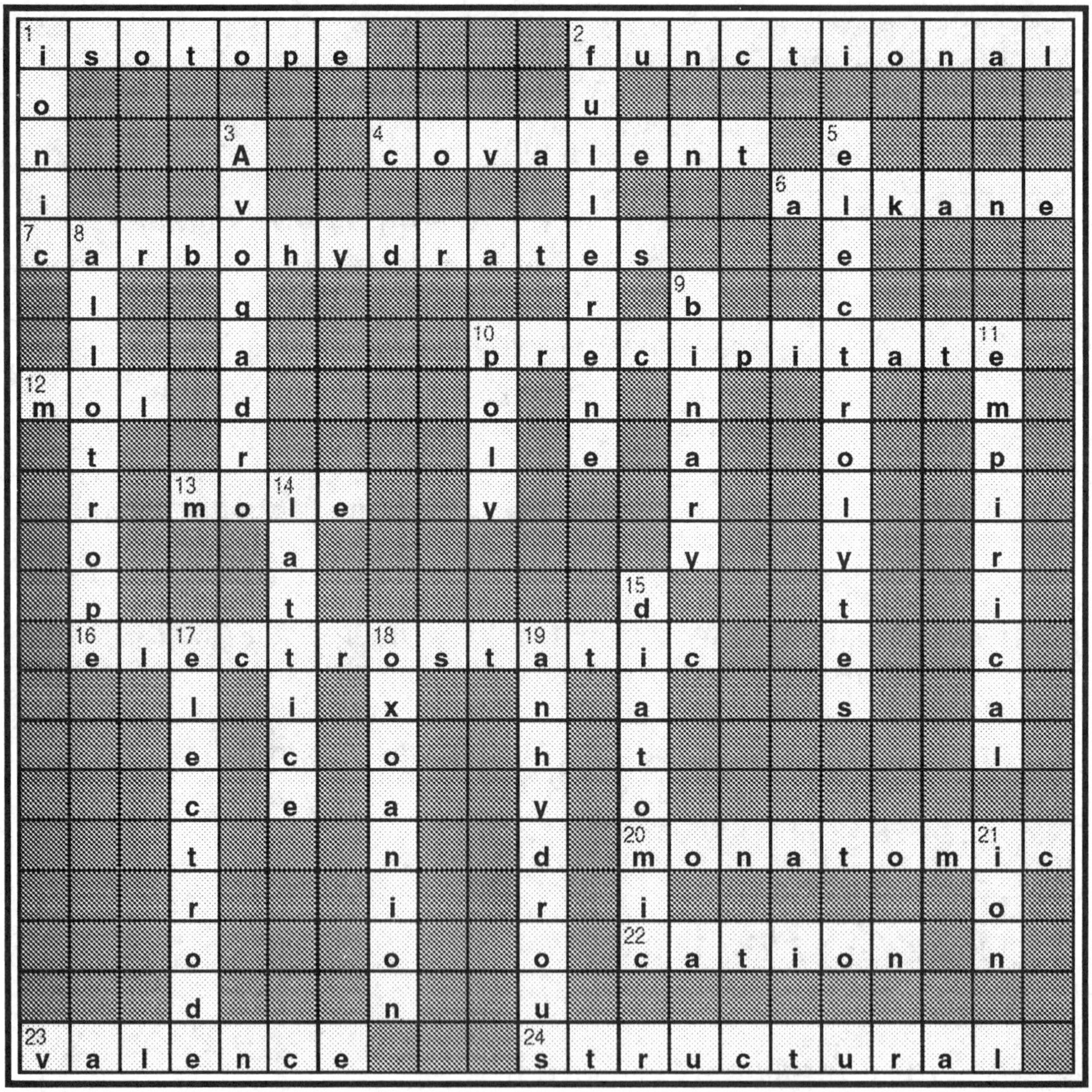

Some Supplementary Readings Relating to This Chapter:

1. Buckyballs, Chem Matters , December 1992, page 7-9

2. Fullerene Bioactivity, C_{60} derivative inhibits AIDS viruses, Chemical and Engineering News , August 2, 1993, page 3.

3. Fullerene' Carbon Molecules Found in Ancient Mystery Mineral, The New York Times , July 10, 1992.

4. New Class of Molecules, The New York Times , May 26, 1992.

5. "Buckyballs" as Filters: Purer Gases May Result, The New York Times , May 20, 1992.

Chapter 5 Chemical Reactions

This chapter discusses how to write and interpret chemical equations whether they result from reactions of molecules or ions. Reactions are classified by type, and materials involved in reactions are put into identifying classes. Major driving forces for reactions are identified.

Focus Your Study Efforts

Various types of reactions can occur.

Objectives:

1. You should be able to identify common types of reactions.
2. You should be able to recognize oxidation-reduction reactions by elements that change oxidation state.

Chemists tend to group reactions by type. There are several types of reactions and these often overlap so classification is not always specific. The general classes of reactions are:

Exchange reactions or Replacement reactions -

These involve two compounds which exchange parts. This group includes acid-base reactions, which are very important. Acid-base reactions are often called neutralization reactions.

It is possible for an exchange reaction to occur between a free element and a compound also. When this situation occurs, the free element replaces the like-type element from the compound. For example, a metal would replace a metal in the compound but not a nonmetal.

There are several common driving forces that make these reactions work well. They include formation of gas, a precipitate or non-reactive material such as water. We sometimes classify reactions which involve the formation of a precipitate (ppt) as a separate group.

Decomposition reactions -

These reactions involve the breakup of larger molecules into smaller molecules or elements. They may be relatively simple such as heating mercury oxide or much more complex such as the heating of ammonium dichromate.

Some materials are predictable in the way they break up. When carbonates are heated, they will decompose to metal oxides and carbon dioxide. Chlorates will form metal chlorides and oxygen. Some oxides will form the metal and oxygen.

Combination reactions or Synthesis reactions -

These reactions involve joining elements to make compounds, or joining small molecules to make larger ones. Oxygen is very often involved in this type of reaction since oxygen has the ability to combine with both metals and nonmetals. The halogens also often undergo this type of reaction.

Oxidation - reduction reactions -

This type of reaction involves an electron exchange between particles. Ion formation will involve redox reactions but redox is not limited to the formation of ions. Combustion reactions are in this category as are the combination reactions. Decomposition reactions may be of this type also.

Because this class of reactions can be so inclusive it is broken into subdivisions such as the combination class outlined above.

This group is associated with the gain and loss of electrons by reactants as products are formed.

Information is provided by Balanced chemical reactions.

Objectives:

1. You should be able to balance chemical equations.
2. You should be able to predict reaction products for common types of reactions.
3. You should be able to identify the conditions that reactants and products exist at.

Chemists find it useful to identify the condition in which reactants and products are present, so you will often see a subscripted letter in parenthesis following the formula. This may be an (s) for solid, (l) for liquid, (g) for gas or (aq) meaning water solution of. Page 149 of your text has examples.

The only valid chemical equation is a correctly balanced one. Be sure that you have balanced the equation before using it in any way.

A balanced equation will assure that several items are conserved. The number of atoms must be conserved, as must the mass between reactants and products. If ions are involved, the equation must conserve charge also.

It is not necessary that the number of moles be conserved. For example, in a decomposition reaction of the type, $CaCO_{3(s)}$ ===> $CaO_{(s)}$ +$CO_{2(g)}$, it is not possible to conserve moles.

Some general guidelines to remember when balancing equations include:

- Write the correct formulas for the reactants. Subscripts are used to make the formulas correct.
- Determine the type of reaction that will occur.This will help you identify the probable products.
- Write the correct formulas for the products. Use subscripts.
- Use coefficients to balance the number of atoms of each kind between the reactants and products. Once the formulas are correctly written subscripts are never changed to effect a balance.
- The oxygen and hydrogen are generally balanced last.
- When finished compare the number of atoms of each kind between reactants and products. They must be the same.

The following combinations are known to react. Determine the type of reaction each will undergo, write a balanced equation for each and write the stoichiometric coefficients for the reactants and products.

(A) Magnesium hydroxide and sulfuric acid
(B) Cobalt and bromine to +3 state of cobalt
(C) Aluminum and tin(IV) phosphate
(D) Cobalt(III) nitrate and potassium carbonate
(E) Heating of aluminum carbonate
(F) Calcium oxide and sulfur dioxide

To solve each the first step is to write correct formulas for the reactants. Then identify the most probable reaction type to occur. This is followed by writing correct formulas for the products. Finally the reaction is atom balanced between reactants and products.

Subscripts are used to make formulas correct. Once this is done the formula is <u>not</u> changed. Coefficients are used to balance atoms between reactants and products. Remember, never try to balance from reactant to product with subscripts.

(A) $Mg^{+2}OH^{-1} + H_2SO_4$ ===> use subscripts for formulas

$Mg(OH)_2 + H_2SO_4$ ===> exchange reaction

this choice based on the fact it is two compounds.

$Mg(OH)_2 + H_2SO_4$ ===>Mg^{+2} $SO_4^{-2} + H^{+1} + OH^{-1}$

Inspection shows the products are correct.

Inspection also shows the H and OH are not balanced.

$Mg(OH)_2 + H_2SO_4$ ===> $MgSO_4$ + HOH

Need to double the H and OH in products

$Mg(OH)_2 + H_2SO_4$ ===>$MgSO_4$ + 2HOH

Stoichiometric coefficients 1:1 ==> 1:2

(B) Co + Br_2 ===> both are elements, so expect a combination type reaction. It would also be a redox reaction. Remember that bromine as an element will be diatomic.

Co + Br_2 ===> $Co^{+3}Br^{-1}$ balance the product with subscripts.

Co + Br_2 ===> $CoBr_3$ product is now correct. We need to balance the bromine. Notice the even odd values on the bromine. When this combination occurs find the lowest subscript multiple and put that number of atoms on each side of the equation. This must be done with coefficients. In this case the multiple is 6. Multiply Br_2 by 3 and $CoBr_3$ by 2.

Co + 3 Br_2 ===> 2 $CoBr_3$ the bromine is balanced but the cobalt is not . We have 2X as many cobalt on product side.

2 Co + 3 Br_2 ===> 2 $CoBr_3$

The stoichiometric ratio is 2:3 ==> 2

What does this mean to a chemist? The answer can be addressed from three levels. The atomic scale, the chemical mole and the gram scale.

Atomic scale: 2 atoms of Co and 3 molecules of Br_2 form 2 molecules of $CoBr_3$.

Mole scale: 2 moles of Co + 3 moles of Br_2 form 2 moles of $CoBr_3$

Gram scale: 117.5 g of Co + 479.4 g Br_2 form 597.2 g $CoBr_3$

(C) $Al^0 + Sn^{+4}PO_4^{-3}$ ===> balance with subscripts

$Al + Sn_3(PO_4)_4$ ===> expect a single exchange reaction which is also a redox. Like type materials will replace like so metal will replace a metal.

$Al + Sn_3(PO_4)_4$ ===> $Al^{+3}PO_4^{-3} + Sn$ both Sn and phosphate do not balance. Which do you start with? It may not matter, but it often works better to start with the larger value, in this case, 4 phosphates.

$Al + Sn_3(PO_4)_4$ ===> $4\ AlPO_4 + Sn$ now balance the Sn.

$Al + Sn_3(PO_4)_4$ ===> $4\ AlPO_4 + 3\ Sn$

Finally balance the Al between reactants and products.

$4\ Al + Sn_3(PO_4)_4$ ===> $4\ AlPO_4 + 3\ Sn$

Stoichiometric coefficients 4:1==> 4:3

(D) $Co^{+3}NO_3^{-1} + K^{+1}CO_3^{-2}$ ===> balance with subscripts

$Co(NO_3)_3 + K_2CO_3$ ===> two compounds so we would predict an exchange reaction.

$Co(NO_3)_3 + K_2CO_3$ ===> $Co^{+3}CO_3^{-2} + K^{+1}NO_3^{-1}$

use subscripts to make the product formulas correct.

$Co(NO_3)_3 + K_2CO_3$ ===> $Co_2(CO_3)_3 + KNO_3$

balance the carbonates with coefficients.

$Co(NO_3)_3 + 3K_2CO_3 ===> Co_2(CO_3)_3 + KNO_3$

this leaves 6K on the reactant side so it needs 6K as products.

$Co(NO_3)_3 + 3K_2CO_3 ===> Co_2(CO_3)_3 + 6KNO_3$

inspection shows the need for 2 Co on the reactant side.

$2Co(NO_3)_3 + 3K_2CO_3 ===> Co_2(CO_3)_3 + 6KNO_3$

Stoichiometric coefficients 2:3 ==> 1:6

(E) $Al^{+3}CO_3^{-2}$ ===> heating of carbonates results in decomposition to a metal oxide and carbon dioxide. First use subscripts to balance the reactant formula.

$Al_2(CO_3)_3$ ===>

$Al_2(CO_3)_3 ===> Al^{+3}O^{-2} + CO_2$

With subscripts balance the product formulas.

$Al_2(CO_3)_3 ===> Al_2O_3 + CO_2$

We now need to balance the carbon

$Al_2(CO_3)_3 ===> Al_2O_3 + 3CO_2$

Stoichiometric coefficients 1 ==> 1:3

(F) $Ca^{+2}O^{-2} + SO_2$ ===> Materials have correct formulas. Since these are two compounds we might expect an exchange reaction. However, both are oxides so exchange produces no net change. A combination reaction is more probable.

$CaO + SO_2 ===> Ca^{+2}SO_3^{-2}$ This gives a balanced product.

$CaO + SO_2 ===> CaSO_3$

This reaction involves a metal oxide which is a base oxide and a nonmetal oxide which is an acid oxide. These will react directly to form a salt, in this case calcium sulfite.

Stoichiometric coefficient 1:1 ==> 1

Water as a solvent has significant effects on reactions.

Objectives:

1. You should be able to recognize electrolyte and non-electrolyte systems.
2. You should be able to recognize when dissociation will occur.
3. You should be able to identify strong and weak electrolytes.
4. You should be able to identify chemicals that make a solution acidic, basic or neutral.
5. You should be able to write ionic equations and identify spectator ions.

A majority of chemical reactions, at least those studied in the beginning course, occur in water. These are known as aqueous systems.

When materials dissolve there is more than one possible outcome. Some materials will break apart into charged particles called ions, in a process known as dissociation. The materials are given the name electrolytes. See the discussion on pages 154 & 155 of the textbook.

Electrolytes in turn may dissociate in varying amounts. Those which produce many ions are identified as strong electrolytes and have the characteristic of conducting an electric current very well. If only a few ions are produced it would be a weak electrolyte.

You often see comments about 100% dissociation. However, this seldom is the case since ions will have some attraction for each other and ion pairs will result. It would be better to think in terms of approaching 100% dissociation.

The weak electrolytes have only a small percentage of their molecules separated into ions. The term "ionize" refers to these materials.

The process of ionization is a dynamic situation. As some particles form ions other ions recombine to form the molecule. This is a condition known as equilibrium. Equilibrium is common to weak electrolytes.

The formation of an insoluble product, known as a precipitate, is a driving force causing some reactions to go to completion. Ions which react to form insoluble substances give us many valuable materials.

Predicting solubility can be very difficult but there are a few general rules which work well for generalization. You should spend the time to learn them. The time learning these guidelines will be saved many times over as you work on writing balanced equations later on. Use figure 5.7 of your text on page 157 for a summary.

An extremely large number of reactions fall into the classification of acid base reactions. These reactions often are very important both industrially and environmentally, such as in “acid rain”.

Chemically, acids and bases can be thought of as chemical opposites and when put together neutralize the properties of each other. Their reactions are known as neutralization reactions.

Acids and bases are electrolytes. In water the major acid ion is H^+, more properly the hydronium ion, H_3O^+. The major base ion is the hydroxide ion, OH^-. If the acid or base dissociates to a large extent we consider the acid or base to be strong.

When acids and bases react the products of the reaction are water and a material called a “salt”. Sodium chloride, common table salt, is one of many materials that a chemist calls a salt.

Completion of an acid-base reaction is difficult to determine since all materials frequently have no color. For that reason indicators, which are highly colored compounds in either an acid or base, are added to identify when the reaction has used up one of the reactants.

When some oxides are dissolved in water an acid or base solution may result. Oxides of metals will interact with the water to produce a basic solution. Thus, metal oxides are called basic oxides. Nonmetal oxides are acid oxides and often produce an acid when added to water. Metal and non-metals oxides are discussed on pages 161 and 162 of the text.

When an acid and base react, it really is only the H^+ from the acid and the OH^- from the base that chemically interact regardless of which acid or base is used. The other ions present in the reactants are also present in the products and chemically have done nothing. These ions are **spectator ions**.

Chemists often write a reaction showing only the parts that are involved in the actual chemistry. These are known as **net ionic equations**.

Write net ionic equations for the following:

(A) Sodium hydroxide and nitric acid
(B) Potassium hydroxide and acetic acid
(C) Lead carbonate and hydrochloric acid
(D) Sodium sulfide and zinc chloride

Several of these reactions involve acids. There are a few common acids which you should know the names and formulas of. Three are given here, nitric, acetic, and hydrochloric.

These will each require a correctly balanced equation. Additionally, you must know if each forms an ion and if the materials are soluble. First determine if the materials are strong or weak electrolytes. You can check solubility charts, such as

Figure 5.7 in the text, to determine something about solubilities. Only after this is done can you determine the net ionic equation.

(A) $NaOH + HNO_3$ ===> $NaNO_3 + HOH$

All sodium compounds are ionic and HNO_3 is a strong acid. Water is molecular.

$Na^{+1}{}_{(aq)} + OH^{-1}{}_{(aq)} + H^{+1}{}_{(aq)} + NO_3{}^{-1}{}_{(aq)}$ ===> $Na^{+1}{}_{(aq)} + NO_3{}^{-1}{}_{(aq)} + HOH_{(l)}$

Both sodium and nitrate are spectator ions since they are present on both sides.

Net ionic equation is $OH^{-1}{}_{(aq)} + H^{+1}{}_{(aq)}$ ==> $HOH_{(l)}$

(B) $KOH + CH_3COOH$ ===> $KCH_3COO + HOH$

All potassium compounds are ionic. Acetic acid is a weak acid.

$K^{+1}{}_{aq)} + OH^{-1}{}_{(aq)} + CH_3COOH_{(aq)}$ ===> $K^{+1}{}_{(aq)} + CH_3COO^{-1}{}_{(aq)} + HOH_{(l)}$

The only spectator ion is the potassium ion.

$OH^{-1}{}_{(aq)} + CH_3COOH_{(aq)}$ ===> $CH_3COO^{-1}{}_{(aq)} + HOH_{(l)}$

Note that charge is conserved. There is an ion with a negative charge in both reactants and products.

(C) $PbCO_3 + HCl$ ===> $PbCl_2 + H_2CO_3$

HCl is a strong acid. Carbonic acid, H_2CO_3, is a weak acid. Both lead compounds are insoluble.

$PbCO_{3(s)} + H^{+1}{}_{(aq)} + Cl^{-1}{}_{(aq)}$ ===> $PbCl_{2(s)} + H_2CO_{3(aq)}$

There are no ions common to both sides so this is the net ionic equation.

(D) $Na_2S + ZnCl_2$ ===> $ZnS + 2\ NaCl$

All these materials are ionic and soluble except the zinc sulfide.

$2\ Na^{+1}_{(aq)} + S^{-2}_{(aq)} + Zn^{+2}_{(aq)} + 2\ Cl^{-1}_{(aq)} \Longrightarrow ZnS_{(s)} + 2\ Na^{+1}_{(aq)} + 2\ Cl^{-1}_{(aq)}$

Sodium and chloride are common to both sides so they are spectator ions.

$S^{-2}_{aq)} + Zn^{+2}_{(aq)} \Longrightarrow ZnS_{(s)}$

When writing net ionic equations it is essential that a conservation of charge be maintained between the reactants and products.

Combination reactions are common between elements or elements and small molecules, such as SO_3. Oxygen is capable of reacting with many metals or nonmetal to form a large class of compound known as oxides.

When nonmetal atoms react with oxygen it is often possible to produce multiple oxides. Nitrogen, phosphorus, sulfur and carbon all produce more than one oxide each.

Decomposition reactions often occur upon heating of materials. This type of reaction is different than combustion so be careful not to confuse them. Carbonates decompose to carbon dioxide and the metal oxide. Some metal oxides will decompose to the metal and oxygen. Metal chlorates will produce the metal chloride and oxygen.

Exchange reactions often involve ions. These reactions can be driven to completion rather than being in equilibrium if a precipitate (ppt), a gas or water is a product of the reaction. All of these effectively remove a material from the reaction system so the reverse reaction cannot easily occur.

Oxidation-reduction reactions are very common.

Objectives:

1. You should be able to recognize various forms of oxidation-reduction reactions.

2. You should be able to distinguish between the materials that change and the processes by which they change.
3. You should be able to explain the differences in the chemistry exhibited by atoms and ions.

A very important class of chemical reactions fits the category of oxidation-reduction. These reactions include such things as battery reactions, burning, metabolism, and corrosion. Oxidation-reduction reactions (redox) will also include reactions from the types discussed above.

Oxidation-reduction terminology can be confusing, so take your time and make sure you have it straight in your mind. There are two important aspects to keep in mind - there are **things**, oxidizing and reducing agents and **processes**, oxidation and reduction.

Electron gain and loss is fundamental to redox, however it may also be defined in terms of gain or loss of other items such as oxygen, halogens and hydrogen. Remember that the latter three are just convenient extensions of electron transfer.

Oxidizing agents gain electrons. Nonmetals, such as oxygen, are examples.When they react they combine with a reducing agent and undergo a reduction reaction. $S + 2e^- ===> S^{-2}$

Reducing agents lose electrons. Metals are good reducing agents. When they react, they combine with an oxidizing agent and undergo an oxidation reaction. $Mg ===> Mg^{+2} + 2e^-$

Oxidizing agents could be elements such as oxygen, ions such as Fe^{+2}, or polyatomic ions such as NO_3^{-1}

Using electrons, write a balanced equation for the following redox reactions.

(A) Lead(IV) oxide and carbon
(B) Zinc metal and copper(II) ion (from dissolving $CuSO_4$)

Redox can be defined in terms of particles other than electrons transferred however electrons will ultimately be involved.

We need to be familiar with some basic definitions:

Reagents (things)

Reducing agents lose electrons (become more positive in charge). They often are metals.
Oxidizing agents gain electrons (become more negative in charge). They often are nonmetals.

Reactions (processes)

Reduction reactions happen to oxidizing agents. They involve the gain of electrons by the oxidizing agent.
Oxidation reactions happen to reducing agents. They involve the loss of electrons by the reducing agent.

Stated another way the reducing agent gets oxidized and the oxidizing agent is reduced.

(A) $PbO_2 + C$ ===>

$Pb^{+4}O_2^{-4} + C^0$ ===> $Pb^0 + C^{+4}O_2^{-4}$

$Pb^{+4} + 4\ e^{-1}$ ===> Pb^0

Reduction reaction with the lead ion being the oxidizing agent.

C^0 ===> $C^{+4} + 4\ e^{-1}$

Oxidation reaction with the carbon being the reducing agent.

(B)	$Zn_{(s)}^{0} + Cu_{(aq)}^{+2}$ ===> $Cu_{(s)}^{0} + Zn_{(aq)}^{+2}$
	Zn^{0} ===> Zn^{+2} 2 e^{-1} oxidation reaction
	$Cu_{(aq)}^{+2}$ $2e^{-1}$ ===> $Cu_{(s)}^{0}$ reduction reaction

Be careful and keep in mind that atoms and ions are two different things. A metal atom can be a reducing agent but the metal ion would be an oxidizing agent.

Ions of some of the transition metals can function as either reducing agents or oxidizing agents. Iron(II) ion is a good example:

Fe^{+2} ===> Fe^{+3} + e^{-1} here it is a reducing agent

Fe^{+2} +2 e^{-1} ===> Fe^{0} in this case it is an oxidizing agent

The chemical environment is what will determine the way the material will react. You must know the chemical system before you can accurately predict how a particular transition metal ion may react.

Symbols found in Chapter 5

Page	Symbol	
149	(s)	As a subscript this means a solid.
149	(g)	As a subscript this means a gas.
149	(1)	As a subscript this means a liquid.
151	M_xO_y	General formula for metal oxides.
154	(aq)	As a subscript this means a water solution.
156	⇔	Double arrow means an equilibrium system.
160	H_3O^+	The hydronium ion, more correct than H^+. Used to indicate an acid is present.
161	OH^-	Hydroxide ion, indicates a base is present.
182	e^-	Electron involved in redox reaction.

Terms in Chapter 5

The following terms are listed for your convenience. They should not become a list for you to memorize. You should have a working definition for each. If you have doubt about any particular term, check its definition in the glossary and/or review it in the chapter. Some of these terms are included in the crossword puzzle later in the chapter.

Reactant	Product	Stoichiometry
Oxides	Combustion	Dissociate
Ion pairs	Aqueous solution	Electrolytes
Strong	Weak	Ionize
Equilibrium	Nonelectrolyte	Neutralize
Acid	Hydronium ion	Base
Basic oxide	Acidic oxide	Acid rain
Spectator ions	Net ionic equation	Precipitate
Salt	Oxidation	Reduction
Oxidizing agent	Reducing agent	Alloy
Corrosion		

Concept Maps for Chemical Compounds

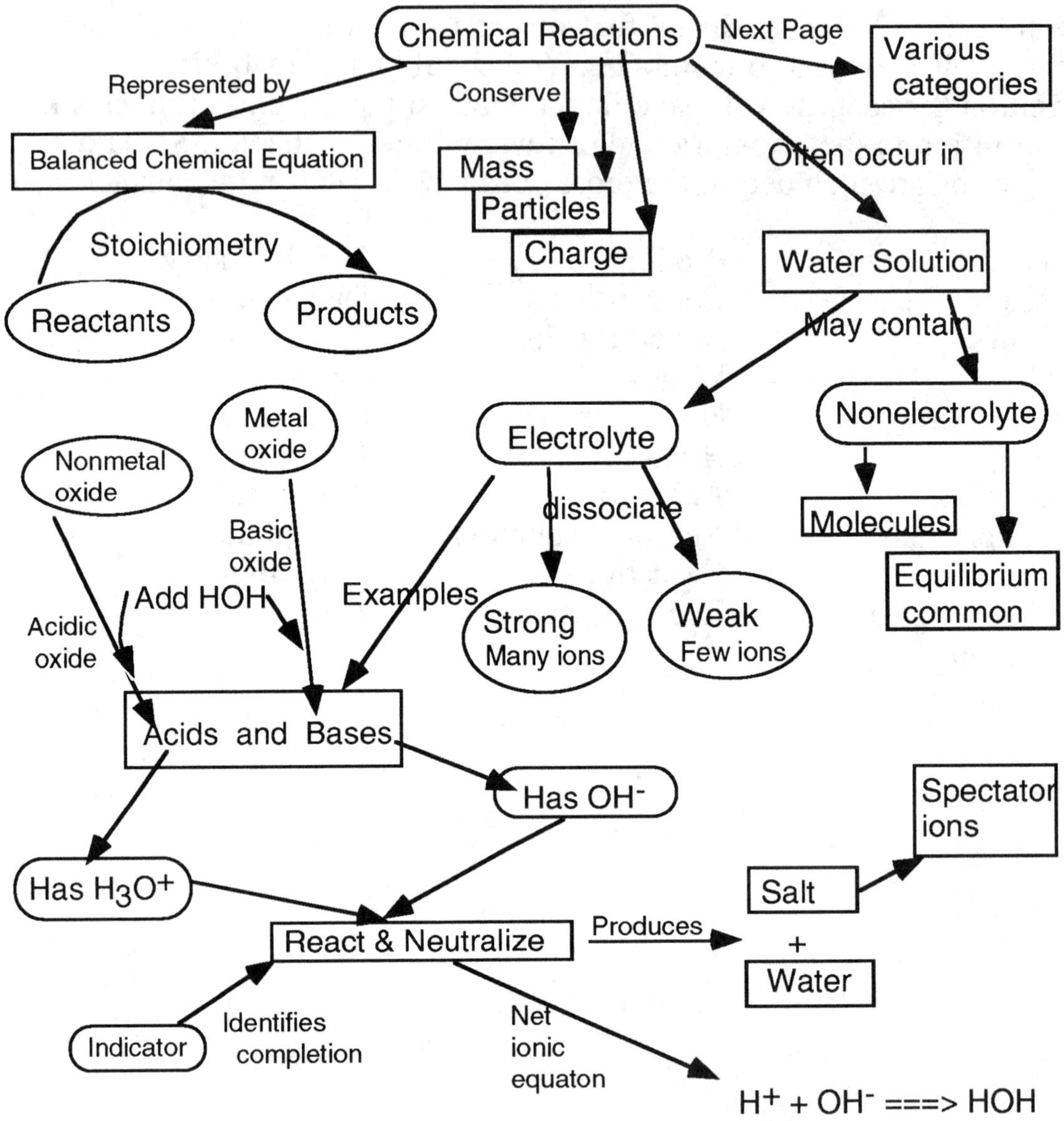

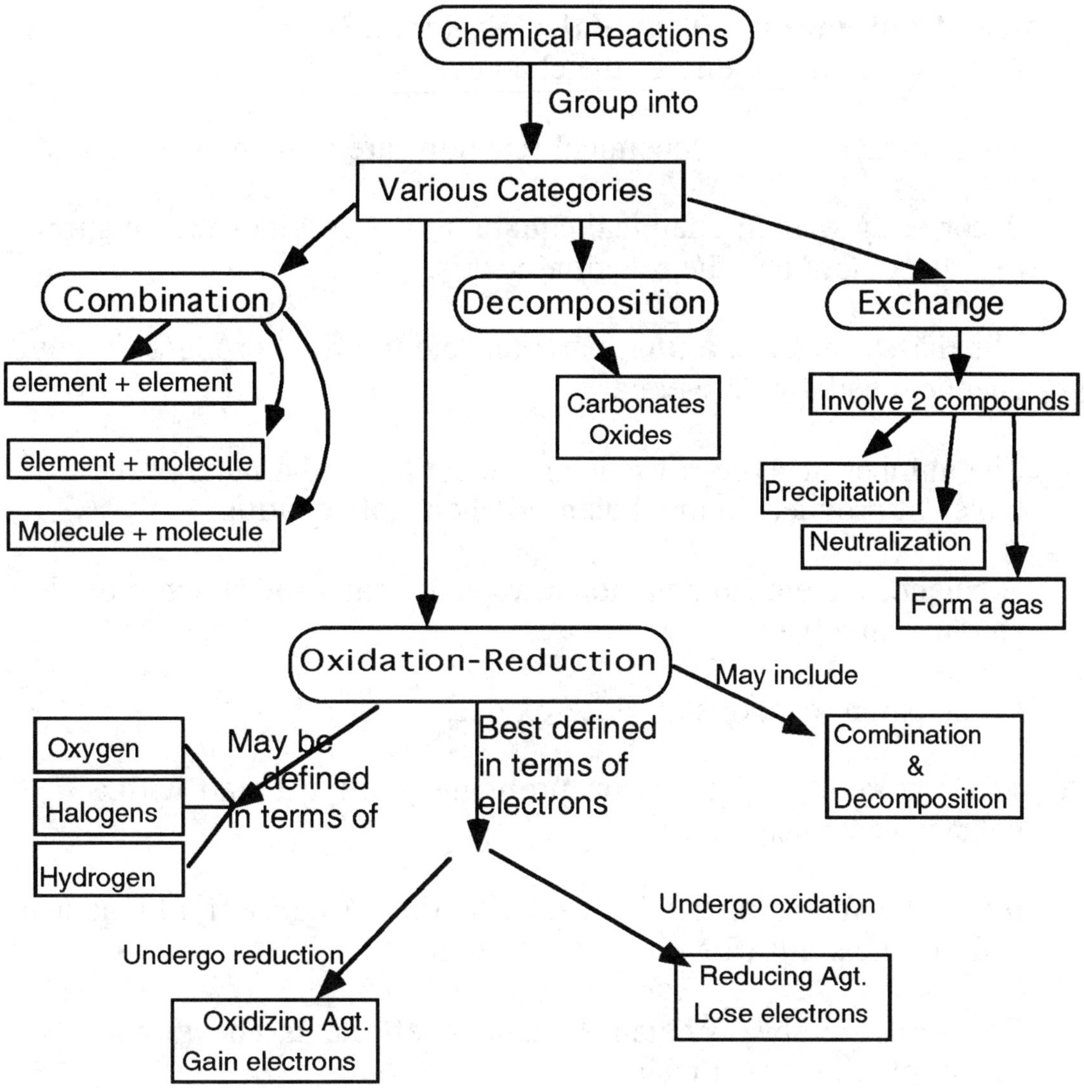
Chemical Reactions
Group into
Various Categories
Combination
Decomposition
Exchange
element + element
element + molecule
Molecule + molecule
Carbonates
Oxides
Involve 2 compounds
Precipitation
Neutralization
Form a gas
Oxidation-Reduction
May include
Combination
&
Decomposition
Oxygen
Halogens
Hydrogen
May be defined in terms of
Best defined in terms of electrons
Undergo reduction
Undergo oxidation
Oxidizing Agt.
Gain electrons
Reducing Agt.
Lose electrons

Chemical Compounds True/False Statements

Answers are at the end of the chapter.

1. Starting materials for chemical reactions are known as products.

2. A correctly written chemical equation also provides information regarding how fast the reaction occurs.

3. The masses of the starting materials and the final products in any chemical equation are equal.

4. The number of moles of reactants must equal the number of moles of products in any balanced chemical equation.

5. A balanced chemical equation always indicates the stoichiometric amounts involved.

6. Combustion reactions produce oxides.

7. When working stoichiometry problems you must start with a balanced chemical equation.

8. Before balancing a chemical equation the formulas of all reactants and products must be correctly written.

9. Equations are always balanced using coefficients - never with the subscripts in the formulas.

10. A strong electrolyte has many molecules but few ions present in the solution.

11. A strong electrolyte dissociates completely.

12. Chemical reactions that do not go to completion are said to be at equilibrium.

13. Nearly all compounds made with Group 1A elements are soluble.

14. Neutralization reactions are exchange reactions.

15. Acids are more likely to react with metals that bases are.

16. Strong acids when placed in water will produce very few H_3O^+ions.

17. A slightly soluble strong base will not have many OH^- ions in solution.

18. Oxides of nonmetals when placed in water will produce a basic solution.

19. All ionic reactions must conserve charge.

20. Ions which do not participate in the chemical reaction are known as spectator ions.

21. In a balanced ionic reaction the number of cations must equal the number of anions present.

22. Decomposition reactions may also be oxidation-reduction reactions.

23. Precipitation type reactions frequently occur in nature.

24. One way to produce a "salt" is to react an acid and a base.

25. Reducing agents gain electrons.

26. Formation of a ppt. (precipitate) tends to drive equilibrium reactions to completion.

27. Metal ores are converted to the metal by reducing the ore

28. Oxidation is the gain of electrons.

29. The ion Fe^{+2} can act as both an oxidizing and reducing agent.

30. Oxidation-reduction reactions are defined only in terms of electron gain and loss.

Testing Your Understanding of Concepts in Chapter 5. Answers are at the end of the chapter.

1. Why is it not necessary to conserve moles in chemical reactions?

2. What essential information is given by a balanced chemical equation?

3. What will be produced in every combustion reaction? What products will be formed in the combustion of a hydrocarbon?

4. What is the distinguishing factor between a strong and a weak electrolyte?

5. Why is water such a good solvent for ionic substances?

6. Equilibrium is associated with weak electrolytes. Why do weak electrolytes exhibit this characteristic?

7. Identify one material that nearly always forms soluble compounds and one material which has almost no soluble compounds.

8. Will any acid undergo neutralization, or just strong acids? Explain.

9. SO_2 contributes to some acid rain. What properties does this compound have that cause it to make rain acidic?

10. When you are told to write a net ionic equation for a given reaction what items appear in this equation?

11. Identify three different types of chemical reactions.

12. What is a salt, and what are two ways a salt could be formed?

13. What is the difference between oxidation and oxidizing agent?

14. Identify three ways that could be used to define oxidation-reduction reactions.

15. What is an extremely common and costly form of oxidation?

Testing You on Some Analytical Concepts.

Answers are at the end of the chapter.

1. Several things are conserved in chemical reactions.Depending on the type of reaction what must be conserved in each specific case?

2. You were given the job of preparing copper carbonate and told to use other compounds but no pure elements. How could this be accomplished?

Crossword - Chemical Reactions

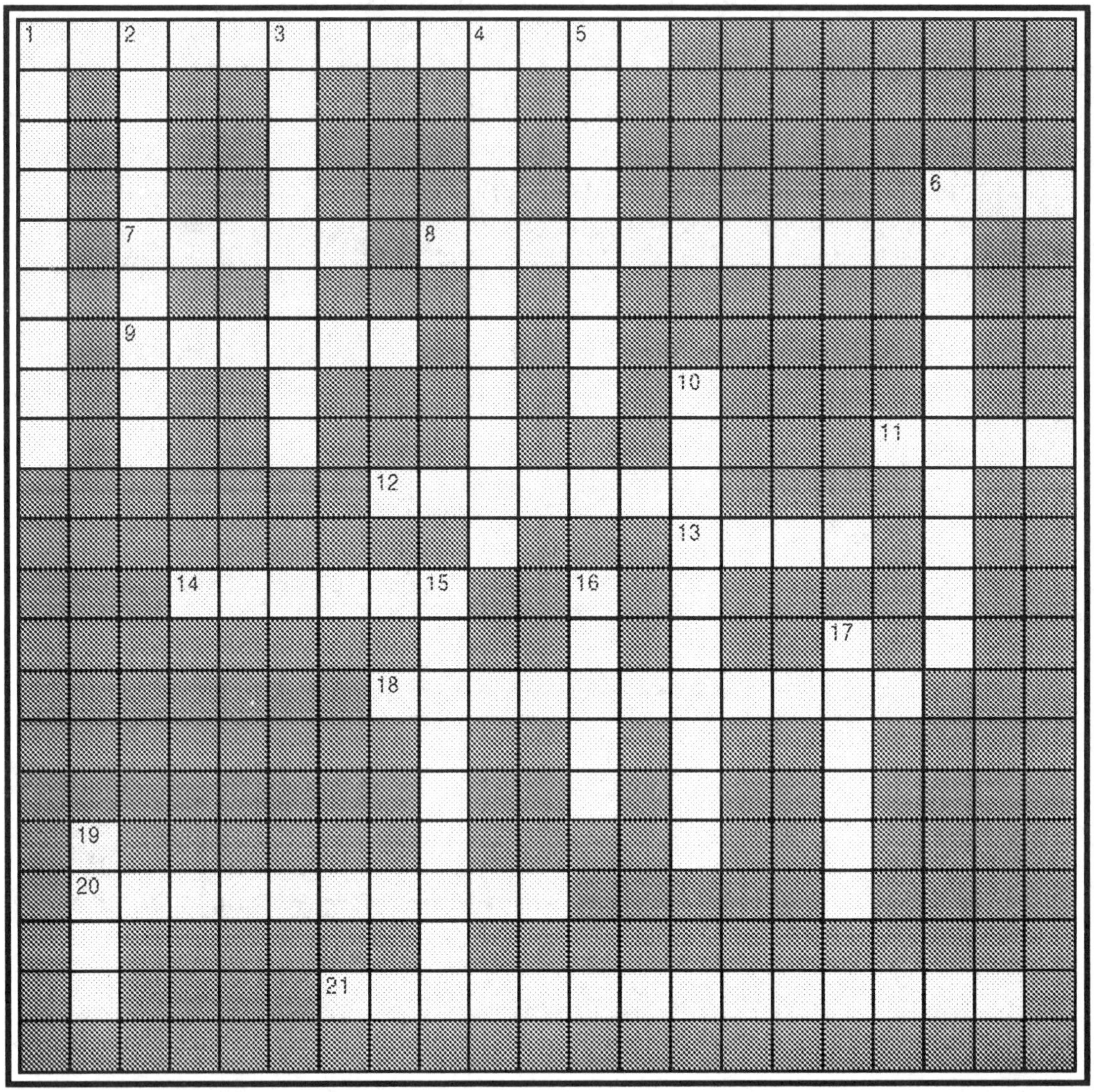

Crossword - Chemical Reactions Clues

ACROSS

1 Refers to quantities involved in reactions.
6 The kind of ionic equation that shows only ions used.
7 A mixture of metals.
8 Material that can conduct an electric current.
9 Happens to nonionic compounds whose molecules form ions in water.
11 Has a concentration of hydroxide ions.
12 Water solution of a material.
13 Will be formed when an acid and base react.
14 Water solutions of oxides of nonmetals are this.
18 Solid material formed as a result of a reaction.
20 Kind of reaction that results from burning.
21 Materials that do not ionize when dissolved in water.

DOWN

1 Ions that do not actively participate in a reaction.
2 This happens to reducing agents.
3 Name for the hydrogen ion in water.
4 Occurs when forward and reverse reactions happen at equal rates.
5 Material that loses electrons in a reaction.
6 Acid and base solutions do this to each other when reacted.
10 Ionic materials do this when put into water.
15 A common and expensive oxidation-reduction reaction.
16 Result of a material reacting with oxygen.
17 A material that forms many ions when put in solution.
19 Produces hydrogen or hydronium ions when put in water.

Ties to Previous Chapters

The basic ideas of elements, atoms compounds and molecules were presented in sections 2.3 and 2.4 of Chapter 2.

The periodic table is used in this chapter in several ways. You were first introduced to this organizational tool in Chapter 3 section 3.6.

This chapter is a direct extension of Chapter 4. If you have trouble with the idea of ions, review sections 4.3 and 4.4. Compound naming can be found in section 4.5 and formula information is in section 4.7.

Answers to True/False Statements:

1. False. The starting materials are the reactants.

2. False. The balanced equation gives indication of quantities only, nothing about reaction speeds

3. True. Mass is conserved in all chemical reactions.

4. False. Moles are not conserved in chemical reactions. Moles vary in mass.

5. True. This is why we are so careful in working only with balanced equations.

6. True. These usually are oxides of carbon and hydrogen which are carbon dioxide and water.

7. True. See # 5 above.

8. True. Once the formulas are written they cannot be changed in the process of balancing a chemical equation.

9. True. The coefficients are only used to make the formulas correct.

10. False. The reverse is true.

11. True. Completely means nearly 100%. There may be some interaction to produce ion pairs.

12. True. This is when a system has the reverse reaction occurring at a rate equal to the forward (forming products) reaction.

13. True. There are very few exceptions to this statement.

14. True. They have an exchange of the cations with the anions they were originally paired with. The outcome is formation of water.

15. True. Bases react with very few metals. Aluminum is an exception.

16. False. Strong means formation of many ions regardless of the type of material.

17. True. A material may be ionic but not dissociated in solution.

18. False. Metal oxides form bases. Nonmetal oxides cause a water solution to be acidic.

19. True. Charge is one of the entities which must be conserved in a chemical change.

20. True. These ions are essential to maintain charge balance but are ions at the start and are still present at the end.

21. False. The total charge must be the same but not the number of ions. Some ions may have a greater charge per particle than others so fewer would be needed.

22. True. Decomposition of a metal oxide would be a good example.

23. True. This is shown in many geologic situations.

24. True. This would produce a salt and water. There are other methods which will also produce a salt. They will be discussed in a later chapter.

25. False. Metals are reducing agents and they form positive ions by the loss of electrons.

26. True. Equilibrium reactions depend on availability of reactants. When a ppt. forms, reactants for the reverse reaction are removed from the solution and are not available to react.

27. True. The ore contains the desired metal in the form of a compound such as an oxide. The removal of the oxygen is reduction.

28. False. Oxidizing agents gain electrons but they are in turn reduced.

29. True. Fe^{+2} could lose 1 electron and act as a reducing agent or it could gain 2 electrons and be an oxidizing agent.

30. False. There are many ways to define this process. It could be in terms of oxygen, halogens or hydrogen as well as electrons.

Answers to Understanding of Concepts Questions:

1. Moles need not be conserved because they vary in mass. A large molecule could be split up by decomposition and produce smaller size molecules but the total number of atoms would still be the same and mass would still be conserved.

2. Every balanced chemical equation identifies the reactants and products. The theoretical amounts of each reactant and produced are known. The chemical makeup of all materials is given.

3. Every combustion reaction will produce an oxide. If the combustion involves a hydrocarbon both carbon dioxide and water (hydrogen oxide) will be formed.

4. Strong and weak electrolytes are distinguished by the number of ions which are formed upon solution. The more ions the stronger the electrolyte.

5. The answer to this goes back to chapter 4. Water has a shape that gives it ends that are slightly negative and positive. These ends can interact with the solute particle ions causing them to separate from the crystal and go into solution.

6. Weak electrolytes have strong interactions between the ions so only a few separate and those that do separate readily interact with each other to reform the molecule. Strong electrolytes on the other hand have only weak interactions between the free ions.

7. Nearly all compounds of sodium, or any of the 1A metals, are soluble. Sulfur, on the other hand, forms almost no soluble compounds. These are examples and others are also correct.

8. Any acid will undergo neutralization by reacting it with a base. The hydrogen ions in the acid are what react. Strong acids just have them more readily available.

9. Sulfur dioxide is a nonmetal oxide. When added to water, nonmetal oxides, produce an acid solution. These materials are acidic oxides. Sulfur dioxide is also soluble.

10. Net ionic equations list only those particles which actively participate in the reaction. Many reactions have particles present that only maintain electrical neutrality in the system and these are left out of net ionic equations. These equations will conserve mass and charge.

11. Chemical reactions are classified in several groups and examples are: oxidation-reduction (which may include some of those which follow), combination, decomposition, combustion and exchange which includes acid base, precipitation and gas formation.

12. A salt is an ionic material which combines a metal ion and a nonmetal ion. Reactions of acids and bases produce salts. The direct reaction of a metals and nonmetals as well as a reaction of metal oxides with nonmetal oxides will also produce salts.

13. Oxidizing agents are things. They gain electrons. Oxidation is a process,reaction type, and this happens to reducing agents.

14. The most comprehensive way to identify redox (oxidation-reduction) is by the gain and loss of electrons. However it can also be defined in terms of oxygen exchange, halogen exchange or hydrogen exchange all of which are very common in organic chemistry.

15. Corrosion is a very common, as well as expensive, oxidation reduction process.

Answers to Testing the Analytical Concepts

1. Every chemical equation must conserve particles and mass. Net ionic equations also conserve charge. Redox reactions must have the same number of electrons gained by oxidizing agents as were lost by reducing agents.

2. Since you want to produce $CuCO_3$ you should start with a material that contains copper ion and is soluble. You also need a soluble compound that contains carbonate. A check of solubility information would show that copper sulfate is soluble and any of the 1A elements have soluble carbonate compounds.
An example answer would be:

$$CuSO_{4(aq)} + Na_2CO_{3(aq)} \Longrightarrow CuCO_{3(s)} + Na_2SO_{4(aq)}$$

Answer to Chapter 5 Crossword

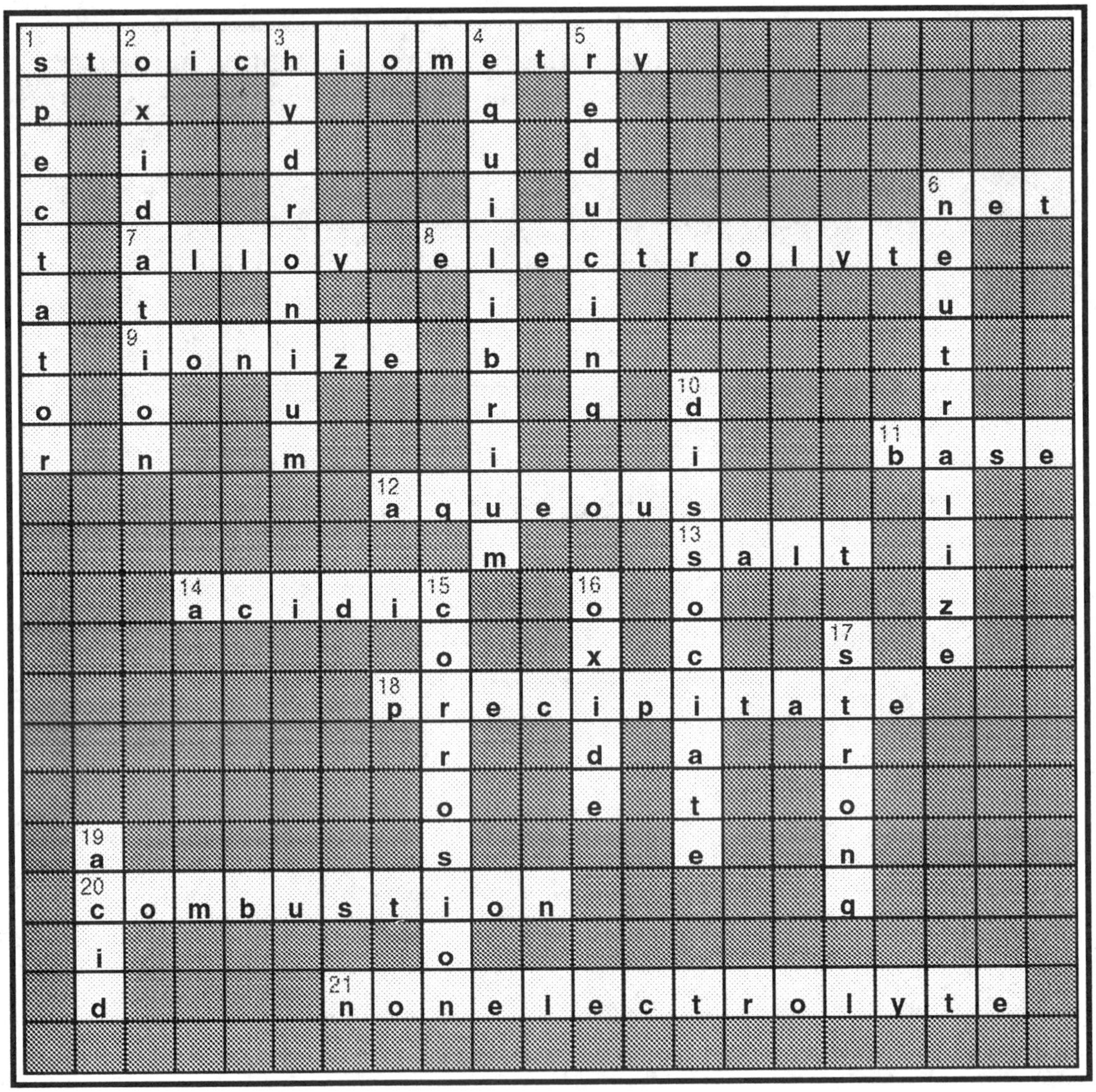

Some Supplementary Readings Relating to This Chapter:

1. Chemists Find Way to Make An "Impossible" Compound. New York Times, March 5, 1993.

2. Precipitated Calcium Carbonate Finds Growing Niche in Paper Market, Chemical And Engineering News, April 29, 1991, pages 9-11.

Chapter 6 Chemical Stoichiometry

The emphasis in this chapter is on quantities. The information learned in the last chapter regarding chemical reactions will be used to determine the amounts used in reactions, the amount of product that should be obtained as well as the yields of reactions

This chapter will also stress the importance of solutions in chemical systems. It will also introduce the concept of solution concentration known as molarity.

Focus Your Study Efforts

Stoichiometry is the study of quantities in chemical reactions.

Objectives:

1. You should be able to define stoichiometery.
2. You should be able to set up and solve stoichiometery problems.
3. You should be able to convert form grams to moles and moles to grams.
4. You should be able to relate the coefficients in chemical equations to their significance as a quantity.

A major effort is made by chemists to determine how much product can be obtained from a given amount of reactants. Along with this is an attempt to get maximum use from expensive reactants by using less expensive materials in excess. To accomplish this the exact quantities, known as stoichiometric amounts, must be determined.

All calculations involving amounts must start with a correctly balanced chemical equation. When working problems do this first.

When working with chemical systems there are two sets of values that need be considered. The amount which chemist's can visualize and measure - grams, and the amounts that are actually used by the chemical system - molecules, ions, or atoms. The factor that relates these two sets of units is the mole.

Solving stoichiometric problems can be divided into a set of sequential steps as outlined below:

- Write the balanced chemical equation. Formulas are made correct with subscripts and the atom conservation between reactants and products is done with coefficients.
- Make use of dimensional analysis to set up the ratios used in the following steps. Dimensional analysis is an excellent aid to assure that you set the problem up correctly.
- Convert chemist's units (grams) to chemical amounts (moles). This is done by dividing the mass involved by the molar mass of the material.
- Relate the moles of the materials involved in the reaction to each other by involving the reaction coefficients. This is why the balanced equation is critical.
- Convert the chemical unit of moles back to the gram unit which the chemist can measure. This is done by multiplying the moles of material by the grams per mole.

At the actual reaction level it is particles that react and anything you can do to bring particles into contact will enhance the speed of the reaction. Remember, that moles are a measure of particles, and a mole of any material has the same number of particles present.

The coefficients in a balanced chemical equation represent the theoretical amounts required for the reaction to occur. This will use up all of the reactants and produce a quantity of product that conserves mass.

If 1.00 mole of copper metal is reacted with a stoichiometric amount of silver nitrate how many moles of silver metal will be produced?

$$Cu_{(s)} + 2\ AgNO_{3(aq)} \Longrightarrow Cu(NO_3)_{2(aq)} + 2\ Ag_{(s)}$$

The stoichiometric amounts from the balanced equation show that two moles of silver nitrate will be required for the reaction and that 2 moles of silver metal will be produced.

If 6.36 grams of copper were used in the above reaction, how many grams of silver metal would be formed?

First determine the moles of copper available:

$$\text{moles Cu} = 6.36\ \text{g} / 63.6\ \text{g mol}^{1-} = 0.10\ \text{moles}$$

Determine the moles of silver this would produce:

$$\text{moles Ag} = 0.10\ \text{mol Cu} \times \frac{2\ \text{mol Ag}}{1\ \text{mol Cu}} = 0.20\ \text{mol Ag}$$

Convert to grams of Ag:

$$\text{g Ag} = 0.20\ \text{mol Ag} \times \frac{107.87\ \text{g Ag}}{\text{mol Ag}} = 21.57\text{g Ag}$$

This problem could be solved in one single set up using dimensional analysis. We need to know grams of Ag, we have the conversion factors to convert grams of Cu to moles of Cu and moles of Ag to grams of Ag. The starting mass of Cu is known. The equation gives us the stoichiometric amounts.

$$\text{g Ag} = 6.36\ \cancel{\text{g Cu}} \times \frac{1\ \cancel{\text{mol Cu}}}{63.6\ \cancel{\text{g Cu}}} \times \frac{2\ \cancel{\text{mol Ag}}}{1\ \cancel{\text{mol Cu}}} \times \frac{107.87\ \text{g Ag}}{1\ \cancel{\text{mol Ag}}}$$

$$= 21.57\ \text{g Ag}$$

This same technique can be used to calculate any quantity in a reaction. How many grams of silver nitrate would have to be dissolved to complete the above reaction? How many grams of copper nitrate could be reclaimed from solution?

$$Cu_{(s)} + 2\ AgNO_{3(aq)} \Longrightarrow Cu(NO_3)_{2(aq)} + 2\ Ag_{(s)}$$

Solving these problems separately using dimensional analysis:

$$g\ AgNO_3 = 6.36\ \cancel{g\ Cu} \times \frac{1\ \cancel{mol\ Cu}}{63.6\ \cancel{g\ Cu}} \times \frac{2\ \cancel{mol\ AgNO_3}}{1\ \cancel{mol\ Cu}} \times$$

$$\frac{169.87\ \cancel{g\ AgNO_3}}{1\ \cancel{mol\ AgNO_3}} = 33.97\ g\ AgNO_3$$

$$g\ Cu(NO_3)_2 = 6.36\ \cancel{g\ Cu} \times \frac{1\ \cancel{mol\ Cu}}{63.6\ \cancel{g\ Cu}} \times \frac{1\ \cancel{mol\ Cu(NO_3)_2}}{1\ \cancel{mol\ Cu}} \times$$

$$\frac{187.6\ g\ Cu(NO_3)_2}{1\ \cancel{mol\ Cu(NO_3)_2}} = 18.8\ g\ Cu(NO_3)_2$$

Another advantage of dimensional analysis is that you can stop any place in the sequence and get a value. If we now wanted to know the moles of silver nitrate we could have terminated the problem at the end of the first line in the first problem above.

There is a relationship between limiting reagents and reaction yields.

Objectives:

1. You should be able to explain why a chemist may not always chose to use stoichiometric amounts.
2. You should be able to calculate the theoretical yield for a given reaction system.
3. You should be able to calculate the percent yield for a given

reaction.

4. You should be able to determine which of two reactants is the limiting reactant.

Sometimes chemists intentionally do not use the stoichiometric amounts when completing a chemical reaction. When this is done, the material which is present in smaller molar amount according to the balanced equation will control the extent of the reaction. This becomes a limiting reagent.

When 40.0 grams of sodium hyroxide is added to 30.0 grams of hydrochloric acid in solution what quantity of salt will be produced?

This problem is different in that the amounts for both reactants are given. It is necessary to determine if these are stoichiometric amounts or if one is a limiting reagent.

As always, we must start with a balanced chemical equation.

$$NaOH_{(s)} + HCl_{(aq)} \text{ ===> } NaCl_{(aq)} + HOH_{(l)}$$

The equation tells us that the stoichiometry is one mole to one mole for the reactants. From the given information we need to determine the moles of each present.

$$\text{Moles of NaOH} = \frac{\text{g NaOH}}{\text{mol wt NaOH}} = \frac{40.0 \text{ g NaOH}}{40.0 \text{ g mol}^{-1}} = 1.00 \text{ mol}$$

$$\text{Moles of HCl} = \frac{\text{g HCl}}{\text{mol wt HCl}} = \frac{30.0 \text{ g HCl}}{36.5 \text{ g mol}^{-1}} = 0.822 \text{ mol}$$

The limiting reagent is HCl so all calculations will be done based on the HCl and we can ignore the NaOH amounts.

$$\text{g NaCl} = 30.0\ \cancel{\text{g HCl}} \times \frac{1\ \cancel{\text{mol HCl}}}{36.5\ \cancel{\text{g HCl}}} \times \frac{1\ \cancel{\text{mol NaCl}}}{1\ \cancel{\text{mole HCl}}} \times \frac{58.5\ \text{g NaCl}}{1\ \cancel{\text{mol NaCl}}}$$
$$= 48.1\ \text{g NaCl}$$

The amount of product we expect to get from a reaction if the stoichiometric amounts are used is known as the theoretical yield.

The actual yield is seldom equal to the theoretical value for several reasons; such as incomplete reactions, loss or reactants or products in the lab procedures and purification, and the production of products from secondary reactions.

Percent yield is a means of determining how well a reaction works. If the actual yield is divided by the theoretical yield and multiplied by the conversion factor, 100%, the percent yield is determined.

The actual amount of NaCl obtained in the reaction from the problem above was 39.8 grams, what is the percent yield for the reaction?

From the above problem we know the theoretical yield is 48.1 g.

$$\text{Percent Yield} = \frac{\text{Actual Yield}}{\text{Theor. Yield}} \times 100\% = \frac{39.8\ \text{g}}{48.1\ \text{g}} \times 100\% = 82.7\%$$

Analytical data is critical to understanding chemical systems.

Objectives:

1. You should be able to discuss the importance of, and need for good analytical data.
2. You should be able to explain what combustion analysis is.

Purity of the reaction product is important. Much effort goes into product analysis to check on purity and prove that a given material is what the chemist thinks it is.

A student was given 1.500 grams of a sample known to be a mixture of KCl and $KClO_3$. When the mixture was heated there was a mass loss of 0.480 grams. What percent of the sample was $KClO_3$?

The first thing to do is write a balanced equation. We know that when metal chlorates are heated, a metal chloride and oxygen are produced.

$$2\ KClO_3 \Longrightarrow 2\ KCl + 3\ O_2$$

To solve the problem we need the mass of $KClO_3$ and mass of the sample which we know to be 1.500 g.

$$\%\ KClO_3 = \frac{\text{Wt } KClO_3}{\text{Wt. of Sample Used}} \times 100\%$$

Using dimensional analysis, the mass of oxygen produced and the stoichiometric coefficients from the equation, we can determine the amount of $KClO_3$ needed to produce the given weight of oxygen.

$$g\ KClO_3 = 0.480\ \cancel{g\ O_2} \times \frac{1\ \cancel{mol\ O_2}}{32.0\ \cancel{g\ O_2}} \times \frac{2\ \cancel{mol\ KClO_3}}{3\ \cancel{mol\ O_2}} \times$$

$$\frac{122.6\ \cancel{g\ KClO_3}}{1\ \cancel{mol\ KClO_3}} = 1.226\ g\ KClO_3$$

We now have both values needed to calculate the % $KClO_3$

$$\% \ KClO_3 = \frac{1.226 \text{ g } KClO_3}{1.500 \text{ g Sample}} \times 100\% = 81.7\%$$

Combustion analysis is a useful means of analyzing several compounds, especially hydrocarbons, and sometimes materials with nitrogen, sulfur or halogens present.

Much of today's analysis work is done using instrumentation.A whole field of chemistry known as instrumental analysis, has been developed to facilitate obtaining quality analytical values.

Using analysis data, the empirical and molecular formulas of materials can be determined. Refer back to chapter 4 in the text for a review of these calculations.

Many chemical reactions occur in solution.

Objectives:

1. You should be able to explain what solutions are.
2. You should be able to describe how to prepare a solution of a given molarity by solution dilution or adding solute to a solvent.
3. You should be able to describe what a standard solution is.
4. You should be able to describe and explain the process of titration.
5. You should be able to explain the difference between endpoint and equivalence point.

A very large percentage of chemical reactions are done in solution. A major solvent is water, which gives us the term aqueous solution. The material placed in the solvent is called the solute. The solvent is usually the material present in greater quantity.

Water is not the only solvent used in chemistry. Other common ones include hydrocarbons, alcohols and other organic liquids.

When working with solutions chemists must know how much material is available for reaction. The choice unit for expressing concentration is molarity.

Molarity, M, is the ratio of moles of solute per liters of solution. Note that it is liters of solution not a liters of solvent.

M = Mol solute/L solution.

It is important that you are able to recognize the way chemists express the concentration. One way is to place the solute formula in brackets and this is used to represent the concentration in moles per liter of solution. For example, $[AgNO_3]$ = 1.00M, refers to a silver nitrate solution of exactly one molar.

Solutions of known molarity are made in special flasks known as volumetric flasks. They have an unique shape and are designed to contain exactly a specific volume of solution at a given temperature. The temperature is usually 20°C and is labeled on the flask.

Solutions are made in the volumetric flask by placing the correct amount of solute in the flask and then adding solvent to the exact volume of the flask. See page 211 in the text.

A student needed 500.0 mL of 1.25 molar solution of the hydrate, barium chloride dihydrate. What mass of the solid is needed? What are the steps to actually preparing the solution?

You would need a 0.500 L volumetric flask and pure $BaCl_2 \cdot 2H_2O$. The compound has a molecular mass of 244.33 grams per mole.

1.25 moles is 244.33 g/mol x 1.25 mol = 305.41 grams.

One liter of 1.25 M solution would require that mass. However the student needs to prepare only 0.500 liter so only half the mass is required, or 152.7 grams.

The weighed amount of barium chloride is placed a small amount of water and dissolved. This is added to the volumetric flask and several small rinsings of the beaker are added to the flask. Water is then added to the flask to the volume line.

A student weighed out three 10.0 gram samples of NaOH and placed them in three volumetric flasks of 0.500 liter, 1.000 liter and 0.100 liter. Water was then added to the line, what solution molarity would the student have in each flask?

$$\text{Molarity} = M = \frac{\text{Moles of solute}}{\text{Liters of Solution}}$$

Since the mass of each sample is the same the number of moles is constant.

$$\text{Moles NaOH} = \frac{\text{Mass of NaOH}}{\text{Mol Wt of NaOH}} = \frac{10.0\text{ g}}{40.0\text{ g mol}^{-1}} = 0.100\text{ moles}$$

$$\text{Molarity} = \frac{0.100\text{ moles}}{0.500\text{L}} = 0.200\text{ M in 500 mL}$$

$$= \frac{0.100\text{ moles}}{1.000\text{ L}} = 0.100\text{ M in 1000 mL}$$

$$= \frac{0.100\text{ moles}}{0.100\text{L}} = 1.00\text{ M in 100 mL}$$

6.36 grams of copper metal was added to 0.300 M silver nitrate solution. What volume of the solution would be needed to react all of the copper?

Start with the balanced equation:

$$Cu_{(s)} + 2\ AgNO_{3(aq)} \Longrightarrow Cu(NO_3)_{2(aq)} + 2\ Ag_{(s)}$$

The equation tells us that one mole of copper requires 2 moles of silver nitrate. The problem tells us that each liter of solution used has 0.300 moles of $AgNO_3$.

Moles of copper = 6.36 g/ 63.6 g mol^{1-} = 0.10 moles

$$\text{Moles of } AgNO_3 \text{ required} = 0.10 \text{ mol Cu} \times \frac{2 \text{ mol } AgNO_3}{1 \text{ mol Cu}}$$

$$= 0.20 \text{ moles } AgNO_3$$

$$\text{Liters } AgNO_3 = 0.20 \text{ mol } AgNO_3 \times \frac{1.00 \text{ L } AgNO_3 \text{ solution}}{0.300 \text{ mol } AgNO_3}$$

= 0.667 L or 667 mL of the silver nitrate solution.

Once the solution is made and its concentration is known exactly, it is a standard solution. Other standard solutions can be made from any given standard solution by dilution or chemical reactions.

If a 1.0 M solution of Na_2SO_4 forms ions what will be the concentration of the sodium ion and sulfate ion? When the equation is written and examined the amounts can be readily identified.

$Na_2SO_{4(aq)}$	$\Longrightarrow$	$2\ Na^{+1}_{(aq)}$	+	$SO_{4(aq)}^{-2}$
1.0 mole		2.0 mole		1.0 mole

A common mistake made with this type of problem is to try and conserve moles. The equation shows us that we end up with three moles of product, but keep in mind those moles are smaller than the original one. The total quantity of material is still the same. This is analogous to cutting a pie into 6 pieces - the number of pieces is greater, but the amount of pie is still the same.

Another situation which often causes misunderstanding involves the dilution of solutions. What you must remember is that at the end the total number of moles must be the same as when you started. All you are doing is spreading the particles further apart. If you increase the volume you must in turn decrease the concentration of the material. The number of moles can be determined from the equation:

$$\text{Number of moles} = \frac{\text{Moles}}{\text{Liter}} \times \text{Liters}$$

If you dilute 0.5 liters of a 2.0 M solution to 2.5 liters, what will the final molarity be? $M_{conc} \times L_{conc} = M_{dilute} \times L_{dilute}$

$$2.0 \text{ M/L} \times 0.5 \text{ L} = ? \times 2.5 \text{ L}$$

$$M_{dilute} = 0.4 \text{ M/L}$$

The process of analysis using solutions involves the quantitative technique known as titration. The standard solution can be added from a buret which is a cylinder usually calibrated in 0.1 mL units.

When analysis involving liquids is done, very often, the reactants and products have no color and it is impossible to know when the reaction is complete. Materials called indicators are added in very small amount to identify when the reaction is complete. Indicators are materials that have intense colors that are different if there is an excess of one reactant or the other. This is illustrated on page 220 of the text.

The point at which a reaction is complete is called the equivalence point. The place where the indicator changes color is known as the endpoint. Every effort must be made to have the endpoint match the equivalence point or else there will be experimental error.

> Standardization of an unknown concentration KOH solution was accomplished by titrating 25.00 mL of 0.1125 M oxalic acid solution with 20.47 mL of the KOH to the indicator endpoint. What is the concentration of the KOH solution?
>
> Again start with a balanced equation:
>
> $$2\ KOH_{(aq)} + H_2C_2O_{4(aq)} \Longrightarrow K_2C_2O_{4(aq)} + 2HOH_{(l)}$$
>
> $$\text{Moles of acid} = 0.1125\ \frac{\text{mol}}{\text{L}} \times 0.02500\ \text{L} = 0.002813\ \text{moles}$$
>
> $$\text{Moles of KOH} = 0.002813\ \text{mol acid} \times \frac{2\ \text{mol KOH}}{1\ \text{mol acid}} = 0.005625$$
>
> $$\text{Molarity of KOH} = \frac{\text{moles KOH}}{0.02047\ \text{L KOH}} = 0.2748\ \text{M}$$

When solution concentrations are determined, a standard solution is made. A standard solution is any solution of known concentration.

The material used as the first standard is the primary standard. Any known solution made from this is also a standard solution but would be called a secondary standard.

A primary standard must have certain characteristics which include: it must be very pure, it must be readily available, it must have a reasonable cost, it must be stable and it must react with the materials to be analyzed.

Some materials that we would like to use as primary standards will not work because they are deliquescent, which means they pick up moisture from the air, making them impure. Strong bases such as NaOH and KOH are examples.

Symbols found in Chapter 6

Page	Symbol	
210	M	Molarity of a solution
211	[ ]	Way to express the concentration in moles per liter This will be used extensively in chapter 8

Terms in Chapter 6

The following terms are listed for your convenience. They should not become a list for you to memorize. You should have a working definition for each. If you have doubt about any particular term check its definition in the glossary and/or review it in the chapter. Some of these terms are included in the crossword puzzle later in the chapter.

Stoichiometry	Stoichiometric factor	Limiting reactant
Excess reactant	Theoretical yield	Actual yield
Percent yield	Volumetric flask	Solvent
Solute	Concentration	Molarity
Standard solution	Titration	Indicator
Buret	Equivalence point	Endpoint
Standardization	Primary standard	Deliquescent

Concept Maps for Chemical Compounds

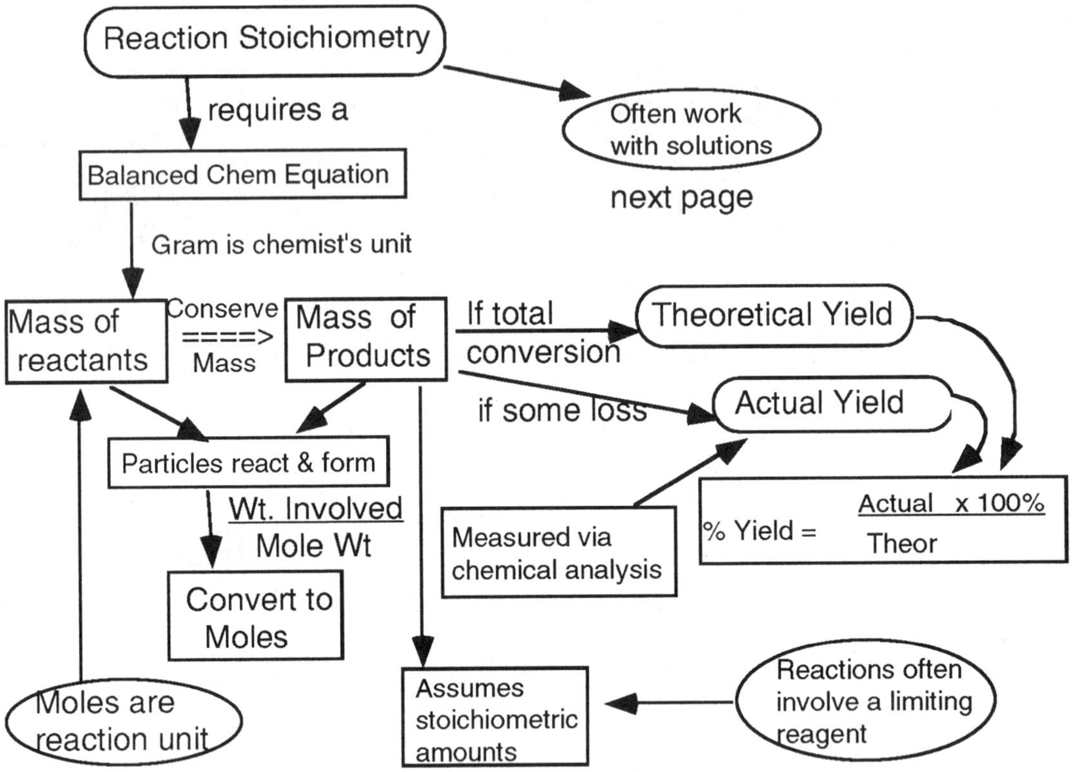

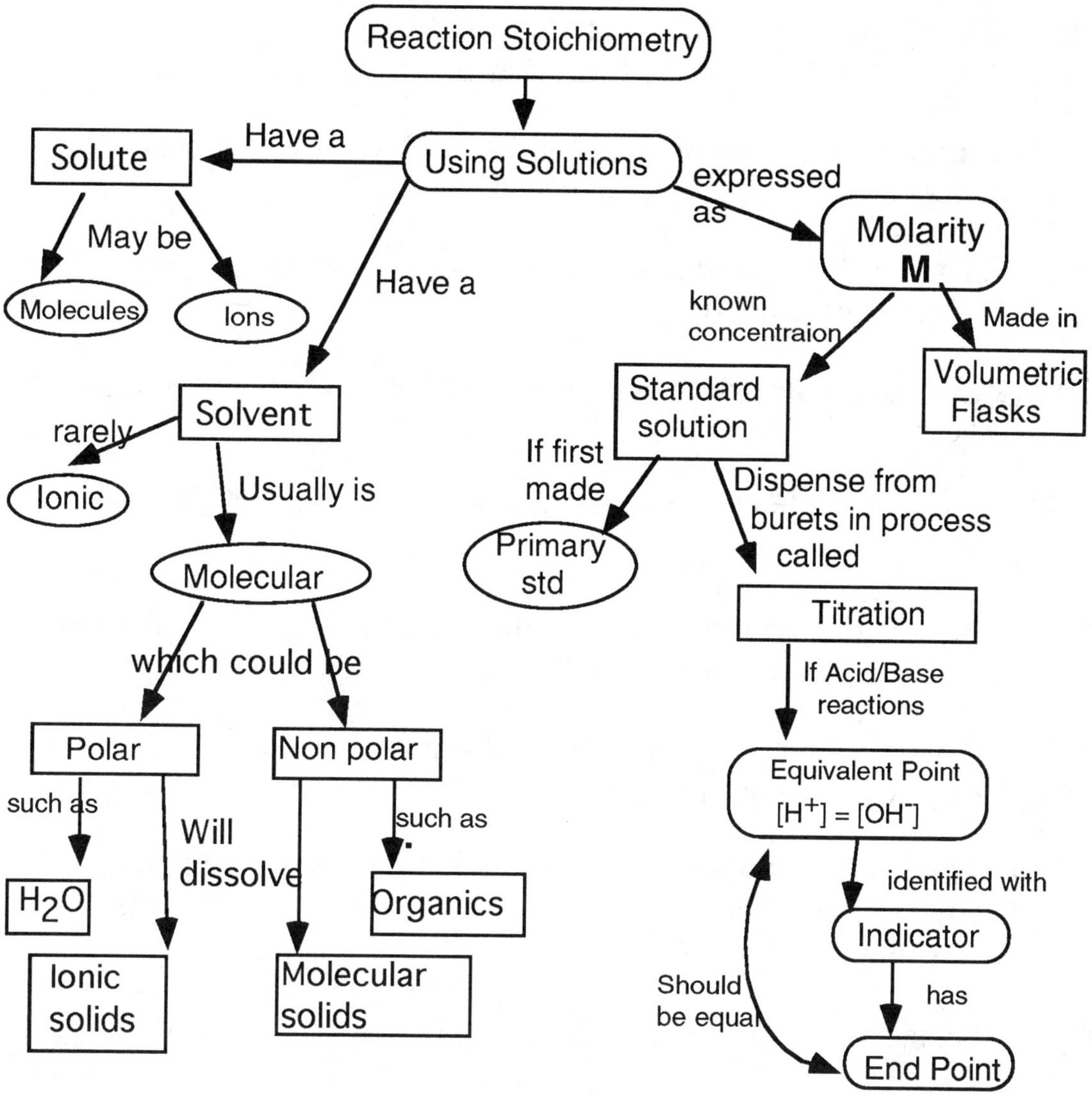
Reaction Stoichiometry
Using Solutions
Have a
Solute
May be
Molecules
Ions
Have a
Solvent
rarely
Ionic
Usually is
Molecular
which could be
Polar
such as
H_2O
Will dissolve
Ionic solids
Non polar
such as
Organics
Molecular solids
expressed as
Molarity
M
known concentraion
Made in
Volumetric Flasks
Standard solution
If first made
Primary std
Dispense from burets in process called
Titration
If Acid/Base reactions
Equivalent Point
$[H^+] = [OH^-]$
identified with
Indicator
has
End Point
Should be equal

Chemical Compounds True/False Statements

Answers are at the end of the chapter.

1. Every balanced chemical equation will have equal numbers of moles of reactants and products.

2. A balanced chemical equation is essential for calculating the correct stoichiometry.

3. If the stoichiometric amounts of reactants are <u>not</u> present, no reaction will occur.

4. Actual yields in chemical reactions seldom match theoretical yields.

5. The material present in lesser than the required molar amount is known as the limiting reagent.

6. Only one material can be the limiting reagent no matter how many reactants are present.

7. Combustion analysis is very common in analytical chemistry. This is true because both CO_2 and H_2O are easily and accurately determined.

8. Standard solutions are usually made in volumetric flasks.

9. Molarity is based on volume of solution rather than thevolume of solvent used.

10. A solution of known concentration is a standard solution.

11. When doing an acid-base titration, the chemist identifies the reactions completion, by using the indicator's end point.

12. Any standard solution is a primary standard.

13. Deliquescent materials can be used for standard solutions but not for primary standard solutions.

14. Titrations are a quantitative method of analysis.

15. Technically, volumetric flasks are accurate at only one temperature.

Testing Your Understanding of Concepts in Chapter 6.
Answers are at the end of the chapter.

1. What is the difference between a stoichiometric coefficient and a stoichiometric factor?

2. Why is the mole so important in stoichiometry problems?

3. Ten grams of compound A was added to ten grams of compound B. They were known to react in a one to one ratio.When the chemist checked the product there was a little of compound B left. Why didn't ten grams of one exactly react with ten grams of the other?

4. Why would a chemist ever intentionally use a limiting reagent rather than exact amounts?

5. Many, if not most, chemical reactions have an actual yield less than the theoretical yield. Why then, do chemists have an interest in the theoretical yield value?

6. We use analysis by combustion to determine the amount of carbon dioxide produced upon buring. We are actually interested in the amount of carbon that was present. How does knowing the

amount of CO_2 produced give any information about the amount of C present?

7. Why is volumetric glassware, such as a volumetric flask, used in making solutions having molarity concentration?

8. What is the difference between M (capital) and *M* (capital italic)?

9. What are standard solutions? Why are they used?

10. Describe the titration process.

11. What is the difference between the equivalence point and the endpoint?

12. Why are deliquescent compounds never used as primary standards, but are often used as secondary standards?

13. When working any chemical problem what is the one of the first things, if not the first thing, you should do?

Testing You on Some Analytical Concepts.
Answers are at the end of the chapter.

1. An oxide of iron, iron(III) oxide can be reduced with carbon monoxide. The following equation indicates what theoretical atom/molecule, mole and gram amounts?

$$3\ CO_{(g)} + Fe_2O_{3(s)} ===> 2\ Fe_{(s)} + 3\ CO_{2(g)}$$

2. Using the above reaction, if a sample of iron oxide weighing 85.0 grams was reduced, what amount of carbon monoxide would be required and what mass of iron would be produced?

3. When 66.0 grams of propane was burned in the presence of 150 grams of oxygen, how much carbon dioxide was added to the environment?

4. If the amount of iron produced in question 2 above was 56.0 grams, what was the percent yield from the process?

5. An organic compound known to contain only the elements of carbon, hydrogen and oxygen was subjected to analysis by combustion. If 0.200 grams of the sample was burned and 0.082 g of water and 0.300 g of CO_2 were produced what is the emperical formula?

6. 5.80 grams of sodium chloride was placed in a 0.250 liter volumetric flask and water was added to the mark. What is the molarity of this solution?

7. What amount of very pure $K_2Cr_2O_7$ is needed to make 2.00 liters of 1.25 M solution?

8. Potassium dichromate is a strong electrolyte. What is the $[K^{+1}]$ in the solution made in problem number 7?

9. What amount of solid magnesium hydroxide would be required to react with 75.0 mL of 0.125 M HCl solution?

10. How many grams of $ZnCl_2$ are present in 1.50 liters of 0.75 molar solution?

Crossword - Chemical Stoichiometry

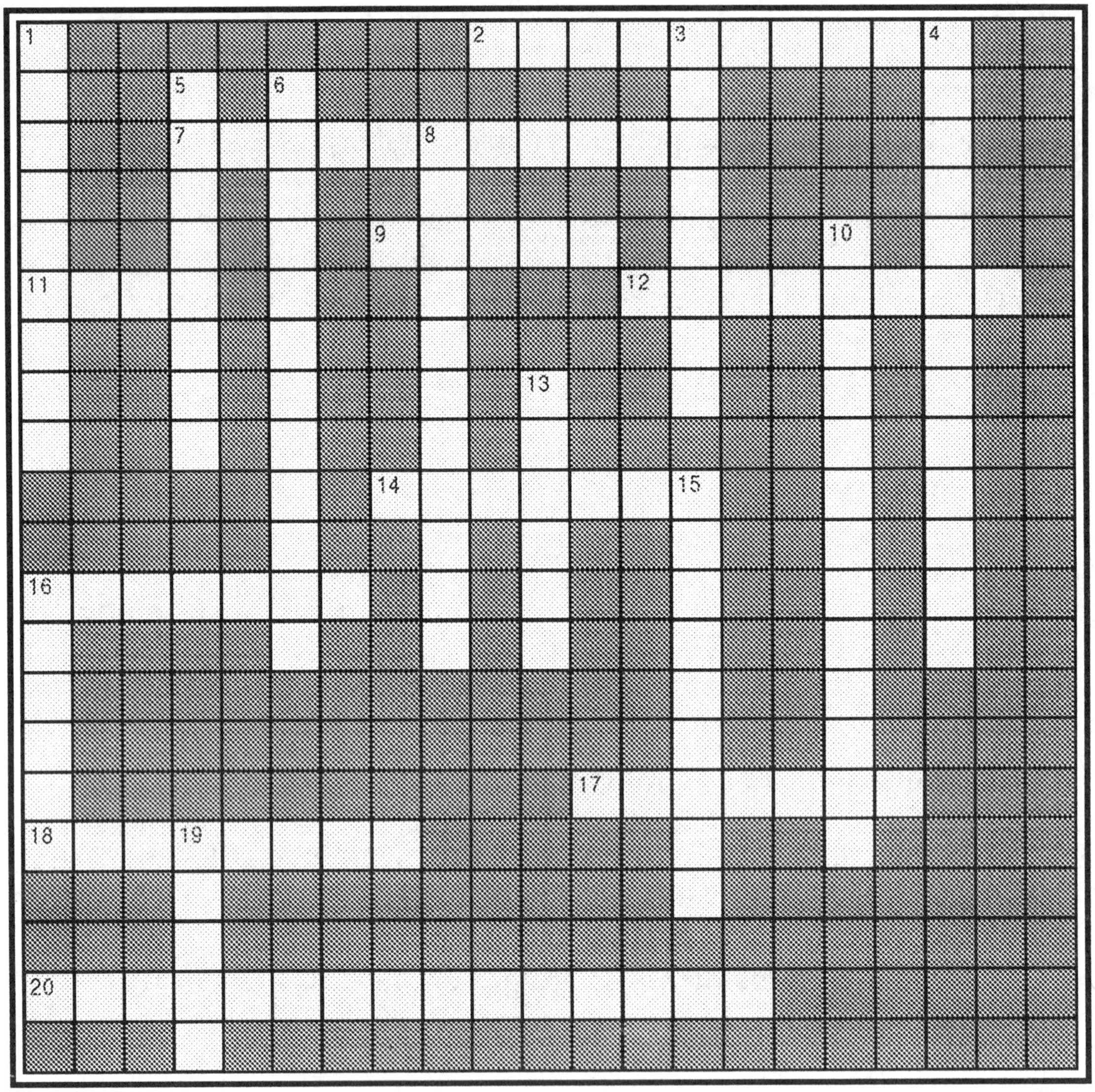

Crossword - Chemical Stoichiometry Clues

ACROSS

2 Type of flask used to make standard solutions.
7 Another name for the expected yield from a reaction.
9 Device used to measure liquid volumes with accuracy.
11 Will be neutralized by a base.
12 The material which, when used up, stops the chemical reaction.
14 A way chemists express the yield of a chemical reaction.
16 The major material in a solution. This is often water.
17 The initial, or first standard solution made is known as this.
18 Where an indicator will change color.
20 Process of determining the concentration of a solution.

DOWN

1 Undergoes a color change and identifies when a reaction is complete.
3 Concentration term favored by chemists and involves moles per liter of solution.
4 Term used to denote the amount of material in solution.
5 Solution of known concentration
6 Material which picks up moisture from the air.
8 This point should be at the endpoint.
10 The study of the quantitative relations between amounts of reactants and products.
13 Reactant which is not completely used up.
15 Process of analysis that uses volumetric measurement.
16 The material which is dissolved to make a solution.
19 Reaction is complete at the equivalence ---- .

Ties to Previous Chapters

This chapter brings several sections of the first year chemistry course together. It also is a chapter that contains concepts you will see used in several of the chapters to follow. This is a key chapter and mastering it will provide dividends in many to follow.

Two sections of chapter 3, 3.5 atomic weights and 3.6 the periodic table are referred to in several instances in this chapter. Chapter 4 is used in many ways, some of which you don't realize, until you give it some thought.You are expected to distinguish molecular from ionic compounds. Writing formulas and naming compounds are more and more left to you. Chapter 5 is used in most problems since the starting point for solving problems is a balanced chemical reaction and that is the essence of chapter 5.

Answers to True/False Statements:

1. False. Moles need not be conserved in chemical reactions. Moles are not all the same mass and it is mass which is conserved.

2. True. The stoichiometry is representative of the amounts exactly needed for the chemical reaction. This can only be known from the correct equation.

3. False. A reaction may occur regardless of what amounts of reactants are present. Unless the stoichiometric amounts are present,howevr, something will be left unreacted.

4. True. Most chemical reactions will have loss of reactant or product. This could occur because of incomplete reactions, loss of material in processing as well as other possibilities.

5. True. The material that does not provide the theoretical amount will be consumed first and when it is gone the reaction has nothing left to react with.

6. True. When any one material is consumed the reaction will stop. Technically it would be possible that all reactants are in stoichiometric amounts and all consumed at the same time.

7. True. Fuels can easily be tested this way.

8. True. If the solutions are based on molarity they are made in volumetric flasks because it is the solution volume which is critical.

9. True. Molarity for example, is moles per liter of solution.

10. True. This is how we can define a standard solution.

11. True. The indicator changes color at its end point. The endpoint should be the same as the equivalence point. This, however, is seldom possible so an effort is made to make the two as close as possible.

12. False. In a sequence of standardizations the first one is the primary standard and others are determined from it so they are secondary and so forth.

13. True. These materials can be standardized by interacting with the primary standard. Once made, the deliquescent factor is not of concern.

14. True. This involves careful measurement of known concentration solutions from a buret.

15. True. These have been calibrated at one temperature which is normally 20°C.

Answers to Understanding of Concepts Questions:

1. The stoichiometric coefficient is the theoretical molar amount involved in the reaction for a given material. A stoichiometric factor is the relationship between two materials in the reaction.

2. Reactions occur because particles react. These particles have different mass values. The mole represents a fixed number of particles.The mole becomes the conversion factor between the mass which is weighed out by the chemist and the particles which actually react.

3. Particles, not grams, react. The same mass of the two different substances did not give the same number of particles since they did not have the same molar mass. B was in excess, or A was the limiting reagent.

4. If a reagent was expensive or difficult to make you would want to be sure all of it was used. One way to help assure that is to use more of the other reactant than is needed.

5. The theoretical yield enables the chemist to predict how much could or should be obtained. By comparing actual with theoretical yields the chemist can adjust the next set of reactions to account for the loss. Knowing the theoretical value also provides a way to check the efficiency of the reaction.

6. The only way the CO_2 can be produced is from the carbon present. Also the percent of C in CO_2 is fixed and if we know the amount of carbon dioxide we can calculate the amount of carbon that was available since 12/44 of CO_2 is C.

7. Concentrations involving molarity are based on having a specific volume of solution, such as one liter. The volumetric flask is designed to allow easy and exact determination of that volume.

8. M (capital) refers to solution molarity and *M* refers to moles of material.

9. A standard solution is any solution whose concentration is known with accuracy. It is used as an analytical tool. When reacted with another material, the reaction enables us to determine the amount of the other item present.

10. Titration is the process of reacting a standard solution with a fixed quantity of another material. The titrant is added from a buret which is a calibrated glass tube which can give volumes to the nearest 0.01 mL. The equivalence point of the reaction is often identified by using an indicator.

11. The equivalence point is where the reaction is exactly complete. The endpoint is where the indicator changes color. It is important that these two points be very close to each other to minimize error.

12. A deliquescent material is one that will pick up moisture from the air becoming impure and changing mass. Primary standards must be very pure and the amount known exactly. However the secondary standard concentration can be determined after it is made so if it picks up moisture it makes no difference.

13. You should write a balanced chemical equation.

Answers to Testing the Analytical Concepts

1. Mole weights:

$CO = 28.0$, $Fe_2O_3 = 159.2$, $Fe = 55.8$, $CO_2 = 44.0$ g/mol

$3\ CO_{(g)}$ +	$Fe_2O_{3(s)}$	===>	$2\ Fe_{(s)}$ +	$3\ CO_{2(g)}$
3 molcules	1 molecule		2 atoms	3 molecules
3 moles	1 mole		2 moles	3 moles
84.0 grams	159.6 grams		111.6 grams	132.0 grams

2. $$g\ CO = 85.0\ g\ Fe_2O_3 \times \frac{1\ mole\ Fe_2O_3}{159.6\ g\ Fe_2O_3} \times \frac{3\ moles\ CO}{1\ mole\ Fe_2O_3} \times \frac{28.0\ g\ CO}{1\ mole\ CO} = 44.7\ g\ CO$$

$$g\ Fe = 85.0\ g\ Fe_2O_3 \times \frac{1\ mole\ Fe_2O_3}{159.6\ g\ Fe_2O_3} \times \frac{2\ mole\ Fe}{1\ mole\ Fe_2O_3} \times \frac{55.8\ g\ Fe}{1\ mole\ Fe} = 59.4\ g\ Fe$$

3. First we need a balanced equation and they we need to determine which is the limiting reagent since we have masses of both reactants.

$$C_3H_{8(g)} + 5\ O_{2(g)} ===> 3\ CO_{2(g)} + 4\ H_2O_{(g)}$$

Mole weight of propane is 44.0 g Mole weight of O_2 is 32.0 g

Moles propane = 66.0 g/44.0 g mol^{-1} = 1.50 mol

Moles O_2 required = 1.50 mol propane $\times \frac{5\ mol\ O_2}{1\ mol\ C_3H_8}$ = 7.5 mol

Moles O_2 available = 150.0 g / 32.0 g mol^{1-} = 4.69 mol
Oxygen is the limiting reagent

$$g\ CO_2 = 150.0g\ O_2 \times \frac{1\ mol\ O_2}{32.0\ g\ O_2} \times \frac{3\ mol\ CO_2}{5\ mol\ O_2} \times \frac{44.0\ g\ CO_2}{1\ mol\ CO_2}$$

$$= 123.8\ g\ CO_2$$

4. $$\%\ yield = \frac{Actual\ Amount}{Theoretical\ Amount} \times 100\% = \frac{56.0\ g}{59.4\ g} \times 100\%$$

$$= 94.3\ \%$$

5. We know the compound is CHO with the subscripts unknown for each element. We also know that in combustion one mole of CO_2 has one mole of C and one mole of H_2O has 2 moles of H. The total mass of sample consists of C,H and O. Once we determine the grams of C and H the grams of oxygen can be obtained by difference.

$$Moles\ of\ H = 0.082\ g\ H_2O \times \frac{1\ mol\ H_2O}{18.0\ g\ H_2O} \times \frac{2\ mol\ H}{1\ mol\ H_2O}$$

$$= 0.0091$$

$$which\ is\ 0.0091\ mol\ H \times \frac{1.008\ g\ H}{1\ mol\ H} = 0.0092\ g\ H$$

$$Moles\ of\ C = 0.300\ g\ CO_2 \times \frac{1\ mol\ CO_2}{44.0\ g\ CO_2} \times \frac{1\ mol\ C}{1\ mol\ CO_2}$$

$$= 0.0068$$

which is 0.0068 mol C x $\frac{12.001\text{g C}}{1\text{ mol C}}$ = 0.082 g C

grams of oxygen = 0.200 g material - 0.0092 g H - 0.082 g C
= 0.1088 g oxygen

moles of oxygen = 0.1088g / 16.0 g mol^{-1} = 0.0068 mol

These mole values can be used as subscripts

$C_{0.0068}H_{0.0091}O_{0.0068}$
divide each subscript by the smallest
$C_1H_{1.34}O_1$ which is equivalent to $C_3H_4O_3$

6. Molarity = $\frac{\text{moles of solute}}{\text{liters of solute}}$ Mole weight of NaCl = 58.5

Moles of NaCl = $\frac{\text{Wt. of Solute}}{\text{Mole Wt Solute}}$ = $\frac{5.80\text{ g NaCl}}{58.5\text{ g mol}^{-1}}$ = 0.0991 mol

M = $\frac{0.0991\text{ mol}}{0.250\text{ L}}$ = 0.396 M

7. We need to rearrange the molarity equation to determine moles of solute. We then need to convert the moles to grams.

Moles of $K_2Cr_2O_7$ = 1.25 M Liter^{-1} x 2.00 Liter = 2.50 moles

Grams of $K_2Cr_2O_7$ = Moles $K_2Cr_2O_7$ x Mole Wt. $K_2Cr_2O_7$

= 2.50 moles x 294.1 g mol^{-1} = 735.2 g

8. Since $K_2Cr_2O_7$ is a strong electrolyte it will be essentially 100% in the form of ions. Each mole of $K_2Cr_2O_7$ will yield 2 moles of K^{1+} ions.

There will be 5.00 moles of K^{+1} present in 1.25 liter
$[K^{+1}] = 4.00$

9. Balanced equation : $Mg(OH)_2 + 2\ HCl \Longrightarrow MgCl_2 + 2\ HOH$

Moles HCl present = 1.25 Mol L^{-1} x 0.075 L = 0.09375 mol

$$\text{g } Mg(OH)_2 = 0.094 \text{ mol HCl} \times \frac{1 \text{ mol } Mg(OH)_2}{2 \text{ mol HCl}} \times \frac{58.3 \text{ g } Mg(OH)_2}{1 \text{ mol } Mg(OH)_2}$$

$$= 2.74 \text{ g } Mg(OH)_2$$

10. We need to determine the total number of moles of $ZnCl_2$ present and convert that to grams of $ZnCl_2$.

$$\text{Moles } ZnCl_2 = \frac{0.75 \text{ mol}}{1.00 \text{ liters}} \times 1.50 \text{ liters} = 1.125 \text{ moles}$$

g $ZnCl_2$ = 1.125 moles x 136.4 g mol^{1-} = 153.5 grams

Answer to Chapter 6 Crossword

1 i									2 v	o	l	u	3 m	e	t	r	i	4 c		
n			5 s		6 d								o					o		
d			7 t	h	e	o	r	8 e	t	i	c	a	l					n		
i			a		l			q					a					c		
c			n		i		9 b	u	r	e	t		r			10 s		e		
11 a	c	i	d		q			i				12 l	i	m	i	t	i	n	g	
t			a		u			v					t			o		t		
o			r		e			a		13 e			y			i		r		
r			d		s			l		x						c		a		
					c		14 p	e	r	c	e	n	15 t			h		t		
					e			n		e			i			i		i		
16 s	o	l	v	e	n	t		c		s			t			o		o		
o					t			e		s			r			m		n		
l													a			e				
u													t			t				
t											17 p	r	i	m	a	r	y			
18 e	n	d	19 p	o	i	n	t						o			y				
			o										n							
			i																	
20 s	t	a	n	d	a	r	d	i	z	a	t	i	o	n						
			t																	

Chapter 7 Chemical Thermodynamics

The basic ideas developed in this chapter relate to energy changes in chemical systems. These may result in state changes or chemical changes of the materials involved.

Focus Your Study Efforts

Energy terminology.

Objectives:

1. You should be able to convert values from one set of energy units to another.
2. You should be able to identify various; classes, types and resources of energy.

This chapter introduces may new terms and symbols. Be sure you have them organized and defined. Several of these refer to the same concept; for example, energy can be expressed in joules or calories. The choice of units will make a big difference in the answer. It is important that you are able to distinguish what appear to be small, but are really, significant differences. Some examples are:

Energy

Classes	- Kinetic and Potential
Types	- Thermal, Chemical, Electrical, Solar, Nuclear and others
Resources	- Ultimate, Primary, Secondary, renewable and nonrenewable

Laws

Conservation of Energy
First Law of Thermodynamics
Hess's Law
Second Law of Thermodynamics

Energy Units

English - calorie and kilocalorie
Food (nutrition) - Calorie (note capital C spelling)
SI - joule and kilojoule

A given reaction releases 473.0 calories of energy. Convert this to units of kilocalories, Calories, joules and kilojoules.

This is a conversion problem and we need conversion units that can be used for dimensional analysis solutions.

kilo means 1000 of
the Calorie is equivalent to the kilocalorie
a calorie is equivalent to 4.184 joules

$$\text{Kcal} = 473.0 \text{ cal} \times \frac{1 \text{ kcal}}{1000 \text{ cal}} = 0.4730 \text{ kcal}$$

Calories = kilocalories = 0.4730 C

$$\text{joules} = 473.0 \text{ cal} \times \frac{4.184 \text{ joule}}{1 \text{ cal}} = 1979 \text{ J}$$

$$\text{kJ} = 1979 \text{ joules} \times \frac{1 \text{ kJ}}{1000 \text{ J}} = 1.979 \text{ kJ}$$

Thermal energy will cause changes in chemical systems.

Objectives:

1. You should be able to distinguish between heat and temperature.
2. You should be able to explain the various changes that the addition of energy to a system can bring about.

Heat and temperature are often confused. The temperature of a material is not related to the quantity present - it measures the kinetic energy of the particles in that quantity. Heat is a means of transferring energy, which always occurs in the direction of the higher to lower temperature object.

Thermal energy content is related to several factors of which the most important are: quantity of material in moles, specific heat capacity of the material, and the temperature the material is at.

If you added 500 joules of energy to each: 100 g of liquid ammonia; and 100 g of concrete, which would have the greater temperature change? How much difference?

Analysis of the data indicates this is a specific heat problem. we need specific heat values for ammonia and concrete. $NH_{3(l)}$ = 4.70 $Jg^{-1}K^{-1}$ and concrete is 0.88 $Jg^{-1}K^{-1}$. Text table 7.1.

The ratio of the specific heat shows it takes 5.34 times as much energy to heat ammonia as concrete. The concrete will warm up more.

This answers both questions but actual values can be obtained.

$$\text{Sp Ht.} = \frac{\text{quantity of heat}}{\text{mass x } \Delta T}$$

$$\Delta T = \frac{\text{quantity of heat}}{\text{mass x Sp Ht}}$$

$$\Delta T_{ammonia} = \frac{\text{500 joules}}{100\text{ g x }4.70\text{ Jg}^{-1}\text{K}^{-1}} = 1.06\text{ K}$$

Concrete would be 1.06K x 5.34 = 5.68 K

A 100 g ball of copper metal was heated to 500°C and dropped into 200 g of liquid ammonia at -80°C. What was the final temperature of the two materials?

We know heat is transferred from the warmer object to the cooler one. We also know energy is conserved so heat lost by the copper = heat gained by the ammonia. The two materials will end up at the same temperature. $q_{copper} = -q_{ammonia}$

$(100g)(0.385Jg^{-1}K^{-1})(T_f-500°C)=-(200g)(4.70Jg^{-1}K^{-1})(T_f-80°C)$
$38.5\ Jg^{-1}K^{-1}(T_f-500°C) = -(940\ Jg^{-1}K^{-1})(T_f-80°C)$
$938.5\ T_f = -55950\ °C$
$T_f = -59.6°C$

Materials have properties, such as melting point and density, which are often called characteristic properties because they help identify one material from others. This chapter introduces several other characteristic properties including: heat capacity, heat of fusion, and enthalpy of formation. Identify others as you work through the chapter.

How much energy is required to convert 200g of water from ice at 0°C to steam at 100°C?

This is a triple problem. We have melting, a state change involving heat of fusion, water warming which is a specific heat problem and evaporation, a state change involving heat of vaporization.

Ice ___melts___ $\text{Liquid}_{0°C}$ ___warms___ $\text{Liquid}_{100°C}$ ___boils___ Steam
Ht of fusion | Sp Ht | Ht of vap.

Heat of fusion = 333 J/g Heat of vaporization = 2260 J/g
Sp Ht = 4.18 j/g°C

Note that the values involving state change are at constant temperature so there is no temperature factor involved.

$J = 200g \times 333J/g + 200\ g \times 4.18J/g°C \times 100°C + 200g \times 2260J/g$
$J = 66600J + 83600\ J + 452000J = 602200\ J$ or 602.2 kJ

Energy added to a system may result in a state change. The change to the vapor state is often much more energy demanding because the atomic/molecular forces must be completely overcome.

Be sure to identify and distinguish the system and surroundings with each situation you study.

Knowledge of thermodynamic symbolism is important.

Objectives:

1. You should be able to explain the symbolism used in the language of thermodynamics.
2. You should be able to use the first law of thermodynamics.
3. You should be able to select thermodynamic data to match the conditions of the reaction.

Usually thermodynamics measures a change, (Δ), in a value. Often the absolute amounts are difficult to measure. Changes, however, are relatively easy to quantify. It is the change that provides the driving force in the chemical system.

Remember that energy conservation is required in any system.

An energy change is in the heat content (q) and work done (w). Because volumes or pressure changes have an effect on the work and heat aspects of a system you will need to be aware if measurements are made with a constant volume (q_v) or constant pressure (q_p) condition.

Thermodynamic values will always have a sign associated with them. The sign indicates what is happening to heat, work, or energy, either in the system or surroundings. It is very important to know if it is the system or surroundings being changed and what the sign direction means for the system and surrounding change.

Negative ΔH values indicate a system is transferring energy from itself to the surroundings and as a result is becoming thermodynamically more stable.

> When iron rusts a reaction occurs as shown in the equation:
>
> $4\ Fe_{(s)} + 3\ O_{2(g)} \text{ ===> } 2\ Fe_2O_{3(S)}$ $\qquad \Delta H = -1648$ kJ
>
> Is this reaction exothermic or endothermic? How much energy change occurs when 10.0 g of iron rusts?
>
> When ΔH is (-) the reaction is exothermic.
>
> From the equation we see 4 moles of iron releases 1648kJ which is 412 kJ per mole.

$$\text{moles of Fe} = \frac{10.0\text{g Fe}}{55.8 \text{ g mol}^{-1}} = 0.179 \text{ mol}$$

$$\text{Energy} = 10.0\text{g Fe} \times \frac{1 \text{ mol Fe}}{55.8 \text{ g Fe}} \times \frac{412 \text{ kJ}}{1 \text{ mole Fe}} = 73.8 \text{ kJ}$$

Mole units are used in thermodynamic systems. Any coefficients you see in reactions are always referring to mole amounts not to atoms or molecules. Think in terms of the macroscopic system when working with thermodynamics.

Energy change values are dependent upon the physical state a system starts from. Always check the state of the reactants before using a table of values. Changes in water from ice involve significantly different energy amounts compared to starting with the liquid state.

Enthalpy changes are dependent upon the quantities reacting.

Objectives:

1. You should be able to apply Hess's Law to find the enthalpy change in a reaction.

Energy changes in chemical reactions are additive. The path of a reaction does not affect the net energy change involved. These are both expressions of the Law of Hess.

Finding the correct reactions to complete a Hess's Law problem may not be straight forward. At times, you will need to have perseverance. Take the reactions a step at a time and in sequence.

With Hess's Law problems some reactions may need to be reversed in order to make the materials cancel as needed. Remember to reverse the sign on the energy when the reaction is reversed.

Some reactions in Hess's Law problems will need a multiplier, either a whole number or factional, to make the system balance. Again remember this also affects the energy change.

How much energy is released in the conversion of $SO_{2(g)}$ to $SO_{3(g)}$?

We know the molar enthalpies of formation for the two gases. We also can rearrange the equations for the formation of the gases to give an equation for the conversion of sulfur dioxide to the trioxide.

$S + O_2 ===> SO_2 \qquad \Delta H = -296.8\ kJmol^{-1}$

$2S + 3O_2 ===> 2\ SO_3 \qquad \Delta H = -395.7\ kJmol^{-1}$

$2SO_2 =====> 2\ S + 2O_2 \qquad \Delta H = +593.6\ kJ$

$2S + 3O_2 ===> 2\ SO_3 \qquad \Delta H = -395.7\ kJ$

When the reaction was reversed the energy sign is reversed.

Net rxn $2\ SO_2 + O_2 ===> 2\ SO_3 \qquad \Delta H = -197.8\ kJ$

Calorimetry is used to measure energy changes.

Objectives:

1. You should be able to describe the connection between enthalpy and entropy changes for a reaction and the Gibbs free energy change.
2. You should be able to describe how one can measure the quantity of heat energy transferred in a reaction by using calorimetry
3. You should be able to use enthalpy and entropy changes to predict whether a reaction is product-favored.

Chemists need a fixed set of conditions to use for comparison purposes. This set of conditions is known as the standard state. This also includes a specific allotropic form for elements such as carbon or tin.

By definition, the ΔH_f of elements is 0. In the case of elements with allotropes only one form has a 0 value.

When thermodynamic terms, like ΔH, are written with a superscript °, such as, $\Delta H°$, you know you are working with a system at standard state conditions.

$\Delta H° = \sum (\Delta H_f \text{ Products}) - \sum (\Delta H_f \text{ Reactants})$ is useful when all heats of formation values are available for the balanced chemical equation. This same format is also valid for entropy and free energy calculations.

The bomb calorimeter has a constant volume but not a constant pressure. This means the energy change is a measure of ΔE not ΔH.

Exothermic reactions result in spreading energy over more bonds, more particles or more random states of matter or some combination of all of these.

The probability toward randomness is a reaction driving force. This is known as entropy, symbolized by S.

The Second Law of Thermodynamics is a statement that the natural tendency of the universe is to become more disordered, to increase in entropy.

Enthalpy changes are more important reaction predictors than entropy changes with two major exceptions. These exceptions are when ΔH is near 0 and/or when the reaction is at a high temperature.

The most useful thermodynamic equation is known as the Gibbs Free Energy equation which relates Enthalpy changes and Entropy changes in a single comprehensive relationship.

> When oxygen reacts with nitrogen monoxide at standard state conditions $\Delta H°$ is -114kJ and $\Delta S°$ is -146 JK^{-1}. What is the value of $\Delta G°$ for the formation of nitrogen dioxide?
>
> We know $\Delta G° = \Delta H° - T\Delta S°$ and T = 298.1K
> We must be careful because ΔH and ΔS have different units.
> We need a balanced equation:
>
> $$2\ NO_{(g)} + O_{2(g)} \Longrightarrow 2\ NO_{2(g)}$$
>
> $$\Delta G° = -114kJ - (298.1K)(-0.146kJ\ K^{-1})$$
> $$= -114kJ + 43.6kJ = -70.4\ kJ$$

The Gibbs Free Energy is a measure of the energy available to do work. It is also a measure of the material as an energy resource.

Notice this chapter did not discuss reactions in terms of how fast they occur, or how complete they may be, or even if they occur spontaneously. These factors are controlled by conditions other than those related to thermodynamics.

Symbols found in Chapter 7

Page	Symbol		
244	c	calorie	An amount of energy equivalent to 4.18 joules
246	C	Calorie	Food calorie Equivalent to 1 kilocalorie
245	J	joule	$kg{\bullet}m^2/s^2$ Energy unit which is preferred
245	kJ	kilojoule	1000 joules of energy
246	K	Kelvin	A temperature unit
247	Δ	change	Means a change in the quantity which follows

247	ΔT	Represents a change in temperature
247	q heat	Represents a quantity of heat
251	q_v	Heat at constant volume
254	q_p	Heat at constant pressure
253	ΔE	Change in energy at constant volume
253	w	Work
255	mol	The chemical mole amount
254	ΔH	Enthalpy change at constant pressure
263	$\Delta H°$	Enthalpy change at standard conditions
263	$\Delta H°_f$	Enthalpy of formation
266	$\sum$ sum	Take the sum of what follows
268	C_{bomb}	Heat capacity of a bomb calorimeter
275	$S°$	Entropy unit J/K•mol
274	ΔS q/T	Change in entropy
282	$\Delta G°$	Change in free energy

Terms in Chapter 7

The following terms are listed for your convenience. They should not become a list for you to memorize. You should have a working definition for each. If you have doubt about any particular term check its definition in the glossary and/or review it in the chapter. Some of these terms are included in the crossword puzzle later in the chapter.

Product-favored Reaction		Thermodynamics
Reactant-favored Reaction		Energy
Potential Energy	Kinetic Energy	Thermal Energy
Chemical Energy	Primary Energy Source	
Secondary Energy Source		calorie
Law of Conservation of Energy		kilocalorie
Calorie	joule	biomass
fossil fuel	Photovoltaic cell	Energy resource
Temperature	Heat	heat capacity
specific heat	Enthalpy change	heat of fusion

heat of vaporization
sublimation
System
Surroundings
Endothermic
Exothermic
First Law of Thermodynamics
Hess's Law
Standard State
Standard enthalpy of reaction
Standard molar enthalpy of formation
Bomb calorimeter
Calorimetry
Nitrogen fixation
Entropy
Second Law of Thermodynamics
Standard conditions
Gibbs free energy

Concept Maps for Thermodynamics

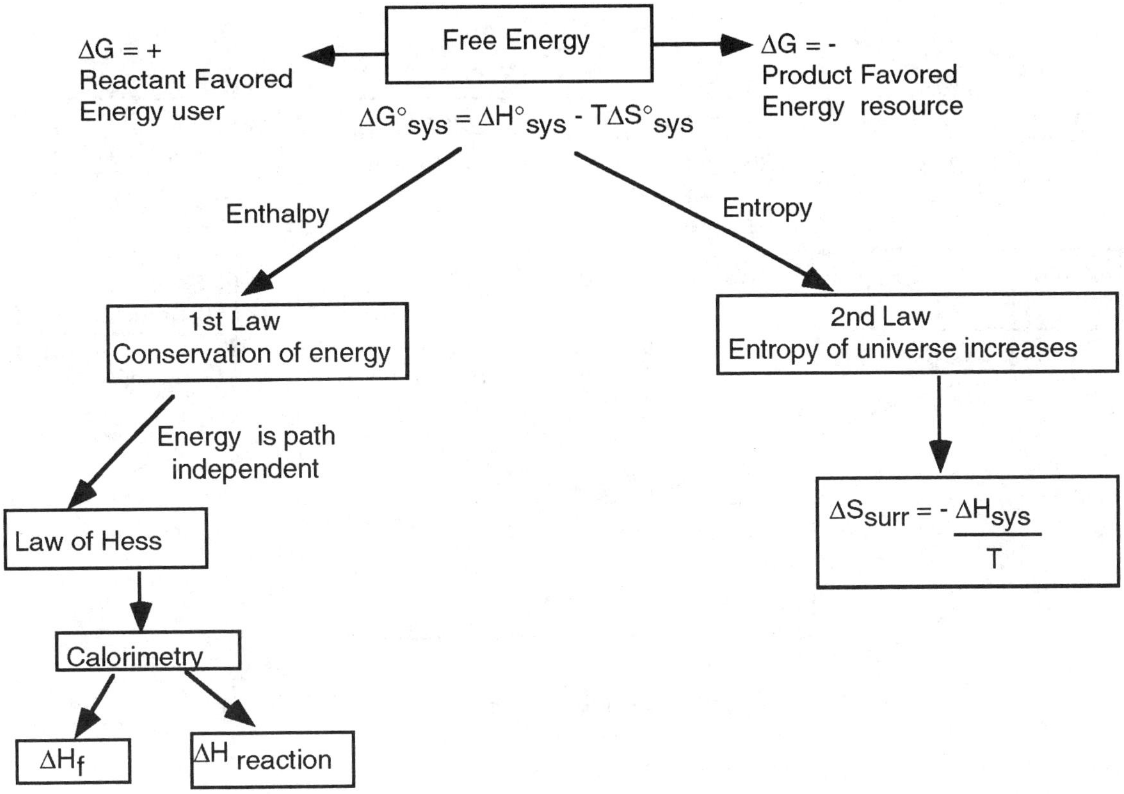

Energy Resources

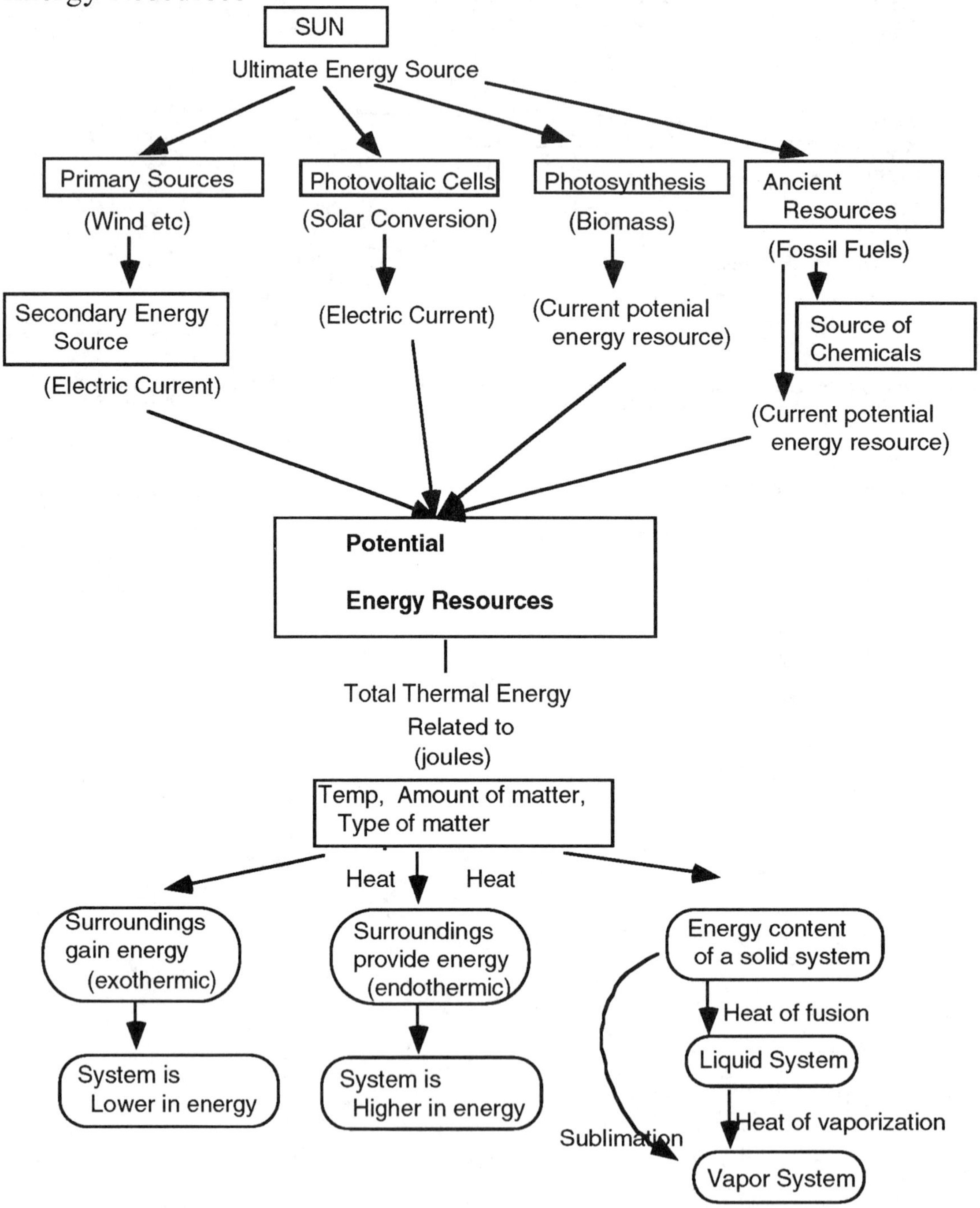

Thermodynamics True/False Statements

Answers are at the end of the chapter.

1. The reactant in smaller molar amount in relation to the chemical equation is known as the limiting reagent.

2. Heat content can be measured by a system's temperature.

3. Fossil fuels and biomass are the same material.

4. Energy resources have a (-) ΔG value.

5. One joule is equivalent to 4.18 calories.

6. Specific heat values vary from material to material and thus can be used as chemical identifiers.

7. Changes in values, such as enthalpy, rather than absolute values of the materials are what chemists usually measure.

8. Heat of fusion values are generally less than heat of vaporization values for any given material.

9. Exothermic systems are generally reactant favored systems.

10. ΔE and ΔH usually have the same numeric value.

11. ΔH values are measured at constant pressure.

12. The more work done by a system at constant ΔE the more heat produced.

13. The path a reaction takes, single step or multiple step, does not determine the overall energy change for a given system.

14. ΔH values for chemical systems are measured on a per mole basis.

15. When a reaction is reversed, the sign of the energy change remains the same.

16. ΔH°_f values are determined only when a reaction is done having elements as the starting materials.

17. Any allotrope of an element can be considered as a standard state condition.

18. The enthalpy of formation of elements is 0.

19. Reaction energies can be measured in a device called a calorimeter.

20. Increasing disorder and an exothermic energy change means a reaction will be product-favored.

21. The Second Law of Thermodynamics is a statement about conservation of energy.

22. Bomb calorimeters measure ΔE, not ΔH.

23. Entropy values become more significant in a system at higher temperatures.

24. All calorimeters will consume some heat from a reaction done in them and this value is the same for every calorimeter.

25. The tendency of the universe is to spontaneously increase organization.

Testing Your Understanding of Concepts in Chapter 7.
Answers are at the end of the chapter.

1. When chemical reactants are brought in contact the exact amounts are not always present. The material totally consumed in the reaction is known as the ____________.

2. If a chemist reacted two chemicals, one of which was inexpensive and the other of which was very expensive,what relative quantities would you recommend be used?

3. A reactant favored system can be driven to produce chemical products with the aid of continuous ________ __________. This type of reaction is energetically classified as __________.

4. Heat is a very common form of energy transfer in chemical systems but other means of energy transfer are common. Some alternative means could be _______ or ______ or ________.

5. The study of chemical science that deals with heat and work is known as _____________.

6. A secondary form of energy resource that is often produced is electrical energy. This form is preferred because it allows ________ to be done.

7. Frequently we hear talk about an "energy shortage". Explain why this is a false statement and what problem is really being referred to.

8. You have two containers of antifreeze solution both at 300K. One container has a volume of 1.0 liter and the other is 10.0 liters in volume. Do they contain the same amount of energy? Would energy be transferred as heat between them?

9. The primary energy resources to us on earth are all the result of the ultimate energy resource, which is the __________.

10. If, in a given system, the surroundings were to do work on the system under study the sign of w in the equation , $\Delta E = q + w$ would be _____ and the systems energy would have ________.

11. When expressing an enthalpy change (ΔH) an exothermic reaction would be identified with energy written as a ________ of the reaction, or by having ΔH expressed with a _________ sign.

12. A chemical change has a ΔH value of -35 kJ. This reaction would be expected to be _____________. However, if a chemist wanted to convert those reaction products back to the original materials what energy conditions would have to be provided?

13. For the reaction:

$$CaCO_{3(s)} \text{ ====> } CaO_{(s)} + CO_{2(g)} \quad \Delta H = 178.3 \text{ kJ}$$

What amount of calcium carbonate is involved ?
How much energy would be required to decompose 25.0 g of calcium carbonate?

14. Using Hess's Law explain why an adequate oxygen supply is important when burning a carbon fuel such as coal.

$C + 1/2\ O_2$ ====> CO	$\Delta H = -110.5$ kj
$CO + 1/2\ O_2$ =====> CO_2	
$C + O_2$ =====> CO_2	$\Delta H = -393.5$ kj

15. Standard state conditions for oxygen include what values ?

16. When are $\Delta H°$ and $\Delta H°_f$ identical?

17. When calorimetry information is collected a certain amount of energy will be absorbed by the calorimeter bomb. This energy is known as the __________ and has units of ___________.

18. By what equation is reaction heat in a calorimeter measured?

19. What constitutes an " energy resource" ?

20. Minimizing the use of energy resources is a major way industry can save money and remain competitive. Identify several ways the industrial establishment attempts to accomplish this.

21. An energy rich chemical bond, such as in a fuel, represents what kind of energy?

22. The spontaneous tendency toward disorder in a system is known as __________.

23. Why do entropy factors become more significant at elevated temperatures?

24. What two factors working together will assure that a reaction will proceed from reactants to products?

25. Arrange the states of matter in order of increasing entropy values.

26. What factors tend to increase entropy values in each of the following cases?
(A) Molecular size (B) Gases with a solvent present.
(C) Pure substance or solution. (D) Ionic sizes.
(E) Ionic charges

27. The single relationship that predicts reaction direction based only on the system is known as _____________.

28. What combination of enthalpy and entropy will always assure a product-driven reaction? Confirm this with the Gibbs free energy equation.

29. Under what conditions can a reaction with a (+) ΔG (reactant-favored) occur?

Testing You on Some Analytical Concepts.
Answers are at the end of the chapter.

1. Two students were holding metal rods in an identical flame. The rods were the same length and both weighed 500 grams. One student held an iron rod and the other held aluminum. If all other factors are equal which student will probably drop the rod first and why?

2. Boiling of ethanol requires adding 204 calories per gram of liquid. How many kJ is required to boil a mole of ethanol?

3. If you had one mole of molten Mg at its melting point of 651°C, how much energy must be removed to cool it to 25°C? The specific heat is 1.05 J/g°C and the heat of fusion is 159 J/g.

4. The specific heat of ethanol is 2.46 J/gK. If 100. grams of ethanol at 24.0°C was mixed with 75.0 grams at an unknown temperature the final temperature was 28.0°C. What was the temperature of the second sample?

5. Magnesium will burn in air to form the oxide, heat and a very bright light.

$$2\,Mg_{(s)} + O_{2(g)} ===> 2\,MgO_{(s)} \quad \Delta H^\circ_{rxn} = 1203\text{ kJ}$$

If a piece of ribbon weighing 0.13 g was burned what amount of heat would be released?

6. Tin can form two oxides, SnO and SnO_2. What is the enthalpy change when tin(II) oxide reacts with oxygen to form tin(IV) oxide?

$$2\ SnO_{(s)} + O_{2(g)} \text{ ===> } 2\ SnO_{2(s)} \qquad \Delta H°_{rxn} = ?$$

The enthalpies of formation for the two oxides are known

$$2\ Sn_{(s)} + O_{2(g)} \text{ ===> } 2\ SnO \qquad \Delta H_f = -572 \text{ kJ/mol}$$
$$Sn_{(s)} + O_{2(g)} \text{ ===> } SnO_{2(s)} \qquad \Delta H_f = -580.7 \text{ kJ/mol}$$

7. If $NO_{(g)}$ were reacted with hydrogen to produce water and $NO_{(g)}$ what free energy change would occur?

Material	NO_2	H_2O	NO	H_2
$\Delta G°$ kJ/mol	+51.8	-237.2	+86.7	0.00

Crossword - Chemical Thermodynamics

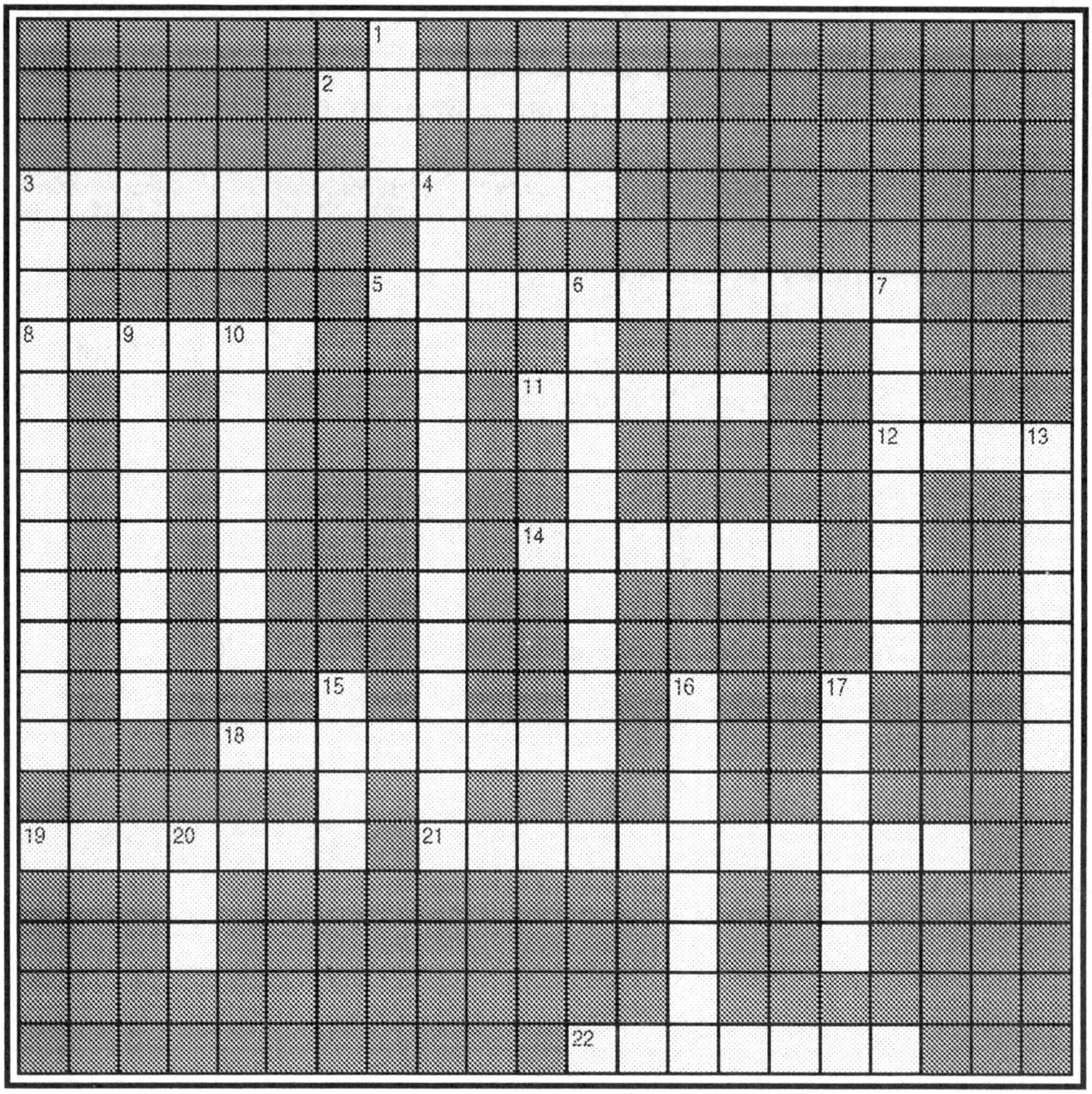

Crossword - Chemical Thermodynamics Clues

ACROSS

2 An energy resource such as wind
3 Technique used to measure heat evolved in a chemical reaction
5 Entity measured as a result of kinetic energy in the particles of a system
8 The substance that is undergoing a reaction.
11 SI unit of energy based on units used in mechanical energy
12 Energy transfer as a result of temperature differences
14 Law that refers to the entropy content of the universe.
18 Heat term that describes energy to change 1 gram of substance 1 degree C
19 Materials such as wood, cornstalks and the like materials
21 Change in a system directly from the solid state to the vapor state
22 Heat term that describes energy to change 1 gram of substance 1 degree C

DOWN

1 Energy available to do work
3 Law that states the total energy of the universe is constant
4 A reaction control based on heat and work
6 Energy change for a product-favored system
7 The heat transferred to or from a system at a constant pressure
9 Reaction conditions that involve 298K, 1 atm and 1 molar solution
10 For a system this generally increases as temperature is increased
13 Energy that is the result of particle motion
15 Law referring to overall enthalpy change of two or more sequential reactions
16 Energy used when a battery starts your automobile
17 Heat term referring to a solid to liquid state change
20 Abbreviation for the chemical unit of quantity

Ties to Previous Chapters

Several terms are used in this chapter which were introduced in chapter 2. If the use of some terminology was not clear a review of parts of chapter 2 may be helpful.

Chemical reactions are used in this chapter and these were discussed in more detail in chapter 5.

You will see several aspects of this chapter again in the next. Spend the time getting a solid understanding of this chapter.

Answers to True/False Statements:

1. True Particles react, when the particles of one are gone the reaction must stop.

2. False. Specific heat of the material is also very important as is the quantity. Both a small amount and a large amount may have the same temperature but they do not cool equally fast.

3. False. Biomass is current material and fossil fuels are the result of long geological changes.

4. True. (-) meaning loss from the system.

5. False. The reverse is true.

6. True. If materials do not have the same specific heat this property can be measured and related to known materials for identification.

7. True. For some materials absolute values are not measurable.

8. True. They often are many times less. Total separation of particles requires more energy than just partial separation.

9. False. Exothermic reactions are product favored.

10. False. ΔE involves work and ΔH does not.

11. True. By definition it is heat transferred at constant pressure.

12. False. There must be a conservation of energy.

13. True. This is confirmed in the Law of Hess.

14. True. This assures a fixed value comparison.

15. False. The sign is also reversed. A reaction favored one way would not be favored in the reverse.

16. True. The subscript, f, refers to formation from the elements.

17. False. Only one allotropic form can be the most stable.

18. True. They are the starting materials, the standards.

19. True. Sometimes called a bomb calorimeter.

20. True. Both of these are favored in nature.

21. False. The second law refers to entropy.

22. True.

23. True. The higher the temperature the greater the motion of the particles and greater chance they will separate.

24. False. Every calorimeter is different so the heat absorbed is not constant. In fact calorimeter constants for a given calorimeter may change over a period of time.

25. False. Disorder is favored.

Answers to Understanding of Concepts Questions:

1. Limiting reagent

2. Make the expensive material the limiting reagent. This would assure that all of it would be consumed. This would result in the lowest reaction cost.

3. Outside intervention
endothermic

4. light; electrical; explosion (and others)

5. thermodynamics

6. Work (Work is much more useful than heat. We want the work our automobile does, not the heat it produces.)

7. The amount of energy is constant. What happens is that the quality of that energy degrades. The major problem is that we take a material which as high potential for use (to do work) and convert it to a form which is much less able to do work.

8. The energy content is different. The important factors are the temperature (both are the same), the specific heat of the material (again they are the same) and the quantity of material. The larger

volume would contain more energy. No energy would be transferred as heat since they are at the same temperature.

9. Sun

10. Positive
increased

11. Product
negative

12. spontaneous (or exothermic)
The reverse system would be endothermic so energy would have to be added.

13. A mole. Thermodynamic values are based on a per mole unit. The mol weight of $CaCO_3$ is 100 grams. 25 grams would be one fourth of a mol and would require 44.6 kj.

14. The Law of Hess states that the total energy is the sum of the individual steps. In this reaction the second step would yield 283.0 kJ. This is the high energy step in the combustion and can only occur if enough oxygen is present.

15. 298K (25°C) , 1 atm of pressure, and since oxygen has an allotropic form known as ozone the gas must be present as O_2 not as O_3.

16. When a material is produced directly from the elements.

17. Heat capacity of the calorimeter
J/K

18. $q_{reaction} = -(q_{bomb} + q_{water})$ water is the usual liquid used in calorimeters since it has a high heat capacity.

19. Materials which have a large (-) ΔG value are high in usable energy content and are our energy resources.

20. Use energy saving equipment.
Design new technologies which use less energy.
Refine the processes used in the manufacturing.

21. Potential energy or an energy resource

22. Entropy

23. As the temperature is raised the materials gain kinetic energy and have greater tendency to change to a more random state. They may even separate from each other and become random individual particles rather than aggregates.

24. More disorder in the particles and dispersal of the energy over more particles or a greater number of chemical bonds.

25. solid < liquid <<< gas

26. A) Larger molecules
B) When the gas separates from the solvent rather than being dissolved.
C) Solutions
D) Larger ionic size
E) Lower ionic charge

27. Gibbs Free Energy

28. Exothermic (ΔH is negative) and increase the system disorder (ΔS is positive). $\Delta G = \Delta H - T\Delta S$

A negative ΔG value is product favored and if the above factors are present the net result is both work to increase the (-) value of ΔG.

29. With a continuous outside influence of sufficient magnitude. Enough energy must be added to offset the energy released during the initial reaction.

Answers to Testing the Analytical Concepts

1. This answer is specific-heat-dependent. The higher the specific heat the longer it takes to heat the rod. Tables give the specific heat of iron as 0.451 J/gK and aluminum is 0.902 J/gK. The iron would heat up about twice as fast and would be dropped first.

2. The mole weight of ethanol is 46.0 grams. One calorie is 4.184 joules.

$$\text{kJ} = 46.0\text{ g} \times \frac{204\text{ cal}}{1\text{ g}} \times \frac{4.184\text{ J}}{1\text{ cal}} \times \frac{1\text{kJ}}{1000\text{ J}} = 39.3\text{kJ}$$

3. The magnesium will be cooled a total of 626°C and this involves the specific heat of the metal. It also must be converted from a liquid to a solid which involves the heat of fusion. The mole weight is 24.0 g.

$$\text{J} = 24.0\text{ g} \times 1.05\text{ J/g°C} \times 626\text{°C} + 24.0\text{ g} \times 159\text{ J/g}$$
$$= 15775\text{ J} + 3816\text{ J} = 19591\text{ joules}$$

4. Heat gained by one sample = heat lost by the other sample
$100\text{ g} \times 2.46\text{ J/gK} \times (28.0°\text{C}-24.0°\text{C}) = 75.0\text{ g} \times 2.46\text{ J/gK} \times (28.0°\text{C} - T_f)$

Since the specific heat is the same that factor will divide out of both sides of the equation.

$400°C = 2100°C - 75T_f$
$-1700°C = 75T_f$
$T_f = 22.7°C$

5. The equation tells us that 2 moles of Mg (48.0 grams) releases 1203 kJ of energy. Since the reaction only uses 0.130 grams of Mg just a fraction of that energy will be released.

$$kJ = 0.130\ g\ Mg \times \frac{2\ mol\ Mg}{48.0\ g\ Mg} \times \frac{1203\ kJ}{2\ mol\ Mg} = 3.26\ kJ$$

6. We need energy for $2\ SnO_{(s)} + O_{2(g)} \Longrightarrow 2\ SnO_{2(s)}$ and we know the energy for the formation of the oxides.

$2\ Sn_{(s)} + O_{2(g)} \Longrightarrow 2\ SnO_{(s)}$ $\Delta H_f = -572\ kJ/mol$
$Sn_{(s)} + O_{2(g)} \Longrightarrow SnO_{2(s)}$ $\Delta H_f = -580.7\ kJ/mol$

If we reverse the first reaction and double the second reaction, by subtraction we will have the reaction we need. When a reaction is reversed the energy change will be reversed so a sign change is needed. When a reaction is doubled the energy will also be doubled.

$2\ SnO_{(s)} \Longrightarrow 2\ Sn_{(s)} + O_{2(g)}$ $\Delta H_f = +1144\ kJ/mol$
$2\ Sn_{(s)} + 2\ O_{2(g)} \Longrightarrow 2\ SnO_{2(s)}$ $\Delta H_f = -1161.4\ kJ/mol$

$2\ SnO_{(s)} + O_{2(g)} \Longrightarrow 2\ SnO_{2(s)}$ $\Delta H_f = -17.4\ kJ$

7. Our reaction with the free energy values is

$$\begin{array}{ccccc} NO_{2(g)} & + H_{2(g)} & ====> & NO_{(g)} & + H_2O_{(l)} \\ +51.8\text{ kJ} & 0.00\text{kJ} & & +86.7\text{ kJ} & -237.2\text{ kJ} \end{array}$$

Energy of reactants is 51.8 kJ and the products is -150.5 kJ. This results in a net change of -202.3 kJ. The products have 202.3 kJ less energy than the reactants.

Answer to Chapter 7 Crossword

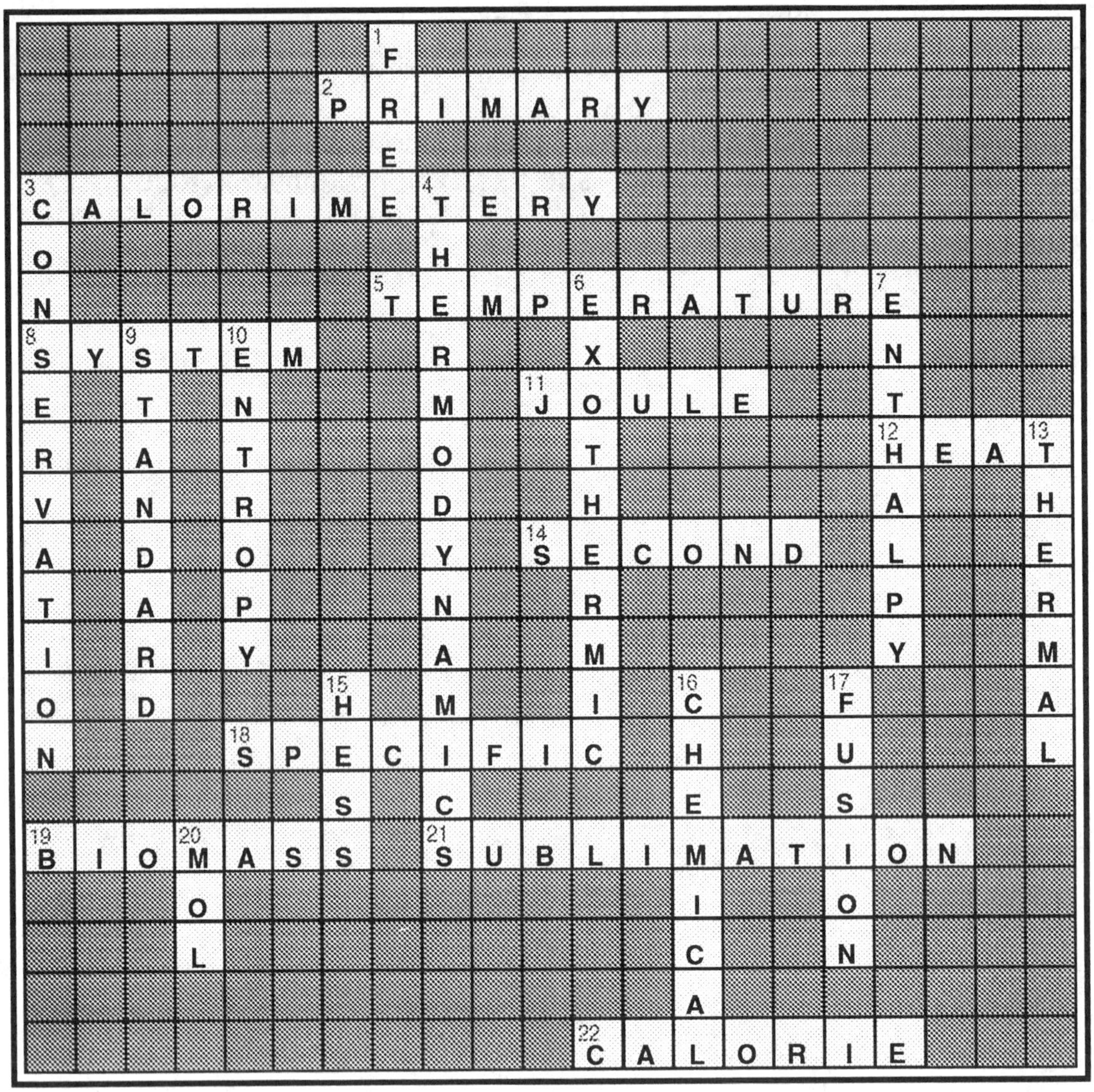

Some supplementary readings relating to this topic:

1. Push for New Materials, Chemicals From Biomass Sparks Active R&D. Chemical and Engineering News , September 10, 1990, page 19

2. Car Cooler, Chem Matters February 1993, page 11.

3. Hot & Cold Packs, Chem Matters , February 1987, page 7

4. Solar Energy, Chem Matters February 1991, page 4

5. Hot Meals, Chem Matters February 1992, page 12

6. Turning a Stew of Old Tires Into Energy, The New York Times, December 27, 1992

Chapter 8 Principles of Reactivity

Chapter 8 continues the theme started in chapter 7. Thermodynamics, provides information regarding the tendency of a reaction to proceed and the energy changes expected. Chapter 8 discusses how complete the reaction will be, how fast it may occur and provides some idea of what sequence of steps the reaction follows.

The information for these two chapters is linked through the concepts of free energy from chapter 7 and equilibrium which is a major part of chapter 8.

Focus Your Study Efforts

Chemists have several questions when a potential reaction is proposed:

Is it likely to occur?
How fast will it be?
How much product will be produced?
Can the reaction rate be modified?
How many steps will the process take?

Several factors affect reaction rates.

Objectives:

1. You should be able to describe the factors that significantly affect the rate of a reaction.

There are several factors that are significant in controlling the speed of a reaction. Temperature, since it has an effect on the energy particles have. The higher temperature, the more energy and more effective the collisions will be. Concentration, including pressure for gases, affect the number of particles that are available to react. Catalysts, including enzymes, function by lowering the activation energy needed. Also the size and shape of the particles will determine how effective collisions will be. The last is known as the steric factor.

Rate means the same thing in chemical reactions as it does in the other uses you associate it with. Chemists measure the change, it could be an increase or decrease, in the amount of material over a given amount of time.

A chemist must be able to measure a change before a kinetic study can be done. Some chemical systems make this very difficult to do and as a result not much may be known about those reactions.

Kinetics is an experimental science. You must have experimental evidence before any valid conclusions can be made about the system.

It is not uncommon to find that experimental data will support more than one conclusion. When this happens more experimental work is required.

Rate laws provide essential information.

Objectives:

1. You should be able to define reaction rate and explain the terms in a rate equation or rate law.
2. You should be able to use rate data to write a rate equation.
3. You should be able to predict how a reaction rate varies as the concentrations change.

The rate equation is of major use and you need to understand the various parts. A generic form of the rate equation is commonly used.

$$\text{Reaction rate} = k\,[A]^m[B]^n$$

This rate equation shows two materials, A and B, affecting the reaction. There may be more or less than two materials depending upon the reaction system. k is a constant for that reaction, the rate constant.

The exponents m and n <u>do not</u> have any correlation to the coefficients in the balanced equation. These must be determined by experimental means.

> A given reaction using CH_3COOCH_3 and H_2O as reactants has a rate constant of 0.075 / minute. What is the initial rate when the concentrations were each 0.250 M ? The reaction is first order in each.
>
> This is a simple complete the rate law equation type of problem. This reaction has both reactants of equal concentration but this does not have to be the case.
>
> Rate = k $[CH_3COOCH_3][H_2O]$
> $= 0.075\ min^{1-}(0.250)(0.250) = 0.00469\ min^{1-}$

k is known as the rate constant. Changes in concentration will not affect it but changes in temperature will alter its value.

> The rate law for a reaction is Rate = k$[CH_3OH][CH_3COOH]$. One data set that was used to establish the rate law is $[CH_3OH]$ = 0.035M, $[CH_3COOH]$ = 0.035 M and the initial rate for the reaction was 0.00350 mol/L·s.
>
> (A) What is the order with respect to each reactant?
> (B) What is the overall order?
> (C) What would happen to the rate if $[CH_3COOH]$ was doubled? If both reactants were doubled?
> (D) What is the value of the rate constant?

This series of problems can be solved in any order since they are independent. The first two are solved by inspection of the rate equation.

(A) The order is the exponent value on each of the terms in the rate law. Both of these are first order.

(B) The overall order is the sum of all exponents in the rate law. In this case 1 + 1 or 2nd. order overall.

(C) Since the order is 1, doubling the CH_3OH concentration would double the rate. Doubling both would make the rate 4 times faster.

(D) Rate = k[CH_3OH][CH_3COOH]

All values are known except k.

$$k = \frac{\text{Rate}}{[CH_3OH][CH_3COOH]} = \frac{0.00350}{(0.035)(0.035)}$$

$$= 2.86$$

The terms rate equation and rate law are often used interchangeably.

After a reaction has been running for a period of time the kinetics may be altered by products formed. To minimize this effect efforts are made to determine the initial rate for the reaction. This initial rate data must be collected over a short time period because as the reaction occurs the amount of material available to react decreases. You must have accurate concentration data to determine the initial rate.

Obtaining good kinetic data involves doing many separate experiments. There are many variables that can change and it is imperative that only one be changed at a time. Several trials are made by varying the concentration of one component with the others being held constant. Then another set of trials is made with a different component varied. Since k is temperature-dependent all experiments must be done at a constant temperature.

Not every component will have an effect on the reaction.

Unfortunately you do not know this until individual reactions are done. Sometimes products may affect the reaction and need be included in the rate law.

The values of the exponents for each component in the rate law are determined by comparing any two experiments when only one variable is changed.

The rate constant value can be determined from any of the individual experiments done on the system once the rate equation is established.

Be sure you understand what the term order means in a given instance. Each component in the rate equation will have a reaction order associated with it and there will also be an overall reaction order which is the sum of the individual component orders.

Isomers of organic molecules, such as cis and trans forms, will give the individual isomers distinct chemistry. You would have to do a separate set of kinetics experiments on each isomer.

Activation energy may control reaction spontaneity.

Objectives:

1. You should be able to explain what activation energy is and what can be done to change overcome it.
2. You should be able to explain how catalysts function.
3. You should be able to describe molecular features that will affect the activation energy needed for a reaction.

A certain amount of energy is required to start any reaction, even those we consider to be spontaneous. This energy is known as the activation energy. This energy may be high or low in value. Often enough energy is present at room conditions to start the reaction so we do not think about anything being added.

Every reaction has this energy barrier that must be overcome. The greater the barrier the higher the activation energy.

> A reaction was found to have an E^* of 90 kJ/mol. This reaction was known to be exothermic. What is known about the reverse reaction activation energy value?
>
> Since the given reaction was exothermic the E^* of the reverse reaction must be greater than 90kJ/mol. The reverse reaction will be endothermic. From the given data we have no way to determine the actual value.

Catalysts function by providing an additional pathway for the reaction to go. If this new path has a lower activation energy the reaction proceeds at a more rapid rate. The catalyst **does not replace** the original mechanism. It **adds another** to it. However if the original mechanism was very slow it may not contribute much to the overall product.

Biological systems make extensive use of catalysts. Biologists usually call catalysts, enzymes.

Shapes and sizes of molecules can be very significant in how fast molecules react. These are called steric factors.

To have a reaction, collisions must occur at the chemically active part of the molecule. Large molecules such as an alcohol like $CH_3CH_2CH_2CH_2CH_2CH_2CH_2$-OH may have many interactions but only those at the OH end are involved in the reaction. We would expect that reagent to react slower than one like CH_3-OH.

If a molecule has a complex shape and the reactive part is shielded by other parts the reaction would be slow.

Reaction mechanisms outline the steps a reaction follows.

Objectives:

1. You should be able to define reaction mechanism.
2. You should be able to identify rate-determining steps.
3. You should be able to define factors that affect chemical stability.

Molecularity is used to describe the number of particles involved in a given reaction step. Unimolecular reactions have only one molecule involved. If two different molecules are involved it is bimolecular.

Although a total reaction may involve several steps to complete it is done one step at a time. Each step is an elementary process.

Reactions occur via a series of simple steps. The chances of several molecules colliding at exactly the same time in exactly the correct orientation is very small.

Another way to enhance a reaction is to get a greater number of the particles at or above the activation energy. Increasing the temperature is one way to do this. A rule of thumb is, for every 10°C rise in temperature, the reaction rate will double. However, remember that thumbs sometimes get hit with hammers so be careful with assumptions.

Reactions occur via a series of steps. Each time a reaction step occurs something is produced. Almost always they are immediately consumed to go to the next step. These short lived materials are called reaction intermediates or intermediate products.

When a reaction occurs by a sequence of steps one of the steps will be slower than the others and control the overall reaction. This slow step is the rate determining step. If a reaction involves a series of 5 steps and four occur in fractions of a second but one takes several minutes the overall reaction will take several minutes.

Catalysts can be involved in a reaction. They may be part of an intermediate product. However, at the completion of the reaction the catalyst is regenerated so it will not appear in the overall reaction. Theoretically a catalyst should never have to be replenished.

A hypothetical reaction is outlined below:

AB + X ===>	AX + B	(rapid)
B + CD ===>	BC + D	(slow)
D + AX ===>	AD + X	(rapid)

(A) What is the net overall reaction?
(B) What is the rate-determining step?
(C) Could X be a catalyst?
(D) If the rate of reaction depended directly on [CD] what is a probable rate law?
(E) Identify any intermediate products.

(A) The net reaction is what remains after products formed and reused in each step are accounted for. In this set of reactions the materials AX,B and D are eliminated.
(B) The rate-determining step is always the slow step. The slow step in this reaction is B + CD ===> BC + D. It makes no difference how fast the steps before and after this are, this one controls the overall reaction rate.
(C) Catalysts, if involved in a reaction, will be regenerated. The material X is possibly a catalyst.
(D) Rate = k[CD] since the data state that the reaction speed depends directly on that material and it is in the rate step.

(E) AX, B and D are all materials formed on the way to a complete reaction. They may have only a very brief lifetime.

Catalysts are divided into two groups. The heterogeneous catalyst is in a different phase than the reactant materials. For example a wire screen may be a catalyst for the reaction of two gases. Homogeneous catalysts are those that have the same phase as the reactants.

It is not easy to predict what may or may not serve as a catalyst. This is an area of chemistry that receives a good deal of study. Often, however, outcomes are the consequence of serendipity rather than planning. A material that works as a catalyst for one reaction may have no effect on one we consider to be very similar.

Chemical stability is due to two different factors. Some materials are stable because of energy constraints, these are thermodynamically stable. Others are stable because other factors; such as size, shape or high activation energies. You should be able to identify which type of stability a material has.

Equilibrium is a measure of how complete a reaction is.

Objectives:

1. You should be able to define what chemical equilibrium is identify factors that cause it to shift.
2. You should be able to use equilibrium constant values to predict if reactions will be reactant-favored or product-favored.
3. You should be able to write equilibrium constant expressions from given balanced chemical equations.
4. You should be able to calculate a value of K using equilibrium concentration values.
5. You should be able to calculate equilibrium concentrations if

the value of K is known.

6. You should be able to use Le Chatelier's principle to predict the effects that concentration or temperature changes will have on the chemcial equilibrium.

When we write a chemical equation we often think that the reactants just form the products. Most chemical reactions are not that clean. In fact, many will undergo a reverse reaction in which the products formed react with each other to reform the original reactants. When this happens we are reaching an equilibrium state.

Equilibrium is when the rate (speed) of the forward reaction to form the products is offset by a reverse reaction of the products reforming the original reactants.

Always keep in mind that at chemical equilibrium the overall reaction is continuing. The reaction does not stop. Equilibrium is a dynamic process.

$CH_3OH + CH_3COOH \Longleftrightarrow CH_3COOCH_3 + HOH$

(all concentrations are in mol/L)

0.500	0.500	0	0	Initial Conc.
-0.200	-0.200	0.200	0.200	Conc.Change

Using the data given above calculate the concentration of all materials at equilibrium. Use that data to calculate the reaction equilibrium constant. Is the reaction product-favored or reactant-favored?

This reaction needs to be solved in the sequence as outlined since each result is required for the one that follows.

We note the reactants both decrease in concentration and both end up at 0.300 M. Products both increase to 0.200 M.

The equilibrium expression has products divided by reactants.

$$K = \frac{[CH_3COOCH_3][HOH]}{[CH_3OH][CH_3COOH]} = \frac{(0.200)(0.200)}{(0.300)(0.300)} = 0.444$$

Product favored reactions have a K value > 1. Since the value of K in this case is < 1 the reaction is reactant favored.

One of the major reasons reaction yields are low is the fact that the system establishes equilibrium and some of the reactants do not get used up.

How far a reaction goes toward completion is identified by a value known as the equilibrium constant. Equilibrium constants greater than one mean a product-favored reaction. The larger the value the more product formed. Very small values indicate little product formed.

Equilibrium constants, K, can have very small or very large values. Numbers such as 10^{+50} or 10^{-50} are known.

The equilibrium constant value is valid only at a given temperature. As temperature changes the relative amounts of reactants and products change and this may have significant effect on the value of K.

The equilibrium expression is equal to the product concentrations raised to their exponents divided by the reactant concentrations raised to their exponents.

The exponents in equilibrium expressions are equal to the coefficients of the materials in the balanced chemical equation. Remember that was not true for kinetic work. This is carefully explained on pages 324-327 of the textbook.

The value of K does have units but they are seldom attached.

If a chemical reaction is reversed the equilibrium constant for that reaction would be the reciprocal. $K_{rev} = 1/K_{fwd}$

The equilibrium expression for the reaction producing sulfur trioxide from sulfur dioxide and oxygen is:

$$2\ SO_2 + O_2 \text{ ===> } 2\ SO_3$$

$$K = \frac{[SO_3]^2}{[SO_2]^2\ [O_2]}$$

If the reaction were reversed and sulfur trioxide were decomposed the expression would be the reciprocal:

$$K = \frac{[SO_2]^2\ [O_2]}{[SO_3]^2}$$

The reaction $2\ NOBr_{(g)} \text{ <==> } Br_{2(g)} + NO_{(g)}$ has an equilibrium constant of 0.16. What is the value of K for the reverse reaction? Which reaction, forward or reverse is thermodynamically favored?

The reverse reaction has a K value which is the reciprocal of the forward reaction K. In this case

$$K = 1/\ 0.16\ = 6.3$$

The reverse reaction has a K greater than 1, it would be the thermodynamically favored.

Solid materials do not show up in the equilibrium expression. As long as there is some solid in the system equilibrium will continue to be established. If we consider that only the very surface of a solid can interact then the amount inside has no influence on the system. The concentration of the solid is is actually included in the K value.

Pure liquids and solvent materials also are not included in equilibrium expressions.

There are several types of equilibrium systems. There are precipitation systems and acid-base systems as specific examples.

Equilibrium and thermodynamics are related. From chapter 7 we found that exothermic reactions favored products. In this chapter we find that K values greater than one favor products. Be careful with both of these however since neither considers the reaction rate.

As chemists, we have the ability to affect equilibrium systems. If we use the reaction of water and carbon monoxide as an example

$$CO_{(g)} + H_2O_{(g)} <===> CO_{2(g)} + H_{2(g)}$$

$$K = \frac{[H_2][CO_2]}{[CO][H_2O]} \quad \text{(temperature constant)}$$

If K must remain a constant, then, if the concentration of any material is changed, other materials must also change. In this system, if we increased the amount of hydrogen then some carbon dioxide would have to be used up and more carbon monoxide and water would be formed. Likewise if some hydrogen were removed more CO and H_2O would react to replenish the hydrogen and at the same time more CO_2 would be formed.

> When the reaction $2\ NH_3 <===> N_2 + 3\ H_2$ reached equilibrium, the concentration of H_2 was 0.30 moles per liter and N_2 was 0.15 moles per liter. The equilibrium constant value is 3.5 x 10^8. What is the ammonia concentration?

To solve this we need to write the equilibrium expression.

$$K = \frac{[N_2][H_2]^3}{[NH_3]^2}$$

It is important to remember that concentration values and exponents are independent in equilibrium type problems.

Rewriting the equation, $[NH_3] = \frac{[H_2]^3[N_2]}{K}$

$$[NH_3]^2 = \frac{(0.15)(0.30)^2}{3.5 \times 10^8} = 3.9 \times 10^{-11}$$

$[NH_3] = 6.2 \times 10^{-6}$ M Very little ammonia remains.

When put under stress any system at equilibrium will react in a way to minimize the stress applied. This is known as Le Chatelier's principle.

Temperature has an effect on the equilibrium. If we think of heat as a reaction entity it may be easier to visualize. If a reaction is endothermic we can think of heat as a reactant. Adding heat in this case would stress the reactant side and drive the reaction to produce more products.

$$N_{2(g)} + O_{2(g)} + \text{heat} <===> 2\ NO_{(g)}$$

Likewise if we heated a reaction that was exothermic, the heat is a product of the reaction, we would force the equilibrium to make more product material.

Symbols found in Chapter 3

Page	Symbol	
	[ ]	This was first introduced in chapter 6.
297	Δ	Delta refers to change in some quantity
300	∞	Proportional to some quantity
300	k	A constant, in this chapter in rate equations
304	cis	Refers to a geometric shape of an isomer
304	trans	Refers to a geometric shape of an isomer
304	=	Represents a double bond between the atoms
306	E^*	Energy of activation
322	$\Leftrightarrow$	Indicates an equilibrium reaction
324	X^*	X is some element symbol that is radioactive
328	K	An equilibrium constant
339	R	The gas constant, this may have several values
339	K_{th}	Equilibrium constant referring to thermodynamics

Terms in Chapter 3

The following terms are listed for your convenience. They should not become a list for you to memorize. You should have a working definition for each. If you have doubt about any particular term check its definition in the glossary and/or review it in the chapter. Some of these terms are included in the crossword puzzle later in the chapter.

Chemical reactivity	Chemical kinetics	Reaction mechanism
Reaction rate	Rate law	Rate constant
Initial rate	Reaction order	Overall order
Cis	Trans	Activation energy
Unimolecular	Bimolecular	Elementary step
Steric factor	Rate-limiting step	Intermediate step(s)
Catalyst	Enzyme	Homogeneous catalyst
Heterogeneous catalyst	Thermodynamic stability	

Kinetically stable	Equilibrium constant	Le Chatelier's principle
Haber Process	Nitrogen fixation	

Concept Maps for Kinetics and Equilibrium

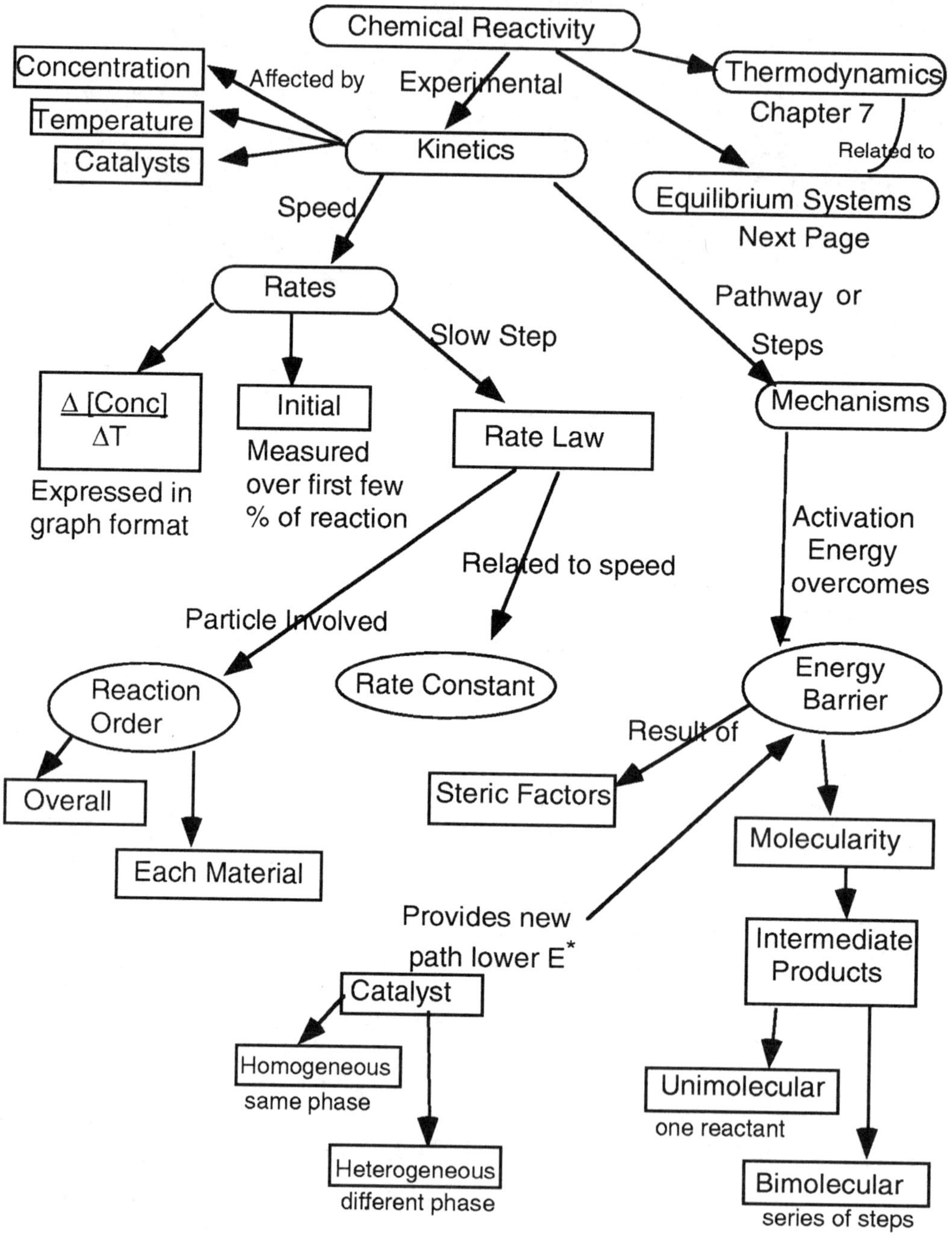

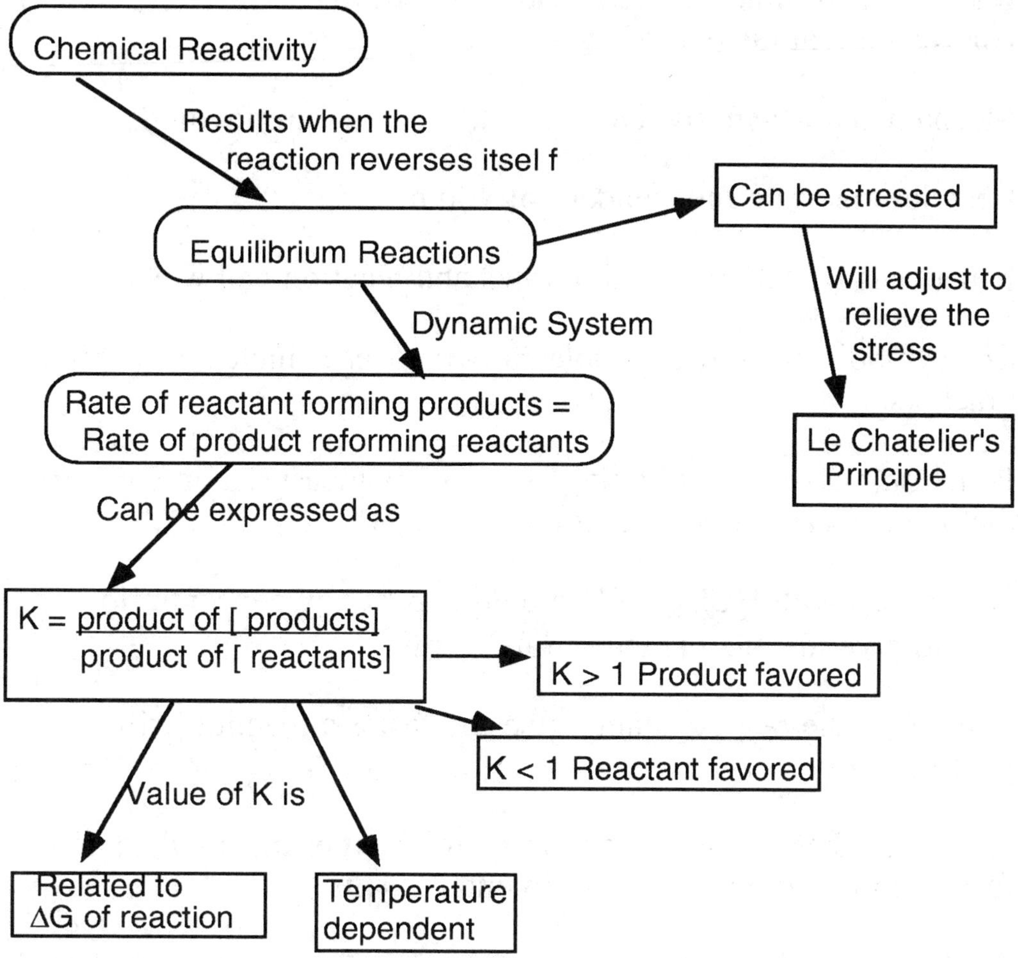
Chemical Reactivity
Results when the reaction reverses itsel f
Equilibrium Reactions
Can be stressed
Will adjust to relieve the stress
Le Chatelier's Principle
Dynamic System
Rate of reactant forming products = Rate of product reforming reactants
Can be expressed as
K = product of [products] / product of [reactants]
K > 1 Product favored
K < 1 Reactant favored
Value of K is
Related to ΔG of reaction
Temperature dependent

Chemical Compounds True/False Statements

Answers are at the end of the chapter.

1. Mechanisms which involve multiple particles are favored.

2. Reaction rates are independent of volume.

3. Kinetics is a study of reaction rates and reaction pathways.

4. The symbolism [ ] means moles of whatever is included within the brackets.

5. Reaction rates are valid only if the data is measured on a material that is decreasing in concentration.

6. $Al_{(s)}$ reacts with $HCl_{(aq)}$. We would expect any concentration of HCl to give identical reaction rate results.

7. The larger the rate constant value the faster a reaction will proceed.

8. Kinetic studies are dependent upon good experimental data and in fact, are invalid without experimental results.

9. Product formation will have no effect on kinetic results so data collected at any time during the reaction is equally valid.

10. Control of variables is critical to good kinetic data.

11. The greater the activation energy the easier it will be for the reaction to occur.

12. Termolecular reactions are more favored than bimolecular ones.

13. The overall reaction order can be determined by inspection of the reaction coefficients.

14. The overall reaction order is 2 for a reaction that has the following rate law. rate = k $[S_2O_8^{-2}][I^{-1}]$

15. The slowest step in a reaction sequence is the rate determining step regardless of when it occurs in the reaction sequence.

16. The molecule below represents a trans isomer.

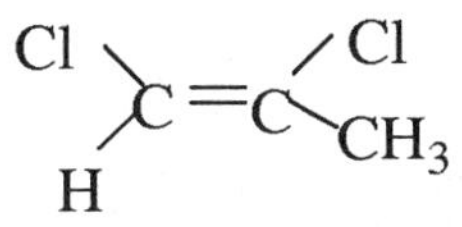

17. E^* is a measure of a reaction's energy barrier.

18. A reaction was run at 10°C and again at 40°C. The second reaction would be about 8 times faster than the first.

19. A reaction has a very large (-) value for ΔG. This assures that the reaction will be very rapid.

20. A reaction mechanism is proposed and it agrees with the experimental rate equation. This assures that this mechanism and only this mechanism is valid.

21. A reaction occurs and evidence is found to suggest that three steps are involved. This means some intermediate products are formed.

22. When a little solid MnO_2 is added to a 10% H_2O_2 solution a rapid decomposition occurs. The MnO_2 is a homogeneous catalyst.

23. Catalysts frequently need replacement. This is due to the fact they wear out.

24. Equilibrium in a chemical reaction occurs only when the amount of product equals the amount of reactant present.

25. An equilibrium constant of 10^3 indicates we have a product favored reaction.

26. According to Le Chatelier's Principle heating a reaction with a (-) ΔH value will favor product formation.

27. In a given reaction one reactant is a gas. Increasing the pressure will speed up the reaction.

Testing Your Understanding of Concepts in Chapter 8.
Answers are at the end of the chapter.

1. The burning of $SO_{2(g)}$ to form $SO_{3(g)}$ is exothermic.

$$2SO_{2(g)} + O_{2(g)} ===> 2SO_{3(g)}$$

Would either heating or increasing the pressure shift the equilibrium, and if so in what direction?

2. Gasoline burns in an automobile engine with a large release of heat. The reaction is very exothermic. Why then do we consider it safe to store the gasoline in the automobile gas tank?

3. Why does a reaction rate change with time?

4. Initial reaction rates are measured over only the first few percent of the reaction. Why do chemists spend the effort to collect data over such a small portion of the total reaction?

5. What is the function of the spark plug in an automobile engine?

6. Why are biological catalysts (enzymes) homogeneous rather than heterogenous?

7. What is the difference between a first order reaction and a unimolecular reaction?

8. What factors need to be considered when planning a reaction? Stated another way, what could be done to facilitate the reaction?

9. The plots in figure 8.1 of your text are of the same data. Why does 8.1a appear to be a straight line while 8.1b is not? What should this make you aware of in data analysis?

10. Explain why reaction rates decline as molecularity increases.

11. Sometimes doubling a reactant's concentration increases the reaction rate and at other times it has no effect. Explain why these differences happen.

12. Why does providing proof of reaction intermediate's existence interest a chemist so much?

13. Referring to figure 8.10 in your text, the catalyzed reaction has several humps in it. What do these humps signify?

14. Discuss what the term stable may mean to a chemist.

15. Once a reaction equilibrium has been established how close to equal are the amounts of reactants and products?

16. Is the value of K_{equil} temperature dependent?

17. The presence of solids is not indicated in an equilibrium expression. Would anything happen to an system at equilibrium if the solid were removed from the reaction area?

18. A given reaction has a large positive value for ΔH. What can be predicted about the K value for the reaction?

Testing You on Some Analytical Concepts.
Answers are at the end of the chapter.

1. When a series of runs were made on a reaction system the following initial rate data was collected:

A + B ====> C + D

Run	[A]	[B]	Initial Rate (mol/L·s)
1	0.020	0.010	3.6×10^{-6}
2	0.020	0.020	14.4×10^{-6}
3	0.020	0.030	32.3×10^{-6}
4	0.040	0.010	7.2×10^{-6}
5	0.060	0.010	10.9×10^{-6}

Determine the rate law and calculate the rate constant. What is the overall order and order with respect to A?

2. When doing a laboratory experiment the following data were collected for the reaction given:

$S_2O_8^{-2} + 3\ I^{-1}$ ====> $2\ SO_4^{-2} + I_3^{-1}$

Run	$[S_2O_8^{-2}]$	$[I^{-1}]$	Initial Rate (mol/L·s)
1	0.038	0.060	1.4×10^{-5}
2	0.076	0.060	2.8×10^{-5}
3	0.060	0.120	4.4×10^{-5}
4	0.038	0.030	0.7×10^{-5}

What is the rate law?

3. Equilibrium data for the reaction in question # 2 above was reported as the following:

$[S_2O_8^{-2}]$	= 0.05	$[I^{-1}]$	= 0.12
$[SO_4^{-2}]$	= 0.75	$[I_3^{-1}]$	= 0.50

Determine the equilibrium constant for the reaction at that temperature. Is this a reactant or product favored reaction?

4. Hydrogen peroxide will decompose to water and oxygen:

$$2\ H_2O_2 \ =====>\ 2\ H_2O + O_2$$

When a little iodide ion from KI is present the following mechanism was found.

slow	$H_2O_2 + I^{-1} ====> H_2O + OI^{-1}$
rapid	$H_2O_2 + OI^{-1} =====> H_2O + I^{-1} + O_2$

What is the net reaction? What is the rate equation?

5. Write equilibrium expressions for each of the following reactions:

$$CH_3OH_{(g)} <===> CO_{(g)} + 2H_2(g)$$
$$CaCO_{3(s)} <===> CaO_{(s)} + CO_{2(g)}$$
$$H_2S_{(g)} + I_{2(s)} <====> 2HI_{(g)} + S_{(g)}$$
$$SO_2 + 1/2\ O_2 <====> SO_3$$

6. When equilibrium was established for the reaction

N_2O_4 $<====>$ $2\ NO_2$ the K value was 4.61×10^{-3}. The molarity of the NO_2 was 0.0113. What was the N_2O_4 molarity?

7. What concentrations would exist at equilibrium if 0.090 mole of HI was placed in a 3.0 liter vessel and heated? The equilibrium constant at the elevated temperature is 3.60×10^{-2}. The reaction is:

$$2\ HI \Longleftrightarrow H_2 + I_2 \qquad \text{(all as gases)}$$

Crossword - Kinetics & Equilibrium

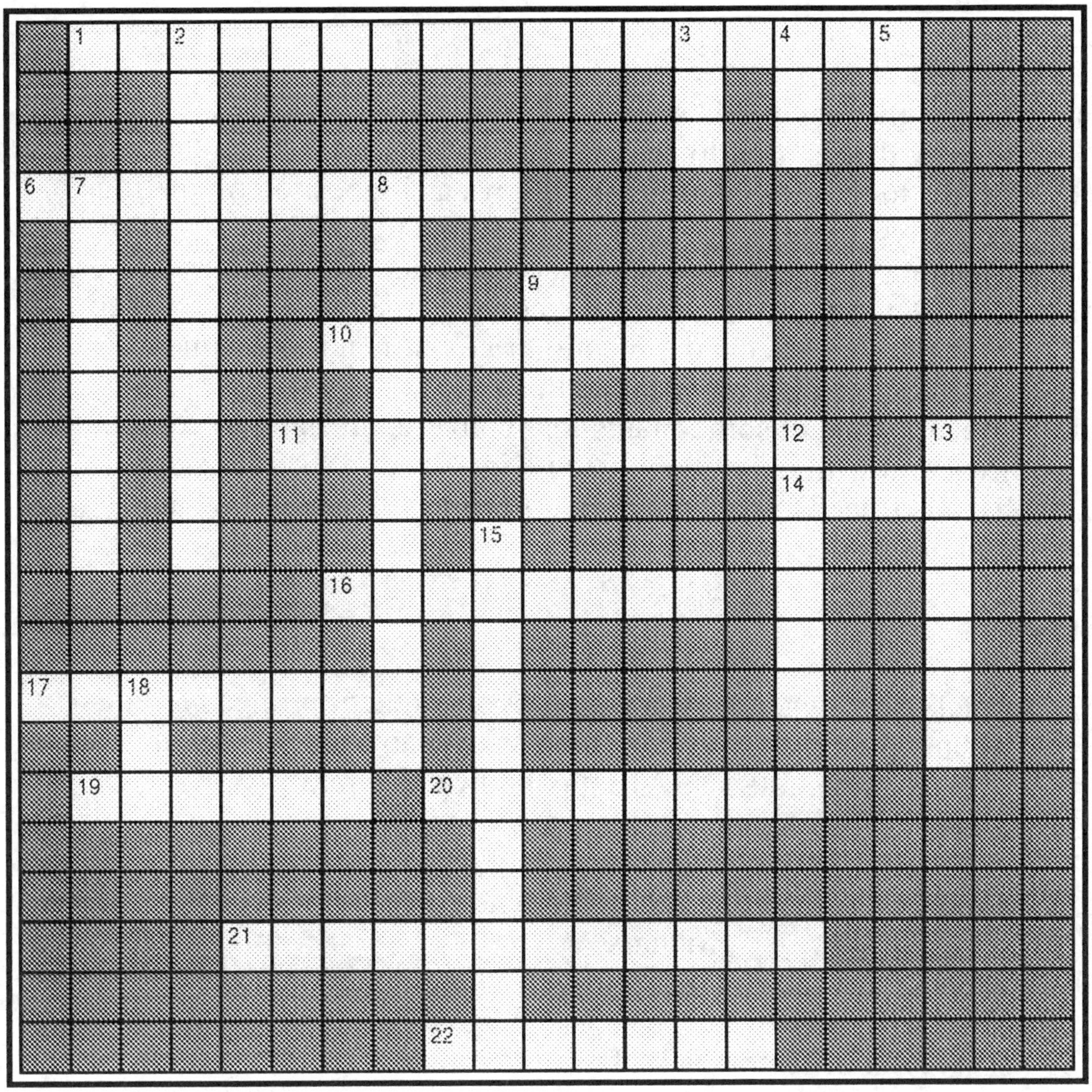

Crossword - Kinetics & Equilibrium Clues

ACROSS

1 Means the material is stable with respect to heat
6 Energy that is used to start a reaction
10 Refers to the steps a reaction follows as it goes to completion
11 A catalyst that has the same phase as the reactants
14 Meaning across from, such as across a double bond
16 The slow step in a reaction is known as the rate-_______ step
17 Material that affects the reaction rate but is not consumed
19 A biological catalyst is called this
20 Process done in some plants that converts nitrogen to compounds
21 Reaction with only one molecule involved in the rate step is
22 This is true of all chemical equilibrium systems. They are ____ and not static.

DOWN

2 Established when forward and reverse reaction rates are equal
3 Meaning on the same side, such as the same side of a double bond
4 The equation, rate = k[A][B] is known as a rate _____
5 Color of the element sulfur
7 Rate = k[A][B], k is the rate _________
8 Products formed but not stable when a reaction occurs are known as ____________ products
9 Process used to convert nitrogen to ammonia on a commercial scale
12 This kind of factor may affect the rate at which a reaction occurs
13 The rate when a reaction first starts is known as this rate
15 A reaction that is favored but occurs at a very slow rate is said to be ______ stable
18 Established when forward and reverse reaction rates are equal

Ties to Previous Chapters

Equilibrium is a thermodynamic property and relates to ΔG (we often use ΔH instead) which is a topic from chapter 7. Be sure you understand the relationship which exists.

It is essential that you have balanced equations when working kinetics or equilibrium problems. The classes of reactions and balancing of equations was covered in chapter 5.

Chapter 6 had some discussion on solutions and molarity. If necessary review section 6.5.

Answers to True/False Statements:

1. False. The fewer particles involved in a step the better the chances for interaction. It becomes a matter of too many just exact collisions having to occur together to be probable.

2. True. Concentration factors are important however.

3. True. This is also identified as speed and mechanism.

4. False. It refers to moles per liter of solution.

5. False. Any item can be measured. You just must account for an increase or decrease with the sign placed on the value.

6. False. Concentration is a major factor in reaction rates.

7. True. The equation shows it is a direct relationship.

8. True. Kinetics is an experimental science.

9. False. It is very common for equilibrium to be established. When this happens the reactants are reformed until no net change occurs. Useful kinetic data is collected before equilibrium occurs.

10. True. It is essential that only one variable change at a time.

11. False. Activation energy is a barrier. The lower it is the better the chance the reaction will occur.

12. False. The fewer particles the more favored. Two particles have a better chance than three of colliding properly.

13. False. Reaction coefficients are not related to reaction order.

14. True. The overall order is the sum of the exponents on each material involved in the rate step. In this case one for each.

15. True. Where the slow step occurs is immaterial to the result.

16. False. The Cl's are on the same side which is cis.

17. True. This term is the symbol for activation energy.

18. True. The generalization is that for each 10°C rise the rate will double. In this case $2^3 = 8$.

19. False. Thermodynamics does not predict reaction rates only if the reaction is possible.

20. False. That mechanism may be correct but it does not rule out other possible mechanisms. All that fit must be tested further.

21. True. After each step some kind of product will result. These may be very transitory in existence.

22. False. These are not the same phase. It is heterogenous.

23. False. They do not wear out. There may be mechanical loss.

24. False. The equal part of equilibrium refers to rates of reaction not to quantities.

25. True. Equilibrium constants >1 mean a product favored reaction is possible.

26. False. Since heat is a product of the reaction adding heat would tend to drive the reaction toward the reactants to relieve stress.

27. True. This is equivalent to increasing the concentration.

Answers to Understanding of Concepts Questions:

1. Since the reaction is exothermic heat is a product. If the reaction system were heated a stress would be placed on the product side. According to Le Chatelier's principle the reaction would shift toward the reactants.

   All of the materials are gases in this reaction and there are three moles of reactant gases and only two moles of product gas. Pressure would have a greater effect on the reactant than the product side. This would result in an equilibrium shift to form more products.

2. The gasoline reacts only when started. Starting the reaction in the engine involves compression and then a high voltage spark. The gasoline has an activation energy which is not available in the gas tank. The gasoline is kinetically stable in the gas tank and unless some activation energy is provided it creates no problem.

3. Chemical reactions are the result of particle collisions. After a period of time more and more reactant particles are used up and this leaves fewer available to collide. Fewer collisions results in ever slowing of the reaction.

4. Most reactions will reach an equilibrium condition. The process of coming to equilibrium is started almost as soon as there are product materials available in the reaction system. Products may alter the conditions of the reaction. The initial reaction has very little of this interference and provides the best information about the original reaction.

5. The spark plug provides the activation energy allowing the gasoline to react with the oxygen in the air. The reaction of gasoline with air is thermodynamically favored but is kinetically stable. The spark provides energy to overcome the initial energy barrier.

6. Biological reactions occur in the cell and the materials are all in solution. Having a catalyst not in solution would be inconvenient and also difficult to maintain in a living system. The catalyst would not be able to work as efficiently.

7. A unimolecular reaction has only one molecule reacting. The first order reaction has a single material present in the rate step but there may be more reactant materials.

8. This question is basic to reaction kinetics. What factors are important to the reaction? Temperature, when increased, almost always increases the rate. Increase the concentration of reactants by putting more in, increase the pressure in the case of a gas or increase surface area of solids by making particles smaller. Making a solution rather than having solids react would also help. Many reactions could be aided by having a catalyst available.

9. The first key to analysis is the time frame involved. Figure 8.1a covers only about the first 10% of the total reaction. This graph is to collect information for the instantaneous rate. Figure 8.1b involves a time which is encompassing most of the reaction. When using graphs for analysis it is important to recognize what the graph is attempting to portray. If one is not carefu, misleading results may be the outcome.

10. Molecularity refers to the number of particles involved in the reaction. Generally the more particles total the more likely more than one would be involved in the rate step. To get more particles to interact at the same time in exactly the correct orientation with the exact energy gets less likely as the number increases.

11. If the material is not involved in the rate step it will have no effect on the reaction rate.

12. If the reaction intermediates can be identified it helps confirm that a particular step is present in the reaction. This is very useful in outlining the reaction mechanism and confirms one possibility out of several possible.

13. Each hump refers to a product produced in one of the reaction steps. They are intermediate products. Most of these are very short lived and seldom are identified.

14. Stable refers to the resistance to change or reaction. A material may be defined as heat stable meaning heating does not destroy it, or it could be thermodynamically stable, meaning the energy of the reaction favors the reactants not the products. Finally, kinetically stable indicates there is a high activation energy which prevents the reaction from occurring even though the reaction energy may be favorable to reaction.

15. This question has no valid answer. The equil in equilibrium refers to rates of forward and reverse reactions not the amounts of reactants and products present. The results at equilibrium could vary from nearly all product to about equal to nearly all reactant.

16. The equilibrium constant is temperature dependent. This relates to the fact that the activation energy for the forward and reverse reactions are not the same. Changing the temperature, would affect one more than the other, which would shift the equilibrium status.

17. At equilibrium the solid present is interacting with the dynamic system. Some solid is going into solution and some is being deposited out. If the solid were removed,the ability for part of the reaction to occur would be lost and there would be a shift in the equilibrium.

18. A positive value for ΔH means the reaction is reactant-favored. The larger the value, the more it favors the reactants. The more the product is favored the larger the value of K. In this case the value of K would be predicted to be small, less than one.

Answers to Testing the Analytical Concepts

1. From the data we can determine that both A and B have an effect on the reaction rate. This tells us the rate law is

$$\text{Rate} = k[A]^m[B]^n$$

The first three runs have a constant A concentration and allow us to evaluate the effect of B. We find that at relative concentration increases for B the rate goes up by a square factor. The value of n is 2. The first and last two runs have B constant and as the concentration of A doubles the rate doubles so the value of m is 1.

The rate law is $\text{Rate} = k[A]^1[B]^2$

Picking data from one run, such as run 2, we can solve for k.

$$k = \frac{\text{rate}}{[A]^1[B]^2} = \frac{14.4 \times 10^{-6}}{(0.020)(0.020)^2}$$

$$= \quad 1.8$$

The overall order is 3rd. and the order with respect to A is 1st.

2. The data indicate that both I^{-1} and $S_2O_8^{-2}$ are in the rate step so the rate law will be rate = $k[I^{-1}]^m[S_2O_8^{-2}]^n$.

Data from runs 1 and 2 show that as $[S_2O_8^{-2}]$ is doubled the rate doubles. The reaction is first order in $S_2O_8^{-2}$. Runs 1 and 4 show that as the $[I^{-1}]$ is halved, the rate is halved so the reaction is first order in I^{-1}.

The complete rate law is rate = $k\,[I^{-1}]^1[S_2O_8^{-2}]^1$

3. The equilibrium expression for the reaction is:

$$K = \frac{[SO_4^{-2}]^2\,[I_3^{-1}]}{[S_2O_8^{-2}][I^{-1}]^3} = \frac{(0.75)^2(0.50)}{(0.05)\,(0.12)^3} = 3250$$

This is a product favored reaction in K is greater than 1.

4. The net reaction is what remains after the materials which are present on both sides of the reaction are canceled out.

slow	$H_2O_2 + \cancel{I^{-1}} \Longrightarrow H_2O + \cancel{OI^{-1}}$
rapid	$H_2O_2 + \cancel{OI^{-1}} \Longrightarrow H_2O + \cancel{I^{-1}} + O_2$
Net	$2\,H_2O_2 \Longrightarrow 2\,H_2O + O_2$

The rate of reaction relates to the materials in the slow step:

$$\text{rate} = k[H_2O_2][I^{-1}]$$

Notice that the rate equation includes a material that is not present in the net reaction.This shows the catalyst effect of the iodide ion.

5. Equilibrium expressions involve the product concentrations being written over the reactant concentrations.

$CH_3OH_{(g)}$ <===> $CO_{(g)}$ + $2H_2(g)$

$$K = \frac{[CO\ ]\ [H_2]^{\ 2}}{[CH_3OH]}$$

$CaCO_{3(s)}$ <===> $CaO_{(s)}$ + $CO_{2(g)}$

Solids do not appear in the equilibrium equation.

$K = [CO_2]$

$H_2S_{(g)} + I_{2(s)}$ <====> $2HI_{(g)} + S_{(s)}$

$$K = \frac{[HI]^2}{[H_2S]}$$

$SO_2 + 1/2\ O_2$ <====> SO_3

$$K = \frac{[\ SO_3]}{[SO_2]\ [O_2]^{1/2}}$$

6. Start with the equilibrium expression which is $K = \frac{[NO_2]^2}{[N_2O_4]}$

and by rearrangement we get $[N_2O_4] = \frac{[NO_2]^2}{K}$

$$[N_2O_4] = \frac{(0.0113)^2}{4.61 \times 10^{-3}} = 0.0277$$

7. From the data given we need to calculate the concentration of the given material, HI, in moles per liter. We then need a table to show what happens to its concentration and the concentration hydrogen and iodine at the elevated temperature.

$$[HI] = \frac{0.090 \text{ mol}}{3.00 \text{ Liters}} = 0.030 \text{ M/L}$$

(concentrations in M/L)

material	HI	H_2	I_2
init conc	0.030	0.00	0.00
conc change	-2x	+x	+x
conc at equil	0.030 - 2x	x	x

Equilibrium Expression is

$$K = \frac{[H_2]\,[I_2]}{[HI]^2} = 3.60 \times 10^{-2} \quad \text{(K value given)}$$

$$3.60 \times 10^{-2} = \frac{(x)(x)}{(0.030-2x)}$$ take the square root of both sides

$$0.190 = \frac{x}{0.030 - 2x}$$ thus $x = 0.190(0.030-2x)$

$x = 0.00569 - 0.38x$ or $1.38x = 0.00569$
$x = 0.00412$ which is the hydrogen and iodine concentration
equilibrium concentration of HI = 0.030 - 0.00412 = 0.0259

Answer to Chapter 8 Crossword

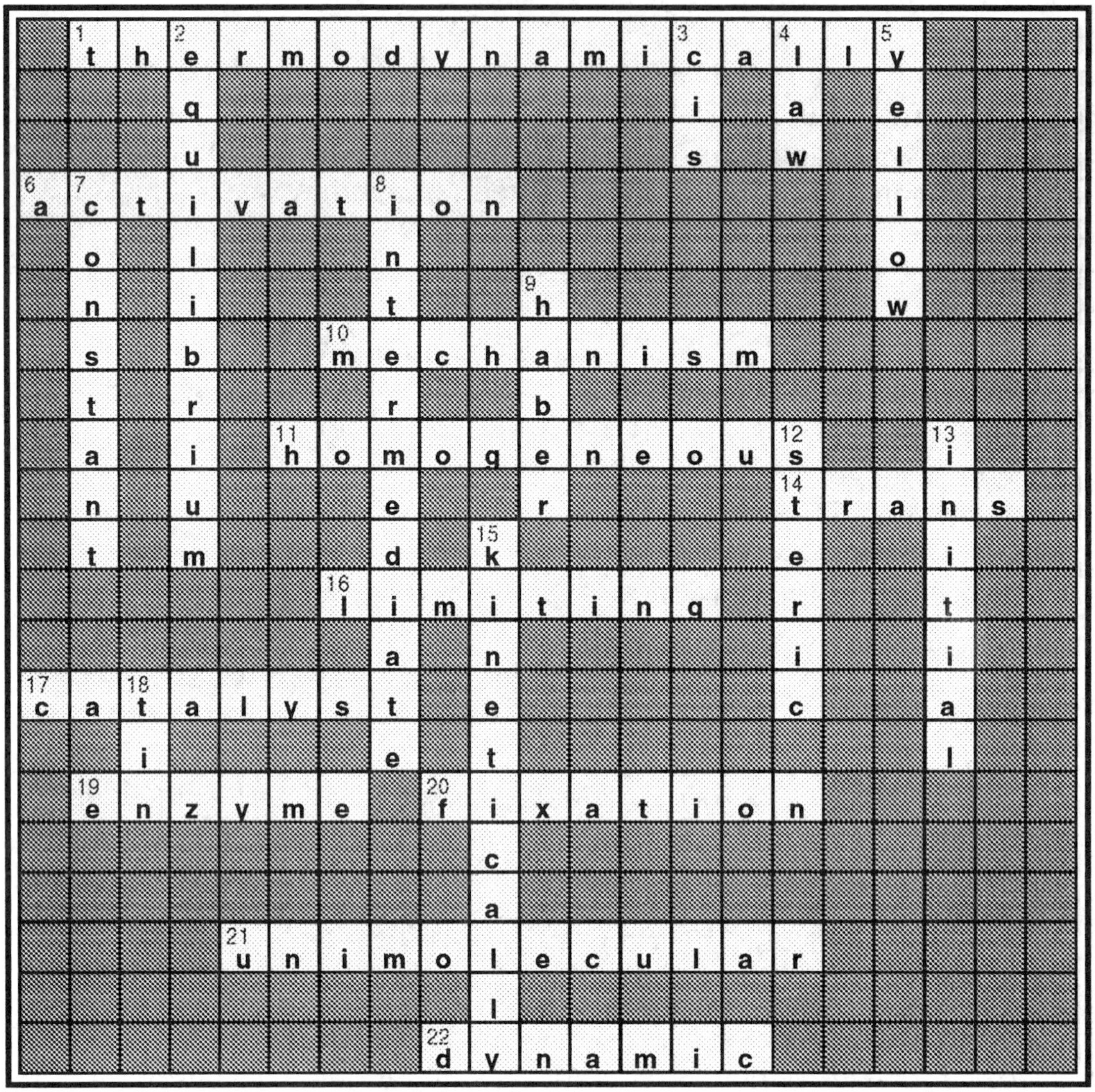

Some Supplementary Readings Relating to This Chapter:

1. A Promising Way to Cut Nitrogen Oxides at Power Plants, The New York Times, June 28, 1992.

2. Hot & Cold Packs, Chem Matters , February 1987, page 7.

3. Poison Ivy, Chem Matters , October 1990 , page 4.

4. The Exploding Tire, Chem Matters, page 12.

5. The Fruits of Ethylene, Chem Matters , April 1989, Page 11.

6. The Smell of Danger, Chem Matters , October 1988, page 9.

7. Concern rises over nuclear waste explosion risk, Chemical and Engineering News , October 29, 1990, page 20

8. Insect Arsenals, Chem Matters , October 1993, page 8

Chapter 9 Electronic Configurations

This chapter is a continuation and extension of the periodic table introduced in chapter 3. The original ideas which led to development of the periodic table were based entirely on observation and collection of facts.

This chapter will provide information about how the periodic table was given a solid theoretical foundation. Several periodic trends are discussed and the family relationships in the periodic table are explained.

Focus Your Study Efforts

All electrons in an atom are not equivalent.

Objectives:

1. You should be able to explain which electrons in an atom will be involved in chemical bond formation.
2. You should be able to use the periodic table to determine the number of valence electrons an atom has.

Electrons are involved in chemical bond formation. Not all electrons are capable of bond formation, however. Chemists have subdivided electrons into two groups:

Valence electrons - those in the outermost energy level and will be involved in bond formation.

Core or Kernel electrons - these electrons do not bond because the energy of interaction with the nucleus of their own atom is very high.

The periodic table is useful in determining how many valence electrons an atom has. For those in any A group it equals the group number.

Just because an electron is in the valence shell does not mean it will be involved in bonding. Those atoms that gain electrons, such as chlorine, may have only one of seven valence electrons bonding.

Electromagnetic radiation may cause atomic changes or be used to study changes in atomic structure.

Objectives:

1. You should be able to predict energy characteristics of the electromagnetic spectrum.
2. You should be able to explain, and calculate, the relationships between wavelength and frequency of electromagnetic radiation.

Electromagnetic radiation covers a span from high energy cosmic rays to low energy radio waves. Visible radiation, light, is a very small part of this spectrum.

Different areas of the radiation spectrum are used by chemists to accomplish different tasks. Since they have different energies different parts of the atom or molecule can be affected.

An important point to remember is that all radiation travels at the same speed, the speed of light, in a vacuum. If either wavelength, λ, or frequency, ν, changes the other will also change. Wavelength and frequency are related through the speed of light constant, c, $c = \lambda \cdot \nu$. The equation shows that as one value gets larger the other value must get smaller.

Infrared radiation may have a frequency of 3.86×10^{14} hertz. What is the wavelength of this radiation? What is the energy content of a quantum of this wavelength?

To solve these problems we must remember that the speed of light is assumed as constant and that wavelength and frequency are related through this constant.

$$c = 2.998 \times 10^{8} \text{ m/s} \qquad c = \lambda \bullet \nu$$

Plank provided evidence that energy is related to frequency by the equation $E = h \bullet \nu$. The value of $h = 6.62 \times 10^{-34}$J s. Looking at the data and formulas we see that the problems are independent and either could be solved first.

$$\lambda = \frac{c}{\nu} = \frac{2.998 \times 10^{8} \text{ m/s}}{3.86 \times 10^{14} \text{ sec}^{-1}} = 7.77 \times 10^{-7} \text{ meters}$$

$$E = h \bullet \nu. = 6.62 \times 10^{-34} \text{J s.} \times 3.86 \times 10^{14} \text{ sec}^{-1}$$
$$= 2.56 \times 10^{-19} \text{ J}$$

Notice that an individual quantum of radiation has an extremely small amount of energy. Only when a large amount of radiation is available is substantial energy present.

Energy is carried by a wave. The more waves passing a point the more energy transmitted. Shorter wavelength, or higher frequency, means a greater amount of energy.

The wavelength is the distance from a point on the wave until it starts to repeat itself. This could be measured at any point. The crest, or top point, of the wave is often used because it is convenient and obvious.

Frequency is measured in reciprocal second units, s^{-1}. The larger the number the more energy. A convenient name given for the unit is hertz, Hz.

Radiation energy may be measured in quantum units.

Objectives:

1. You should be able to explain the quantum concept as it relates to the electromagnetic spectrum.
2. You should be able to why some spectra are "line spectra" and others are continuous.

An energy "quantum" refers to a minimum, or discrete, quantity. The smallest packet. This idea became the basis for the quantum theory of radiation.This theory can be used to quantify the relationship between the frequency of radiation and the energy content: $E = h\nu$.

h is a proportionality constant named after Max Planck.

The photoelectric effect is a statement of the fact that a certain minimum energy content is required to cause any effect in removal of electrons from a system. It is not dependent upon how much radiation is applied, just the energy content. However, once the minimum energy value is surpassed the intensity will have an effect.

Photons are quanta. They are massless radiation that have both particle-like and wave-like properties.

Electromagnetic radiation is not a wave or particle. It is an entity with properties that, at present, we do not have the ability to understand or explain in complete detail.

Radiation may be the result of continuous or discrete energy quantities. A broad band radiation source, such as the sun or an incandescent light bulb, emits energy over the entire visible spectrum and if this light is separated we can see a continuous spectrum of all visible colors. In contrast, if you take a colored light source such as a sodium vapor lamp (some yellow street lights have this) or a neon sign, analysis of the radiation shows only specific color bands in the visible region. This results in a discrete line spectrum.

Any material that will not transmit all wavelengths of light equally will separate a light source into its base components. A prism or diffraction grating is capable of this.

Models are used to explain the atom and properties of atoms.

Objectives:

1. You should be able to compare and contrast the Bohr model of the atom to the quantum mechanical model.
2. You should be able to describe why the Bohr model was significant in the development of atomic models.

Ever since the idea of an atom has been considered there has been a model to explain and/or picture it. Remember, the interpretation we have is a model, based on the best current data.

The current model is known as the Quantum Mechanical Model. A significant aspect of it is that a probability factor is incorporated.

Using the model chemists attempt to draw pictures and make comparisons to things we can relate to in terms of our everyday experiences. When we talk about shells or represent electrons as lines in a box it is probably more representation than real.

Much of the initial background for the current model is the result of work done by Neils Bohr in 1913. When Bohr looked at the spectra generated when gases of pure elements were subjected to high energy he saw a series of lines. He used the bright line spectrum from hydrogen to develop a model.

Line spectra are the result of electrons moving from a ground state to an excited state, and upon returning to the ground state, releasing energy which creates the bright line. From the fact there were lines, not a continuous spectrum, Bohr concluded there were specific regions in space where electrons could reside, represented by their principle quantum numbers. These had a fixed amount of energy, a quantum. Electrons could be excited only if energy applied matched the quantum steps of the principle quantum numbers.

The Bohr model was soon replaced, but it was extremely important because it introduced and validated the idea of quantization of energy.

de Broglie expanded the Bohr concept by suggesting matter, like light, could have both particle and wave properties.

The energy an electron has is very small. When an attempt is made to locate it, the energy of the searching medium is close to the content of the electron. The result is, when the probe "finds" the electron it also imparts energy to it. The net effect is that you only know where the electron <u>was</u>. This dilemma was summarized by Heisenberg in what is known as the **uncertainty principle**.

The current atomic model has a probability factor.

Objectives:

1. You should be able to explain the use of the quantum mechanical model of the atom to represent energy and probable location of electrons.

We now describe electrons in terms of probability rather than position. The probability of an electron being found is greater in some places than others which is what our model is designed to show. Electrons occupy a region of space not just a place or path.

The Quantum Mechanical Model is based on the wave-particle duality and probability aspects. Electrons are described in terms of regions in space with a boundary surface representing a probability (usually 90%) of finding the electron there.

Wave functions are solutions to the mathematical model that defines the areas of highest electron probability. The solution to the wave equation defines the shells, principle quantum numbers, n; the subshells which define electron density patterns in s,p,d and f orbitals.

Boundary surface diagrams, such as Figure 9-10 in your text, are the result of solving the wave equations.

Quantum numbers and orbitals represent regions in space. Those space regions are there all the time even if electrons are not present in them. For example, a hydrogen atom in the ground state has an electron in the 1s orbital, but when energy is applied the electron moves to other spacial regions, that are already there, and does not create them.

Electrons are involved in many atomic properties.

Objectives:

1. You should be able to explain the how electron transitions result in line spectra.
2. You should be able to explain how magnetic properties relate to unpaired electrons in atoms.
3. You should be able to describe how the effect represented by the Pauli principle affects the shape of the periodic table.
4. You should be able to how electron configurations can be read from the atoms position in the periodic table.
5. You should be able to write Lewis dot structures for the elements in the A groups of the periodic table.

The line spectra resulting from electron transitions, such as shown in Fig.9.8 of your text, are the result of transitions occurring simultaneously in huge numbers of atoms. Each hydrogen has only one electron and it can undergo only one transition at a time. However, many hydrogen atoms are excited at once and their electrons are excited by differing amounts of energy and this results in all the possible spectral lines being visible.

Electrons can have two interactions with a magnetic field. The spins can be aligned with the field and opposite to the field. These alignments are quantized and called **electron spin**.

When electron spins are opposite they are "paired" and this favors electrons being in the same orbital. If they are not paired the term used is "parallel" and the interactions are less favored and electrons will occupy different orbitals.

When electrons are not paired they can interact with external magnetic fields and this gives the atom (element) magnetic properties. If all shells are filled the electrons are all paired and the atom will not interact with a magnetic field. The later is known as diamagnetic. Paramagnetic character is present in atoms which have unpaired electrons and interaction with the external magnetic field does occur. If this magnetic character is retained, it is said to be ferromagnetic. Ferromagnetism is common only to a few elements for example; iron.

The Pauli exclusion principle states that no two electrons in a given atom can have the same exact quantum number representation. Only two electrons, each having a different spin, can occupy the same orbital. This restriction places a limit on the number of atoms which are present in each principle quantum level. The value is $2n^2$ for any given principle quantum number. For example the third energy level can have up to 18 electrons.

The atom arrangement in the periodic table is detailed by the Pauli principle.

The representation used to place electrons in atoms or ions is known as the electron configuration. The configurations start with a principle quantum number which is divided into subshells and these in tern are subdivided into sections called orbitals. Electrons will always occupy the lowest energy position available.

The energy values between principle quantum numbers are not the same. This is demonstrated in Fig. 9.8 of your text which shows that the transition between the first and second energy level has energy in the ultraviolet region but transitions between the second and third energy levels are in the lower energy visible region.

The ground state is when all the electrons in an atom are in the lowest energy positions available.

Electrons all have negative charge. Like charges repel, so when electrons are arranged in the atom they do so to minimize interactions. One way to minimize the interactions is to never pair electrons in an orbital until all orbitals are first filled with an unpaired electron. The is expressed as Hund's rule.

Electron configurations can provide useful information, such as showing the valence shell electrons. Writing the electron configuration, especially with large atomic number elements, can be tedious and prone to error. This can be simplified by using the noble gas notation for the nearest completed energy level and then filling in from there. For example strontium has 38 electrons and only two of them are beyond the noble gas configuration for krypton so the Sr configuration can be shortened to $[Kr]5s^2$.

Draw the electron configuration for calcium and bromine. (A) Show them using the electron in a box method. (B) Identify the core electrons, the valence electrons and the noble gas notation.

Starting with a standard configuration so we have a reference to check back against we can write spdf patterns:

Ca # 20 $1s^2 2s^2 2p^6 3s^2 3p^6 4s^2$
Br # 35 $1s^2 2s^2 2p^6 3s^2 3p^6 3d^{10} 4s^2 4p^5$

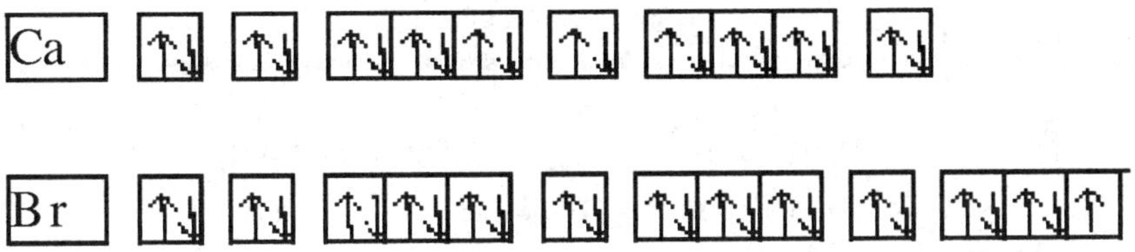

1s 2s 2p 3s 3p 4s 4p

Core	Valence	Short hand
$1s^2 2s^2 2p^6 3s^2 3p^6$	$4s^2$	$[Ar]4s^2$
$1s^2 2s^2 2p^6 3s^2 3p^6 3d^{10}$	$4s^2 4p^5$	$[Ar]\ 4s^2 4p^5$

The electrons of most interest to chemists are those in the valence shell. The other electrons, called core electrons or kernel electrons, are much too tightly bound to the nucleus to chemically interact with any other atom.

A convenient method of showing the valence electrons for the s and p type elements is with a Lewis dot symbol, or Lewis electron dot structure. With this structure the symbol represents the nucleus and the core electrons and a dot (or some other symbol) is used for each electron in the valence shell. For example; fluorine, magnesium and phosphorus, would have the following structures:

$:\ddot{\underset{\bullet}{F}}:$ $\qquad$ $\cdot Mg \cdot$ $\qquad$ $:\dot{\underset{\bullet}{P}}\cdot$

The periodic table is an expression of electron arrangements.

Objectives:

1. You should be able to use the periodic table to write electron configurations for main group and transition elements.
2. You should be able to use the periodic trends to predict trends in size, ion formation reactivity of the elements.
3. You should be able to list some properties of elements from the various groups of the periodic table.

All elements in the same family will have exactly the same number of dots. This makes it very easy to remember how to write them. For example the 1A elements; H,Li,Na,K,Cs and Fr all would be just the symbol with one dot.

$Na \cdot$

When the s and p block elements gain or lose electrons they will do so until they have the same number of electrons as the nearest noble gas element. For example, K will lose one electron to have the same number as argon, and chlorine will gain one electron and have as many as argon. Those materials, which have the same number of electrons, are **isoelectronic** with each other.

The prefix **iso** appears many times in chemistry. Remember it means, the same as, such as species with the same number of electrons.

Write symbols for materials with Lewis dot structures isoelectronic to Ca^{+2} :

Isoelectronic means the materials must have the same electron configuration as Ca^{+2} . Ca has 20 electrons so the ion has only 18.

K^{+1}	Ar	Cl^{-1}	S^{-2}	P^{-3}

Electron configurations for the transition elements are not as predictable as those for the s and p block elements. With the transition elements the filled and half-filled d or f orbitals have enhanced stability. For this reason when the d or f orbital is within one electron of being filled or half-filled, an electron is "borrowed" from the s orbital leaving it also half filled. Copper for example is $[Ar]3d^{10}4s^1$ rather than $[Ar]3d^9 4s^2$.

Some orbital electron energy levels are very close in energy content. This makes it possible for electrons to be taken from one orbital rather than another as might be expected. This is important to the electron configurations noted in the comment just above and also in the formation of transition metal ions.

Periodic trends for properties of the elements are common. Entire books have been written about various trends. There are several common, easy to visualize trends you should know. They include: atomic radius, ionic radius, ionization energies and electronegativity.

Progressing down a family (column) in the periodic table causes energy levels to be added. This results in an increase in atomic radii.

Progressing from left to right in a row (period) results in atomic radii gradually decreasing. The electrons are all being added to the same energy level at the same time more protons are added to the nucleus. This increase in the attractive force of the nucleus results in a slight size decrease.

The next element after a noble gas element always has a large increase in radius because the electron is forced into a new, larger energy level. When ions are formed the tendency is to gain or lose electrons to attain the nearest noble gas electron configuration. For cations (positive ions) formed by the loss of electrons the ionic radius decreases dramatically in size from the neutral atom. Anions (negative ions) are larger than the neutral atom because more electrons are interacting with the same number of protons and the effect is less attraction per electron.

Ionization energy refers to producing cations only, not anions. These energies are always measured with the elements in the gas phase.

The removal of each subsequent electron becomes more difficult. As electrons are removed and the number of protons remains constant, the effective pull of the nucleus increases. The second ionization energy is always greater than the first.

When forming cations it is very difficult to remove electrons that are not in the valence energy level. When that happens the ionization energy shows a dramatic increase.

Going down any particular family results in a larger atom because energy levels are being added. The interaction of the nucleus for electrons obeys Coulomb's Law which means the force is inverse to the distance squared. This results in a decrease in ionization energies as you go down any family because the size increases.

Compounds of hydrogen are known as hydrides. If the hydrogen reacts with a metal it forms the anionic hydrides since it has a negative charge and has gained an electron from the metal. With nonmetals it forms molecules known as molecular hydrides.

Since hydrogen, H_2, is such a small molecule it can be absorbed into the lattice holes of many metals. This set of hydrides are known as interstitial metallic hydrides. There is no real chemical reaction, just a physical absorption of H_2 gas.

The Group 1A elements are named Alkali Metals. They all have similar chemical and physical properties. They all are very reactive chemically. They are different, however, in how they react with oxygen. You may get an oxide such as Li_2O, or a peroxide like Na_2O_2, or a superoxide KO_2. Notice that the percentage of oxygen in the compounds is increasing.

Group 2A elements, known as the Alkaline Earth metals, are less reactive than the corresponding 1A elements because two electrons must be lost rather than just one.

Some of the Group 3A elements have nonmetallic properties. This group has much more varied chemical and physical properties than either the 1A or 2A metals. Some of the elements, such as aluminum, show characteristics of either metals or nonmetals, depending on the chemical environment. As a group, the metallic properties increase as you progress down the family.

The elements in Group 4A are about half nonmetals and half metals. Carbon and silicon are very common elements and their importance to us can not be overestimated.

Group 5A elements react to form oxides. Some, such as nitrogen, form many different oxides. Nitrogen and phosphorus are used to produce two very important industrial acids.

Elements in Group 6A are reasonably reactive nonmetals, yet two, oxygen and sulfur, are found free in nature.

The halogens, Group 7A, are all reactive elements. Compounds of these elements are important sources of many materials in nature.

The noble gas elements, Group 8, are known for their lack of chemical reactivity.

Symbols found in Chapter 9

Page	Symbol	
352	ν	frequency of radiation
352	λ	wavelength of radiation
355	h	Planck's constant
353	c	the speed of light constant
354	hz	hertz, name given to the frequency unit s^{-1}
354	MHz	megahertz unit which is 10^6 hertz
354	s^{-1}	reciprocal centimeters, frequency unit
360	n	principle quantum number symbol
366	spdf	symbols given to orbitals
368	$\uparrow\downarrow$	represent spin on electrons
368	$2n^2$	maximum number of electrons in that energy level
383	pm	picometer 10^{-12} meters, used for size of atoms

Terms in Chapter 9

The following terms are listed for your convenience. They should not become a list for you to memorize. You should have a working definition for each. If you have doubt about any particular term check its definition in the glossary and/or review it in the chapter. Some of these terms are included in the crossword puzzle later in the chapter.

valence electrons	electromagnetic radiation	
wavelength	frequency	hertz
quantum	Planck's constant	quantum theory
photoelectric effect	photon	continuous spectrum
line emission spectrum	excited state	ground state
principle quantum number		uncertainty principle
probability	wave equation	boundary surface
orbital	shell	subshell
Pauli exclusion principle		paired electrons
parallel spin	electron configuration	Hund's rule

noble gas notation	core electrons	octet rule
Kernel	Lewis dot symbols	s-block electron
p-block electron	isoelectronic	transition element
inner transition element	diamagnetic	paramagnetic
ferromagnetic	atomic radius	ionic radii
ionization energy	anionic hydride	
molecular hydride	interstitial metallic hydride	
peroxide ion	superoxide	flux
synthesis gas	ore	gangue

Concept Maps for Electronic Configurations

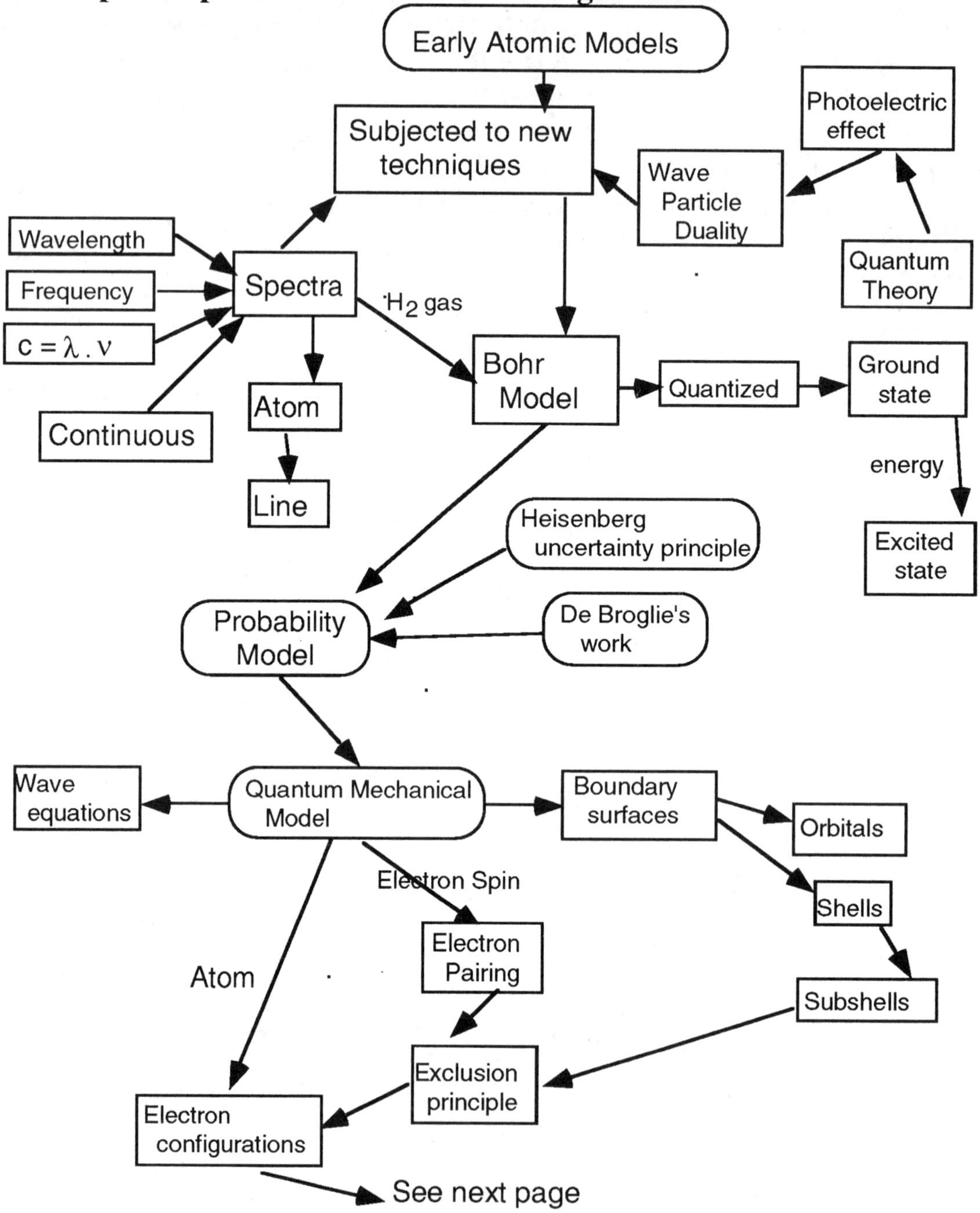

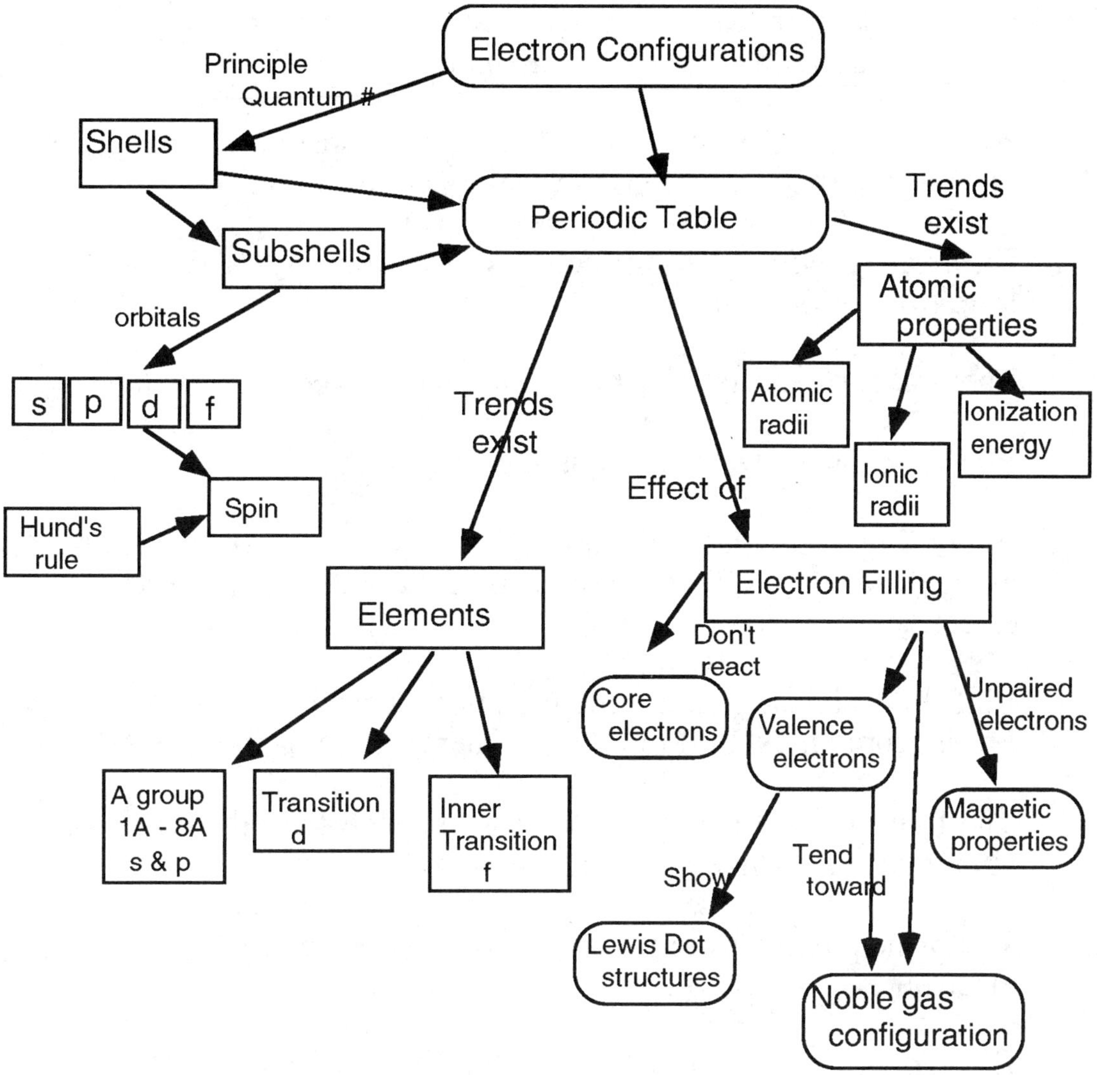
Electron Configurations
Principle Quantum #
Shells
Periodic Table
Subshells
Trends exist
Atomic properties
orbitals
s
p
d
f
Spin
Hund's rule
Trends exist
Effect of
Atomic radii
Ionic radii
Ionization energy
Elements
Electron Filling
Don't react
Core electrons
Valence electrons
Unpaired electrons
Magnetic properties
A group 1A - 8A s & p
Transition d
Inner Transition f
Show
Tend toward
Lewis Dot structures
Noble gas configuration

Electronic Configurations True/False Statements

Answers are at the end of the chapter.

1. According to the equation $c = \lambda \bullet \nu$, as the wavelength increases, the frequency must increase to maintain c as a constant.

2. Electromagnetic radiation includes only infrared, visible and ultraviolet radiation.

3. The unit for hertz is s^{-1} which is related to frequency values.

4. Plank's constant relates frequency and energy.

5. According to the photoelectric effect any amount of radiation energy causes an effect in metals that show the photoelectric effect.

6. Line emission spectra are the result of electrons changing energy levels from an excited to a lower energy state or the ground state.

7. The four visible lines in the hydrogen spectra are the result of a simultaneous change of a single electron.

8. The principle quantum number is the major determinant of electron energy.

9. Excited electronic states are possible for all atoms.

10. Modern atomic theory refers to the electron location in terms of probability.

11. Diagrams or models for electron densities set the maximum size for the orbitals.

12. Electronic subshells have a maximum capacity which varies with the principle quantum number.
13. Energy is required to get electrons to pair in the subshells.

14. Electrons occupy energy levels in a specific order filling lower energy positions before higher energy orbitals.

15. Once electrons fill an orbital position they will never change except upon ionization.

16. Only valence electrons are involved in chemical bonding.

17. All 5A elements will have the same electron dot structure.

18. Both d and f orbital filling elements are known as transition elements.

19. Atoms with unpaired valence shell electrons are diamagnetic.

20. Any time an atomic number increases the size of the atom increases.

21. Cation radii are always smaller than the same atom's atomic radii.

22. Transition elements have less variation in ionization energy than do the representative elements in the same period.

23. The third ionization energy for the element strontium is much greater than the second ionization energy.

24. All hydrides have hydrogen with a negative charge.

25. Ores are minerals with a quantity of an element that makes it profitable to extract.

Testing Your Understanding of Concepts in Chapter 9.
Answers are at the end of the chapter.

1. Describe the relationship that exists between frequency and wavelength.

2. Why, when a specific sample, is radiated with high intensity low energy radiation, there is no photoelectric effect, but when radiated with low intensity radiation of higher energy, a current flows.

3. Describe how a line spectrum is created, for example from a sodium vapor lamp.

4. What is the principle quantum number for valence electron(s) in (A) Cs and (B) Ca?

5. Explain the relationship between the Heisenberg uncertainty principle and our description of electronic structure.

6. How many subshells are present for an element in the 5th period of the periodic table?

7. What is a boundary surface? What does it relate to?

8. A particular ion is found to be paramagnetic. What does that tell us about the ion?

9. Which electronic configuration best represents Si and why?

$$1s^22s^22p^63s^23p_x{}^2p_y{}^0p_z{}^0 \quad \text{or} \quad 1s^22s^22p^63s^23p_x{}^1p_y{}^1p_z{}^0$$

10. Explain why cation radii are smaller than the starting atomic radii but anion radii are larger than the starting radii.

11. Why are ionization energies more useful for metallic elements than the nonmetallic elements?

12. How are interstitial metallic hydrides different than the ionic or molecular hydrides?

13. Why is discussion of the oxides of the 1A metals more complex than the 2A metal oxides?

14. Why do only some noble gas elements form compounds?

Testing You on Some Analytical Concepts.
Answers are at the end of the chapter.

1. Green light has a wavelength of 550 nm. Calculate the energy of a photon in joules.

2. Give an orbital diagram for titanium.

3. Which atom is larger in each set: (a) Sb or P ? (b) Si or Na ?

4. Which particle is larger in each set: (a) Co^{+2} or Co^{+3} ? (b) K or K^{+1} ? (c) F or F^{-1} ?

Crossword - Electronic Configurations

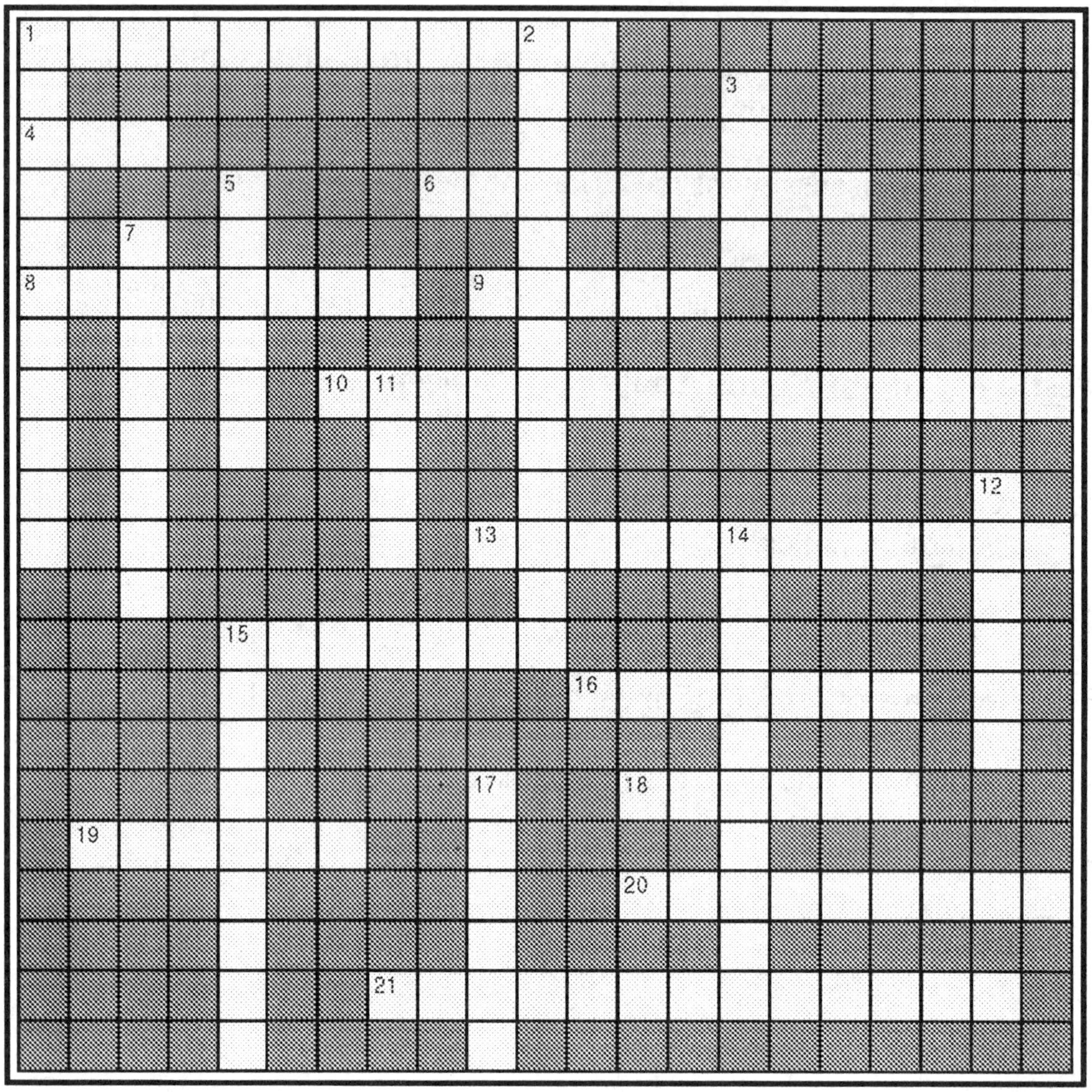

Crossword - Electronic Configurations Clues

ACROSS

1 Property of an atom with unpaired electrons.
4 A mineral that is profitable to extract an element from.
6 Radiation energy is directly related to this measurement.
8 Type of surface diagram for electron densities.
9 Name for frequency unit.
10 Term for the radiation spectrum from X-rays to radio waves.
13 Hydrides that have neutral hydrogen atoms combined with metals.
15 Hydride that is the result of hydrogen reacting with active metals.
16 These electrons are the ones involved in chemical bond formation.
18 Electronic state where all electrons are in the lowest energy possible.
19 Constant that relates energy to frequency.
20 Quantum number that represents the periods in the periodic table.
21 This effect is present when light causes electrons to be ejected from a metal.

DOWN

1 The quantum mechanical model of the atom identifies electron location in terms of _______.
2 Atoms or ions with the same number of electrons.
3 This rule summarizes how subshells are filled.
5 Atomic or ionic sizes are given in terms of their ______.
7 These are designated by one of the letters s,p,d or f.
11 This kind of spectra is the result of electrons dropping from an excited to ground state.
12 Impurity material in mined minerals.
14 This is the kind of oxide that potassium would form.
15 Group of transition elements that are all radioactive.
17 This is what "particles" of light are called.

Ties to Previous Chapters

This chapter builds on periodic properties started in chapter 3. The material in chapter 3 was presented with no theoretical background but based on experimental observations. This chapter presents the theoretical foundation for the periodic table. It may be worth your time to review the general information from chapter 3.

This chapter also extends the information base on the structure of the atom started in chapter 3.

Answers to True/False Statements:

1. False. Since the product is a constant, if one increases the other must decrease. So the frequency must decrease.

2. False. What we call electromagnetic radiation includes everything from radio waves to gamma radiation.

3. True. Both parts of the statement are correct. The symbol is reciprocal seconds and it does refer to frequency.

4. True. A check of the equation shows that Plank's constant this is correct.

5. False. Every material has a threshold energy value which must be present before electrons are released. This does vary from element to element. After this energy is reached the intensity of the radiation will affect the amount of photoelectric effect noted.

6. True. Energy is absorbed when the electron is excited and this is released when the electron returns to the ground state.

7. False. Hydrogen has only one electron and it can only undergo one transition at a time. However, a sample will have a very large number of atoms and they undergo varying changes with the four lines being the net result.

8. True. The principle quantum number is the determinant of the gross energy. The subshells just subdivide the major part.

9. True. The energy to cause the excitation will vary to a great extent.

10. True. This relates to the Heisenberg Uncertainty Principle.

11. False. The maximum size is unknown. The model represents a specific percentage, usually 90%.

12. False. The subshells always hold the same number: s =2, p =6, d =10, and f =14.

13. True. This is shown by the fact that a subshell, such as the p, will fill each of the three sections with a single electron before any will have two. This is expressed as Hund's rule.

14. True. The lowest available energy position has the greater stability. Filling is always from lower to higher energy content.

15. False. This is shown with elements such as copper where an s type orbital will have 2 electrons in nickel but only one in copper.

16. True. The other electrons require too much energy to be activated for bonding. They would have to come from lower energy orbitals.

17. True. Electron dot structures show only the valence electrons and all elements in the same column have the same number.

18. True. Those with f orbitals are often also referred to as the inner transition.

19. False. The unpaired electrons can interact with a magnetic field and these are paramagnetic not diamagnetic.

20. False. Within a given period there is a decrease in size as the atomic number is increasing.

21. True. The formation of the cation removes an electron. Those electrons that remain receive more interaction from the protons which were not changed. Also removal of electrons may decrease the number of orbitals being used.

22. True. For some periods the change is very small.

23. True. This is because the removal of the third electron must involve a whole new energy level which is closer to the nucleus.

24. False. Some like the molecular hydrides have a + charge on hydrogen and the interstitial metal hydrides are not charged.

25. True. This is a definition.

Answers to Understanding of Concepts Questions:

1. The product of these two factors always leads to a constant value, the speed of light. As one of these values increases the other must decrease in value. The longer the wavelength the lower the frequency.

2. Every metal has a threshold energy which must be met before the electrons can be released from the nuclear control. Once the minimum energy is met the number of electrons released is a function of the amount of effective energy present

3. Energy is supplied to element which excites electrons from the ground state to excited states. Different electrons will be promoted to various excited states. Excited states are not stable and the electrons will return to the ground state and release the energy used to excite them. This release of energy is what causes the line spectra.

4. For Cs it is the 6th and for Ca it is the 4th.

5. The Heisenberg uncertainty principle states that both the position and momentum of an electron cannot be known because the energy needed to locate it is sufficient to change its position. You only know where it was. This results in describing electrons in terms of probability locations.

6. The fifth period will have a single s, three p, five d and seven f orbitals.

7. The boundary surface represents a fixed percentage of space for finding an electron. This often is 90%. It is the result of mathematical calculations. To have 100% certainty the electron could theoretically be found any place in the universe.

8. The presence of unpaired electrons results in the material having paramagnetic properties. It would interact with an applied external field.

9. According to Hund's rule electrons will not pair until forced to. This means the second configuration is preferred.

10. When a cation forms electrons are lost from the valence shell. Often the valence shell is emptied and this decreases the size. Anions, however, gain electrons in the valence shell so they expand.

11. The ionization energy is defined as the energy to <u>remove</u> an electron from the gas phase. Metals lose electrons but the nonmetals tend to gain so the term does not normally apply.

12. The anionic and molecular hydrides have the hydrogen bonded. The interstitial hydrides have hydrogen molecules trapped in the materials lattice.

13. The 1A metals form different kinds of oxides, such as peroxides and superoxides. This is not normal for the 2A metals when reacting with oxygen.

14. Only the noble gases with large atomic radii have electrons far enough from the nucleus and low enough ionization energy to react.

Answers to Testing the Analytical Concepts

1. $$E_{quantum} = h \cdot \nu = h \times \frac{c}{\lambda}$$

$$= 6.626 \times 10^{-34}\ \text{J}\cdot\text{s} \times \frac{2.998 \times 10^{8}\ \text{m/s}}{550\ \text{nm} \times 1 \times 10^{-9}\ \text{m/nm}}$$

$$= \frac{6.626 \times 10^{-34}\ \text{J} \times 2.998 \times 10^{8} \times 10^{9}}{550} = 3.61 \times 10^{-19}\ \text{J}$$

2. Ti has 22 electrons:

$1s^2 2s^2 2p^6 3s^2 3p^6 3d^2 4s^2$ (d electrons not paired)

$[Ar]3d^2 4s^2$

3. (a) Sb is larger it has two more energy levels than P.
   (b) Na, when going across a period the atoms decrease in size.

4. (a) Co^{+2} would be larger since there are no electrons taken from the d orbital. The third electron comes from the d orbital.
   (b) K is larger. Atoms are larger than their cations.
   (c) F^{-1} is larger. The anion is larger than the atom.

Answer to Chapter 9 Crossword

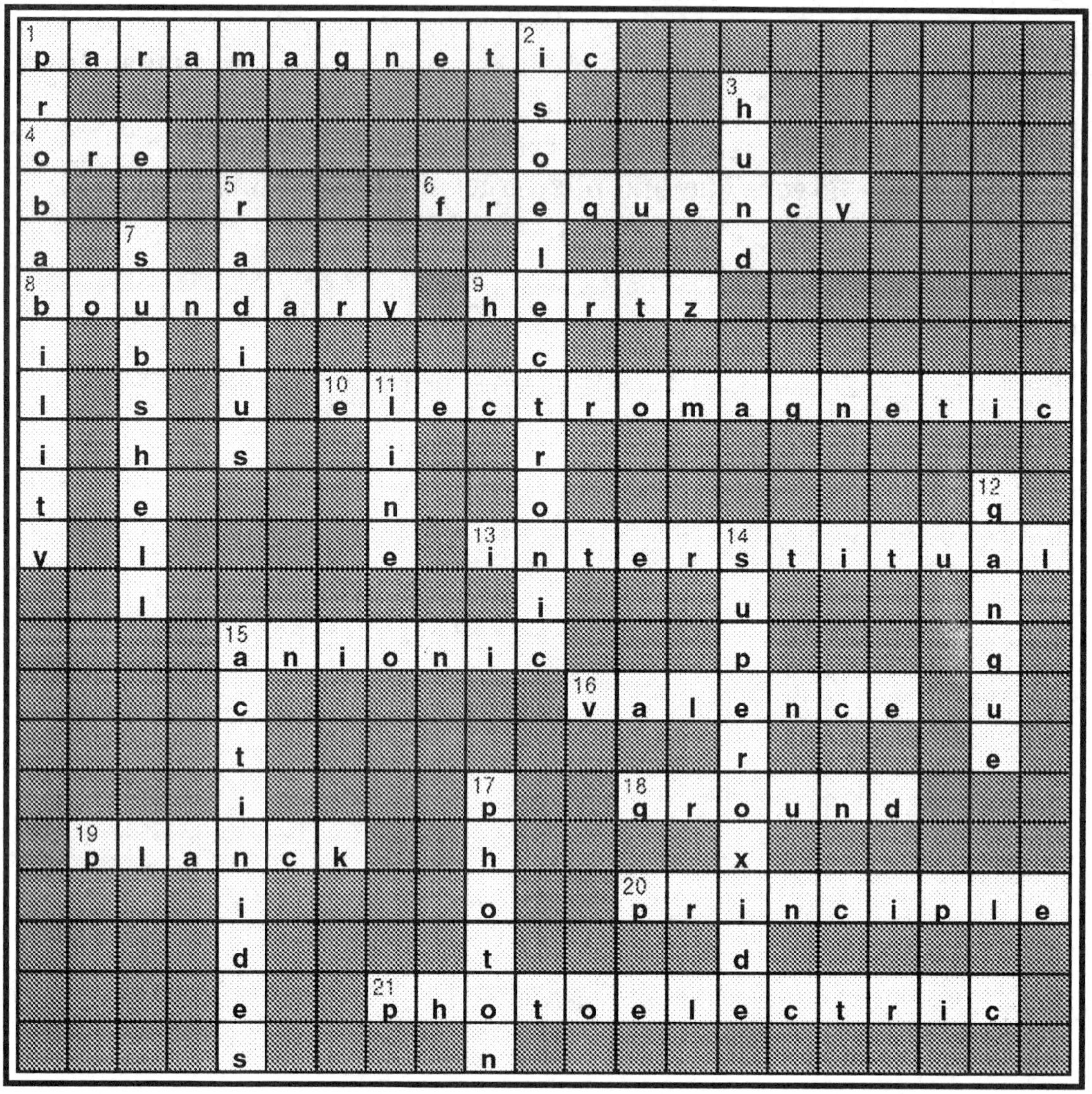

Some Supplementary Readings Relating to This Chapter:

1. Beyond Uranium, Chemical and Engineering News , November 19, 1990, page 18.

2. Three-Dimensional Periodic Table of Elements Proposed, Chemical and Engineering News ,January 1,1990, page 18.

3. Naming Heavy Elements: 107 to 109 attract least controversy, Chemical & Engineering News , September 14, 1992, page 4.

Chapter 10 Covalent Bonding

This chapter introduces you to a very common bond found in chemical compounds, the covalent bond. This bond involves sharing of electrons between atoms in the molecule. Covalent bonding is by far the predominant bond in organic compounds and most nonmetal to nonmetal compounds. Several types of covalent bonds are discussed and the terms important to the bond variations are detailed. Finally the bonding between metal ions and molecules which form coordination compounds are explained.

Focus Your Study Efforts

Electronegativity can be used to predict bond types.

Objectives:

1. You should be able to use electronegativity values to predict the type of bond that will form between atoms.
2. You should be able to explain what a covalent bond is.

Electronegativity measures the ability of an atom to attract the shared electron bond pair. The Pauling electronegativity scale has values from 0 to 4. The greater the value the more electronegative the element.

When elements bond, and the closer their electronegativity values, the more equally shared the bonding electrons are.

Covalent bonds share valence electrons **between** the two atomic nuclei.

Electron sharing is not always equal between the atoms. This results in a bond series being possible. This series is diagramed in the following table.

Covalent	Polar Covalent	Ionic
Identical atoms	Some difference in electronegativity	Large difference in electronegativity
I_2	CO	NaCl
Equal sharing	Unequal sharing	Electron transfer

Sharing of valence electrons allows each atom to reach a noble gas configuration. This structure is very stable and enough electrons are involved so that both atoms attain it.

> Classify the polar character of the Zn-Cl bond in $ZnCl_2$.
>
> The electronegativity of Zn is 1.6 and that of Cl is 3.0. This difference of 1.4 is significant and a shift of electrons occurs toward the Cl atoms. There will be a polarity present.

Lewis structures are used to represent electron patterns.

> **Objectives:**
> 1. You should be able to use, and write, Lewis structures to represent different types of covalent bonds in molecules and polyatomic ions.
> 2. You should be able to recognize the limitations of the Lewis dot system.

Not all valence shell electrons need be involved in bonding. Those that do participate in the bond are labeled as bonding electrons. Those that are not being shared are called lone pair or non-bonding electrons.

Lewis structures are a convenient, effective way to show valence electrons. When writing these structures the **symbols** refer to the nucleus and core electrons. Valence electrons are written, as dots, singly up to 4, and then pairing starts.

Lewis structures of molecules use a "-" to represent a bonding pair of electrons. The lone pairs are identified with dots.

You should not try to imply a molecular shape from Lewis structures.

> Draw the structure of a hydrocarbon molecule with 3 carbon atoms and has the general formula of C_nH_{2n-2} .
>
> The general formula indicates a molecule with a triple bond. From the general formula we can conclude the hydrocarbons formula to be C_3H_4. A reasonable structure would be:
>
> 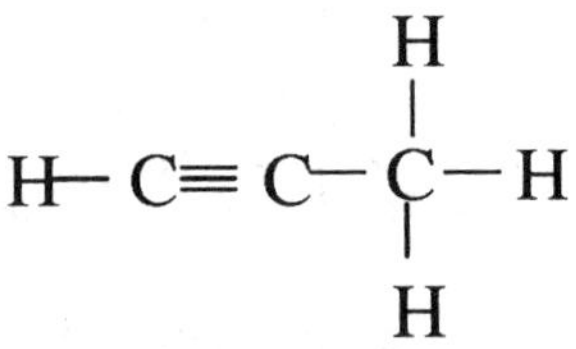
>

Lewis structures work well to predict the number of bonds formed but the positioning of electrons is a convenience only. Electrons do not have a fixed position as two dimensions may imply.

Hydrogen can only form one bond. It will never be between two other atoms in a molecule.

When done writing a Lewis dot structure the number of electrons present as dashes or dots must equal the sum of electrons present on the starting elements. Many structures are shown pages 411-424 of the text.

Carbon has unique bonding characteristics.

Objectives:

1. You should be able to describe catenation and discuss its importance in organic chemistry.

2. You should be able to recognize the different types of C-C bonding in hydrocarbon molecules.
3. You should be able to determine the presence of multiple bonding in hydrocarbons based on the hydrocarbon name.
4. You should be able to write Lewis structures for multiple bonds.

Carbon is the only element that bonds with itself to any extent. The length of carbon chains seems limitless. This ability to bond with other like atoms is known as **catenation**.

Carbon also has the ability to share more than a single pair of electrons between atoms. Four electron bonds, known as double bonds, or six electron bonds, known as triple bonds, are common.

Multiple bonds are common only to the elements of C, N and O.

If double or triple C to C bonds are present the compound is said to be unsaturated.

Hydrocarbons are found in several series. They are outlined in the table:

Alkanes	Alkenes	Alkynes	Cycloalkanes	Aromatic
C_nH_{2n+2}	C_nH_{2n}	C_nH_{2n-2}	C_nH_{2n}	
saturated	unsaturated	unsaturated	saturated	unsaturated
single bond	double bond	triple bond	single bond	alternate
C_2H_6	C_2H_4	C_2H_2	C_3H_6	C_6H_6
-ane	-ene	-yne	cyclo-	

Naming of hydrocarbons is based on reference to the number of carbons. The suffix is changed as a result of bonding: ane if all single bonds, -ene if a double bond is present and -yne if a triple bond is in the molecule. Ring compounds start with a prefix of cyclo- and then the name and series based on bond type.

Draw the structural formula for cyclopentene.

The prefix cyclo means we have a ring. The pent indicates 5 carbons and the ene suffix tells us we have a double bond present.

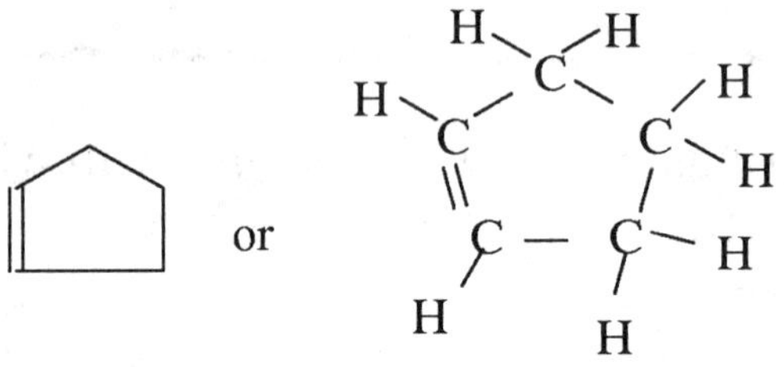

Multiple bonds result when there are not enough electrons to complete the octet of each atom. You can only have as many valence electrons in the molecule as are present in the starting atoms.

Hydrocarbons containing only single C to C bonds are said to be saturated. These materials are not very reactive and undergo substitution reactions - something replaces a hydrogen.

Hydrocarbons with one or more multiple bonds, either double or triple, are unsaturated. These molecules undergo addition reactions at the multiple bond. If hydrogen is added the reaction is known as hydrogenation.

The periodic table encourages us to think in terms of similarity, but we must use caution and not expect exact likeness. The chemistry of silicon and carbon illustrate this. Silicon does not produce an SiO_2 analog to CO_2. In part this is due to a d-orbital with energy close to other orbitals and this opens other bonding options.

The guidelines for writing Lewis structures are detailed on pages 416 and 417 of your text, study them carefully. A couple of useful hints to remember about these structures are: Think symmetrical - if in doubt about atom arrangement in the molecule start with a symmetrical pattern. Hydrogen must always be at the end of the line. The central atom is most likely the only atom that can accept more than 8 electrons.

Covalent bonds may have some uncommon characteristics.

Objectives:

1. You should be able to identify free radicals and explain their high degree of reactivity.
2. You should be able to explain why there are exceptions to the octet rule.
3. You should be able to explain the factors that affect bond characteristics, such as bond length and bond angle.
4. You should be able to use bond energies to estimate enthalpies of reaction.
5. You should be able to explain the significance of bond breaking and bond formation in chemical reactions.
6. You should be able to recognize the conditions that require writing of resonance structures.

Normally covalent bonds are the result of sharing electrons which are provided by both atoms. Sometimes both electrons come from the same atom. When this happens a special name is given - coordinate covalent. Once a bond forms there is no difference between a coordinate covalent and a regular covalent bond.

The representative elements attempt to fill the valence octet. This is not always possible however. Occasionally the octet will have less than eight electrons. When that happens a reactive molecule is the result.

It is possible for representative atoms in the third and higher period to accept more than 8 electrons. This can occur because d-orbitals are available at about the same energy as s and p orbitals.

In a few instances a molecule may have a single unpaired electron. These molecules are called **free radicals** and they are very reactive. Frequently they will react with a like molecule forming a **dimer.**

Bonds have two very important properties - the **bond length** and **bond energy**. Values given for these lengths and energies are average values. Bond length is determined by the atom sizes and the number of electrons involved in the bond. Larger atoms have a longer bond. Bonds shorten as they go from single to double to triple.

Bond energy is dependent upon how close the atoms are (their size), the number of electrons involved and the effectiveness of the nucleus to electron interactions. Triple bonds have more energy than double bonds and these are larger than single bonds. A triple bond, however, does not have three times as much energy as a single bond.

Bond energies are always positive values. They are expressed as the amount of energy needed to break the bond. They are always based on the material in the gas state.

Referring to the following structure.

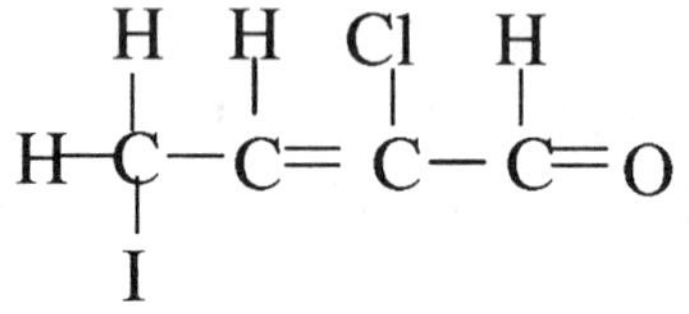

Which bond is longest and which is the shortest in the above molecule?

From table 10.2 in your text the longest bond is C to I at 210 pm and the shortest is C=O at 122 pm.

Which bond is the strongest and which is the weakest in the above molecule?

From table 10.3 in your text the strongest bond is C=O at 745 kJ and the weakest would be C-I at 238 kJ.

Chemical reactions involve a series of bond breakings and bond formations. A reaction will be exothermic if the energy released in forming new bonds exceeds that needed to break the existing bonds.

Bond energy values can be used to estimate the energy change in a chemical reaction. When doing this it is a good idea to draw all molecules showing all bonds involved so that all are accounted for.

Using bond energy data from table 10.3 in your text determine the energy change that occurs in the reaction of:

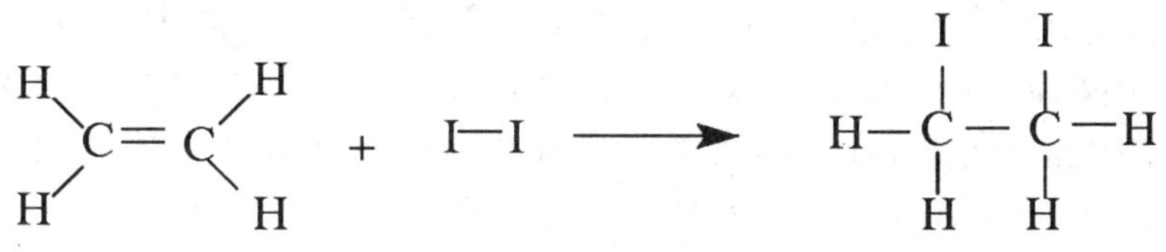

$\Delta H_{rxn} = \Sigma(\text{bonds broken}) - \Sigma(\text{bonds formed})$

Notice that only a few bonds are involved in the reaction. These are the only ones we need consider. We do break a C=C double bond and reform a C- C single bond.

Break C=C and I-I
Form C-I and C-I and C-C

$\Delta H = [C{=}C + I\text{-}I] - [\,2(C\text{-}I) + C\text{-}C]$
$= 611 + 151 - [\,2(238) + 347]$
$= 762 - 803 = -41$ kJ/ mol

When molecules or ions can be written in more than one structure based on electron positions (not changing the atoms arrangement) a situation known as resonance exists. Each structure is known as a resonance structure. In reality there are not several structures. As chemists our ability to explain and visualize is just incomplete. See text section 10.6.

Objectives:

1. You should be able to calculate oxidation numbers for atoms in molecules or polyatomic ions.
2. You should be able to calculate formal charges for atoms in molecules or polyatomic ions.
3. You should be able to use formal charges to choose the most likely resonance structure.

Chemists need a means of keeping track of electrons in molecules. This is accomplished through use of oxidation numbers.The oxidation number concept was developed by chemists to aid us in keeping track of electrons. These numbers frequently are used to determine charge changes on atoms in oxidation reduction reactions.

The rules for assigning oxidation numbers are summarized in your text on page 437. Be sure to review them as this concept is a valuable tool in working with oxidation reduction reactions.

What is the oxidation number of Mn in a potassium permanganate molecule, $KMnO_4$?

Since K is a 1A metal we know it has a value of +1. The oxygen is not present as a peroxide or bonded with F, so it has a value of -2 for each atom or a total of -8. The molecule must be neutral so the Mn oxidation number is +7.

Formal charge calculations are used to estimate the charge distribution in molecules. There are several points to keep in mind when calculating them. The sum of the formal charges must be the same as that on the ion or molecule. Adjacent, like non-zero, formal charges are not compatible. Small formal charges are preferred. The most electronegative atom should have the negative formal charge.

What are the formal charges on Br and O in the BrO_3^- ion?

Electron dot systems involve 7 valence electrons from Br and 6 from each of the oxygens and 1 electron from the charge. This totals 26.

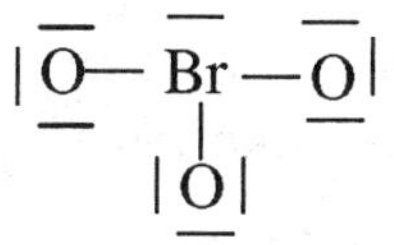

Formal charge = # valence e^- - # nonbond e^- - 1/2 bond e^-

Bromine formal charge = 7 - 2 - 3 = +2
Oxygen formal charge = 6 - 6 -1 = -1 or -3 total

Net for the ion = -3 + (=2) = -1 This equals the charge on the ion so our system checks.

Objectives:

1. You should be able to identify coordination compounds and identify which parts are Lewis acids and Lewis bases.

Coordination compounds are made from Lewis acids, electron pair acceptors, such as metal ions and Lewis bases, which are electron pair donors, such as the ligands.

Metal complexes may be neutral or charged. The net result is dependent upon the charge of the metal ion, charge on the ligand and the number of ligands.

The number of ligands in the coordination sphere varies with the size of the ligand and the number of times it can bond to the central atom

Coordination compounds are very useful and are found in many places. They are common to biological systems and also have many industrial uses.

Symbols found in Chapter 10

Page	Symbol	
414	LPG	Liquid petroleum gas
414	△ □	Cycloalkanes- various molecules are possible
416	$(SiO_2)_n$	Refers to silicon oxide base unit
417	-C=O with H below the C	Aldehyde functional group
418	CFC's	Chlorofluorocarbons
421	C_nH_{2n-2}	General formula for the alkyne hydrocarbon series
427	A°	Angstrom unit, 10^{-8} cm.
427	D	Bond energy symbol
433		Benzene

Terms in Chapter 10

The following terms are listed for your convenience. They should not become a list for you to memorize. You should have a working definition for each. If you have doubt about any particular term check its definition in the glossary and/or review it in the chapter. Some of these terms are included in the crossword puzzle later in the chapter.

Covalent bond	Lewis structure	Bond pair
Lone pair	Catenation	Cycloalkanes
Saturated	Double bond	Triple bond

Alkenes	Unsaturated	Addition reaction
Substitution reaction	Hydrogenation	Cycloalkenes
-ene	-yne	Coordinate covalent
Free radical	Dimer	Bond length
Resonance hybrid	Resonance structure	Delocalized
Nonpolar	Polar covalent	Electronegativity
Electropositive	Oxidation number	Formal charge
Metal complex	Coordination compound	
Coordination sphere	Complex ion	Lewis acid
Lewis base	Ligand	Coordination number
Monodentate	Chelating ligand	

Concept Maps for Covalent Bonding

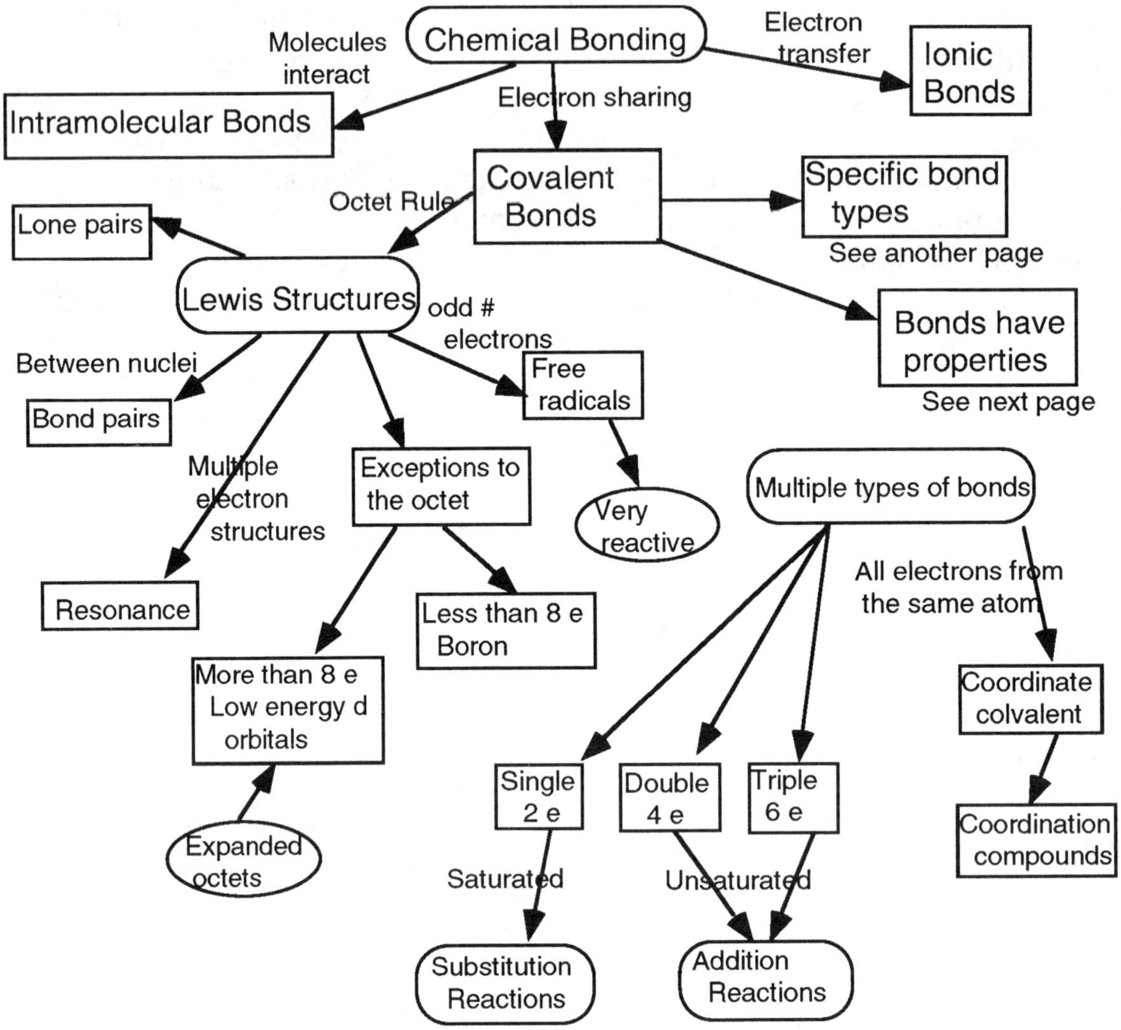

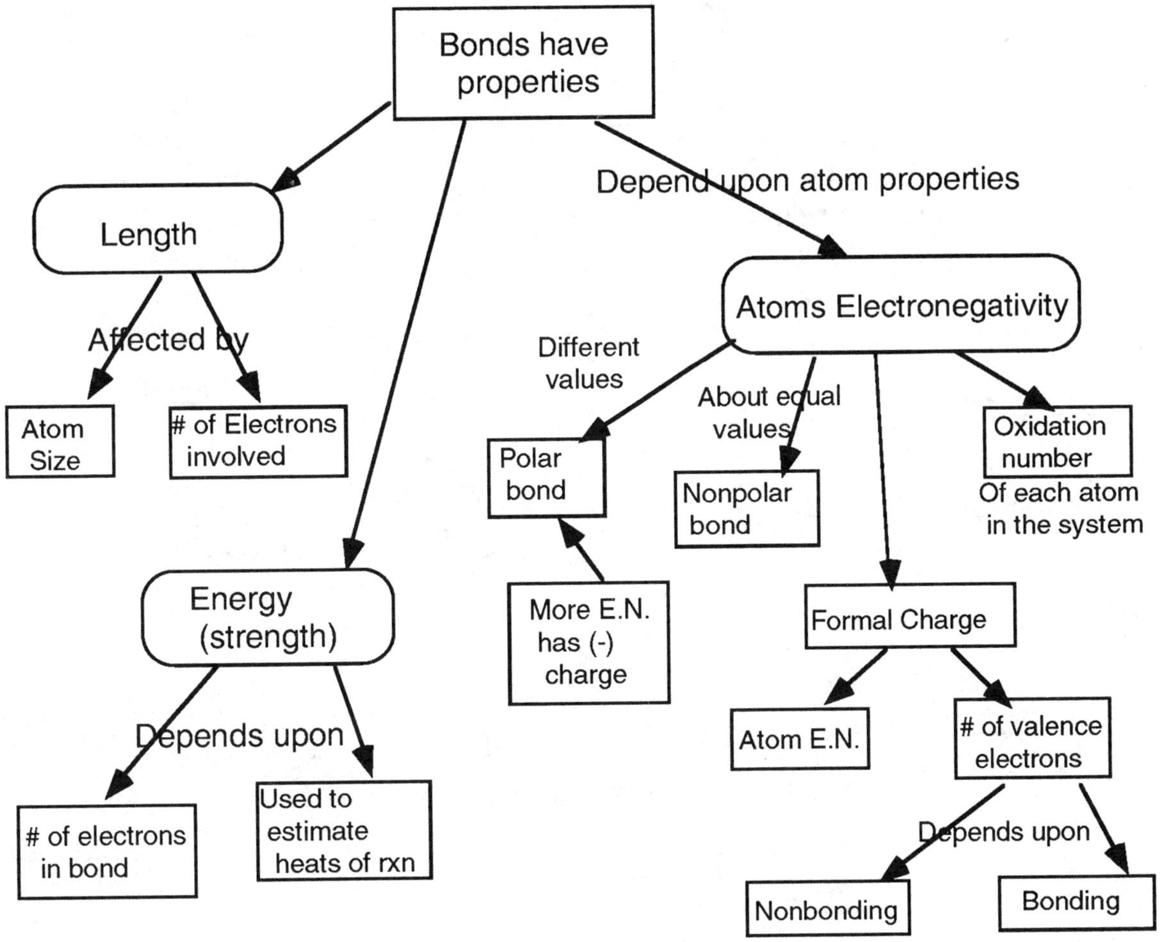
Bonds have properties
Length
Affected by
Atom Size
# of Electrons involved
Energy (strength)
Depends upon
# of electrons in bond
Used to estimate heats of rxn
Depend upon atom properties
Atoms Electronegativity
Different values
Polar bond
More E.N. has (-) charge
About equal values
Nonpolar bond
Formal Charge
Atom E.N.
# of valence electrons
Depends upon
Nonbonding
Bonding
Oxidation number
Of each atom in the system

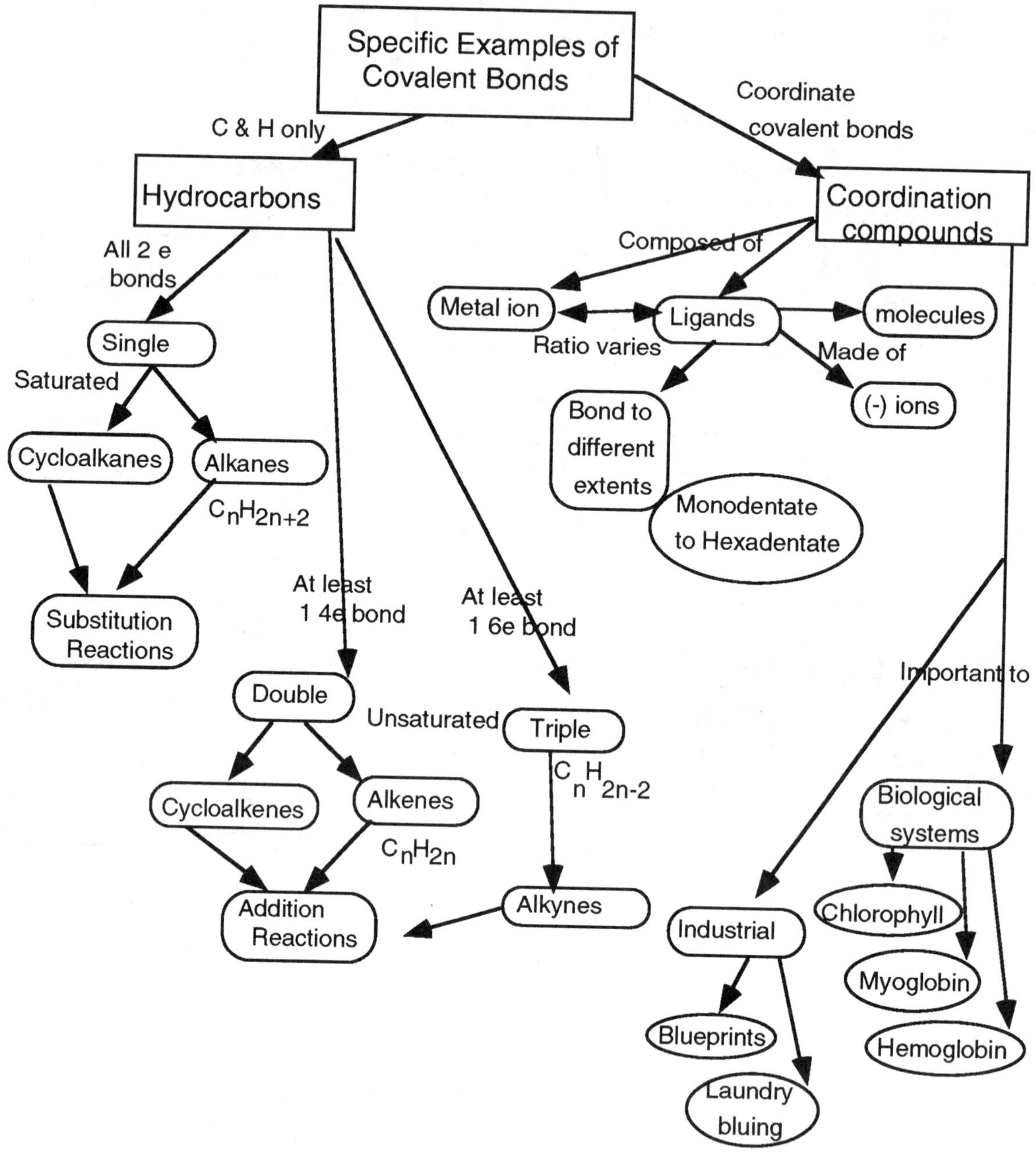
Specific Examples of Covalent Bonds
C & H only
Coordinate covalent bonds
Hydrocarbons
Coordination compounds
All 2 e bonds
Composed of
Metal ion
Ligands
molecules
Ratio varies
Made of
Single
Saturated
Bond to different extents
(-) ions
Cycloalkanes
Alkanes
Monodentate to Hexadentate
C_nH_{2n+2}
At least 1 4e bond
At least 1 6e bond
Substitution Reactions
Important to
Double
Unsaturated
Triple
C_nH_{2n-2}
Cycloalkenes
Alkenes
Biological systems
C_nH_{2n}
Addition Reactions
Alkynes
Industrial
Chlorophyll
Myoglobin
Blueprints
Hemoglobin
Laundry bluing

Covalent Bonding True/False Statements

Answers are at the end of the chapter.

1. Bonds that involve sharing of electrons are known as covalent bonds.

2. Lewis structures can not be used to represent double bonds.

3. A "-" between symbols represents a pair of shared electrons.

4. Hydrogen can only be a terminal atom in a molecule.

5. The octet rule must be followed for all Lewis structures.

6. Only valence electrons are used in Lewis structures.

7. The ability of an atom to bond many times with like atoms is called catenation.

8. Saturated hydrocarbons must have the general formula C_nH_{2n+2}.

9. Multiple bonds are formed when there are not enough electrons to make complete octets for all atoms.

10. Unsaturated hydrocarbons undergo substitution reactions as the prominent reaction type.

11. Hydrogenation is the process of adding hydrogen to multiple bonds.

12. The elements in all periods except period 1 can have more than eight valence electrons.

13. Free radicals have one unpaired electron on the molecule and this results in a very reactive material that often forms a dimer.

14. The element Se would have a longer bond than oxygen.

15. If we know bond energies we can calculate the heat of reaction for gas phase systems.

16. Resonance structures are the way we represent various electron arrangements in a given molecule.

17. Polar covalent bonding is more common if the bonding elements have nearly identical values of electronegativity.

18. Oxidation numbers are experimentally determined charge values for atoms in a molecule.

19. In a given reaction a specific element has its oxidation number decreased. This involves a reduction of the element.

20. Like adjacent formal charges on atoms in a molecule or ion results in a much less stable system.

21. The coordinate covalent bond is very common in coordination compounds.

22. Ligands are Lewis acids.

23. A coordination number of six means six ligands are attached to the central metal ion.

Testing Your Understanding of Concepts in Chapter 10.
Answers are at the end of the chapter.

1. What is the difference between a covalent bond and a polar covalent bond?

2. Why aren't all valence electrons bonding electrons?

3. How does the chemical formula give an indication of what atom should be the central atom in the molecule?

4. The general formula for alkanes is C_nH_{2n+2}, and these hydrocarbons are saturated. The cycloalkanes are also saturated, yet they do not have this general formula. How can this be explained?

5. When writing Lewis structures what gives an indication that multiple bonds are required in the molecule?

6. Why are alkenes more reactive than alkanes?

7. Why do coordination compounds have coordinate covalent bonds rather than covalent bonds?

8. If a molecule forms a dimer with a like molecule what is probably present in the system to bring about this characteristic?

9. What factors are significant in determining bond length?

10. With reference to bonds, under what conditions is a reaction exothermic?

11. What are the two very important aspects of bond formation that facilitate an exothermic reaction?

12. What is the meaning of the term "resonance hybrid" ?

13. What is the relationship between electronegativity and bond polarity?

14. The oxidation number of Cl in ClO_4^{1-} is +7. Is this a reasonable value?

15. What is the value of a formal charge number over an oxidation number?

16. How is it possible for a metal ion to have a coordination number of 6 but have only three ligands?

Testing You on Some Analytical Concepts.
Answers are at the end of the chapter.

1. Does the molecule N_2H_4 have an unpaired electron?

2. Draw a Lewis structure for $SbCl_3$ and $SbCl_5$.

3. In the molecule 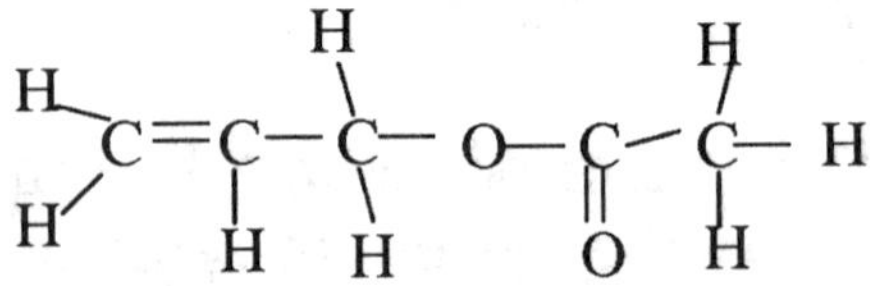 which bond is the shortest and which is the strongest?

4. If the reaction $CO_{(g)} + Br_{2(g)}$ ===> $COBr_{2(g)}$ were to occur, based on bond energies, what enthalpy change would result?

5. Assign oxidation numbers to the elements in $K_2S_2O_8$.

6. Which molecule of those listed below would show the greatest polarity. CH_4 CHI_3 $CHBr_3$

Crossword - Chemical Elements

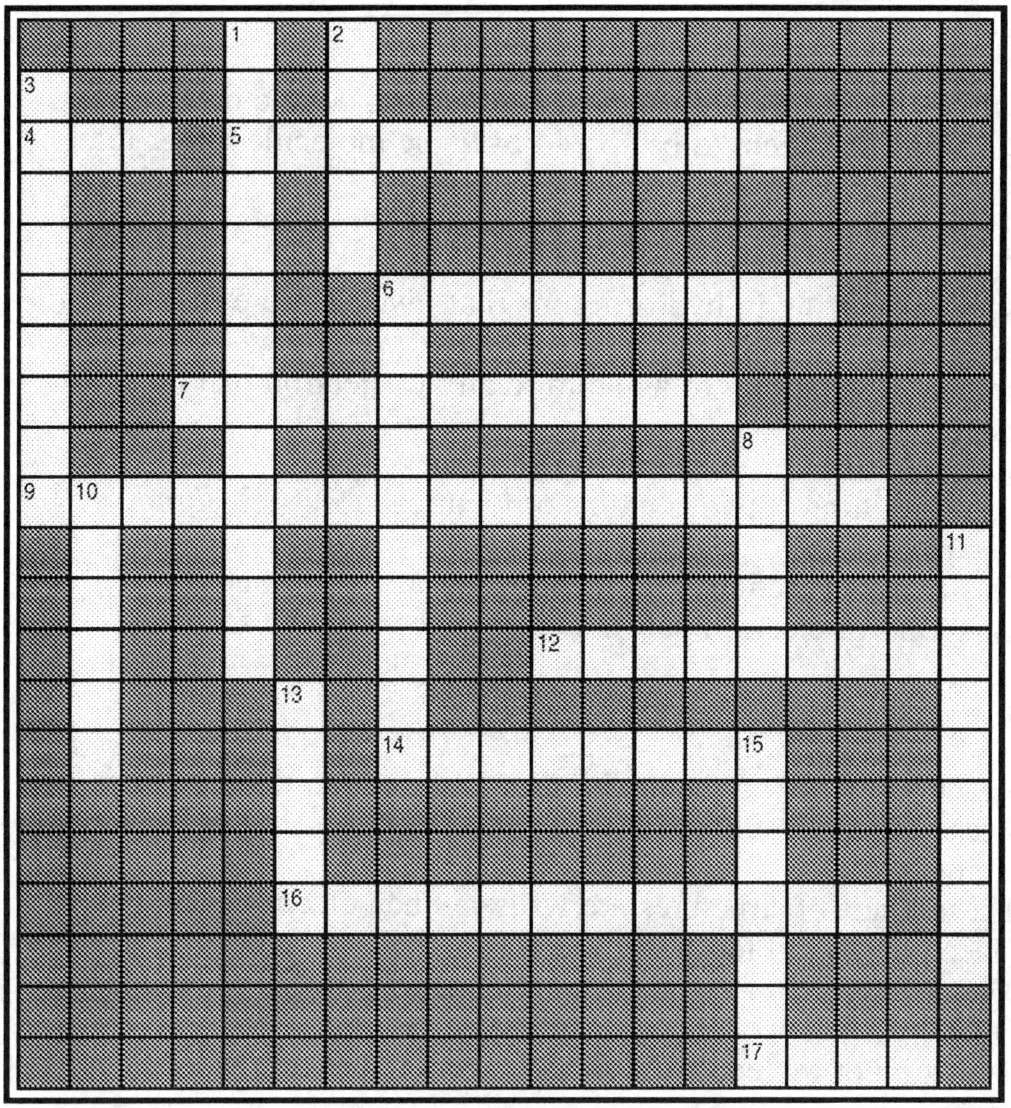

Crossword - Chemical Elements Clues

ACROSS

4 Suffix given to hydrocarbons that contain a double bond.
5 Electrons that seem to be spread over the molecule rather than confined to one space.
6 These ligands are attaching to the central ion in more than one place.
7 Bonds occurring in hydrocarbons that involve more than 2 electrons.
9 Refers to the ability of an atom to attract electron pairs toward itself.
12 Molecule that contains only single bonds.
14 Bond resulting between atoms of identical ability to attract electrons.
16 This type of reaction is common to unsaturated bonds.
17 This kind of electron pair is not involved in bond formation.

DOWN

1 Reaction involving the addition of hydrogen to double bonds.
2 These bonds are the result of unequal electron sharing between atoms.
3 These kinds of structures are written when electrons can be placed in several locations.
6 The ability of an atom to bond repeatedly with its own kind of atom.
8 This kind of molecule often occurs when like free radicals contact each other.
10 Ions or molecules bonded to a central metal ion.
11 A ligand that can bond twice.
13 This kind of acid involves acceptance of electrons.
15 Item with an unpaired electron is known as a free _______.

Ties to Previous Chapters

The ideas of bonding have been seen in several chapters. Each previous chapter has looked at a different type of bond but all have one thing in common, they represent energy.

Chapter 4 introduced ions and ionic bonds. Chapter 7 discussed energy in general but also that bonds were energy.

Answers to True/False Statements:

1. True. Sharing may be covalent which means electrons come from both atoms or coordinate covalent in which case both electrons for the bond come from a single atom.

2. False. A double bond would be represented by 4 electrons or two dashes between the atoms.

3. True. Bonding electrons are represented this way. Lone pairs are usually identified by the dots.

4. True. Hydrogen can only involve two electrons, which is one bond.

5. False. Several materials have an expanded set and a few involve only six electrons.

6. True. Lewis structures are used for electrons in the bonding orbital which are the valence electrons.

7. True. Carbon does this very well. No other atoms does this to any extent. A few, such as S and P, can catenate a few atoms.

8. False. The alkanes do but the cycloalkanes do not.

9. True. Each multiple bond gains two electrons for the system because two atoms make use of them.

10. False. Unsaturated bonds undergo addition reactions.

11. True. This could be either a double or triple bond.

12. False. Period 2 cannot either. Valence shell expansion must be possible and that starts in period 3.

13. True. Single electrons tend to pair so molecules with an unpaired electron would react with another very rapidly.

14. True. Se has more energy levels. Bond length is from nucleus to nucleus and two energy levels enlarge the size.

15. True. A reaction is a series of bond breakings and bond formations. Energy is used in breaking bonds and released when bonds are formed. The difference is the reaction energy. By definition this is true only in the gas phase.

16. True. The electrons belong to the molecular bonding system. For convenience we write a separate structure for all possible different electron locations. Remember the atoms do not move.

17. False. If the electronegativity values are almost equal neither atom has more effect and the electrons are equally shared.

18. False. Oxidation numbers are values invented by chemists to keep track of electrons in the molecule or ion. They are not experimental values.

19. True. Decreasing the oxidation number happens because electrons are gained. Materials that gain electrons are oxidizing agents and undergo reduction reactions.

20. True. The charges should be different so there is minimal repulsion. Like adjacent charges would result in much more repulsive interaction.

21. True. The metal ion has no available valence shell electrons so the ligand provides both electrons for each bond.

22. False. Ligands donate electrons. We define a Lewis base as electron donors.

23. False. The coordination number is the number of bonds attached. Some ligands can attach more than one time, EDTA for example attaches six times.

Answers to Understanding of Concepts Questions:

1. Both of these type bonds involve electron sharing. The polar bond has the sharing distorted so the more electronegative element experiences the electron density the most.

2. The atom only uses as many electrons as necessary to complete the valence shell. Usually only 1,2 or 3 electrons are needed for bonding.

3. The lone atom is usually the central atom. The more electronegative atoms are usually not the central atom and hydrogen must always be at the end.

4. The ring structure removes the need for two end hydrogens since the C-C bond takes their place. Saturation occurs with two fewer hydrogens.

5. If there are not enough electrons to complete all the octets the need for multiple bonds is evident. Lone pairs become bond pairs and actually count for filling two coordination groups.

6. The double bond can undergo substitution reactions. This does not require replacement of an already bonded element.

7. The central atom has no electrons available to share in the bonds. The ligand must provide both electrons for the bond pair.

8. There must be an unpaired electron in the molecule.

9. The size of the atoms involved and the number of electrons in each bond both are important in bond length. Large atoms make for longer bond lengths while multiple bonds result in shorter bond lengths.

10. If the energy released in forming new bonds exceeds energy used in breaking existing bonds then the reaction is exothermic.

11. The more bonds formed, and the stronger those bonds are, the more exothermic the reaction.

12. When several bonding structures are possible for a system none are probably correct. The electrons are delocalized over the entire bonding system and a single structure is written to represent that.

13. The greater the difference in electronegativity values between the bonding atoms the greater the polarity. This will eventually lead to ionization.

14. Probably not. For an element to lose seven electrons the energy involved would be extremely large. More probably it means that the chlorine atoms have lost some electron control to the oxygens. Oxidation numbers are an invention of chemists for our convenience.

15. The formal charge values are more realistic. They take into account the character of the electrons such as bonding or nonbonding.

16. Some ligands can attach more than once to the central atom. In this case there could be three bidentate ligands.

Answers to Testing the Analytical Concepts

1. The total number of valence electrons present include: 10 from the two nitrogen atoms, and 4 from the four hydrogens. This gives a total of 14 and is an even number. Each nitrogen will bond to two hydrogens and the other nitrogen and have a lone pair of electrons.

2. Sb is in group 5A with 5 valence electrons. Cl is in 7A with 7 valence electrons. In the case of $SbCl_3$ there will be a total of 26 electrons and for the $SbCl_5$ there will be 40 electrons.

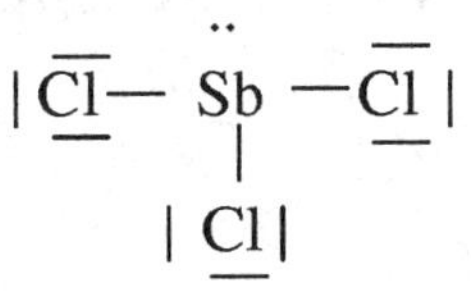

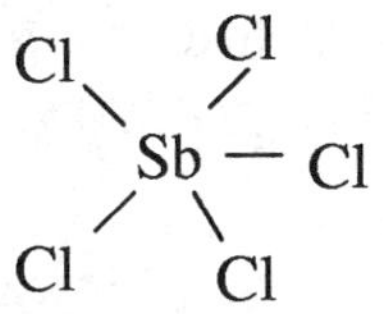

3. Table 10.2, from your text, provides bond length data and table 10.3 gives us bond energy information. Inspection of the system can provide us with some realistic ideas however. We would expect the double bonds to be stronger and we would expect the small hydrogen atom to have a short bond. The data substantiate this, with the C=O bond having an energy of 745 kJ/mol and C=C has 611 kJ/mol. The C-H bond is the shortest at 110 pm.

4. This reaction involves breaking the Br-Br bond and in turn formation of two C-Br bonds. We also must break the triple bond between carbon and oxygen and reform a double bond.

Break bonds		Form bonds	
C-O triple	1075 kJ/mol	C-O double	745 kJ/mol
Br-Br (2 at 192)	384 kJ/mol	C-Br (2 at 276)	552 kJ/mol
net to break	1459	net released	1297

The reaction is endothermic by 262 kJ.

5. Oxidation numbers for oxygen and potassium are known to be -2 and +1 respectively. This gives a net -16 on oxygen and a net +2 on potassium. The two sulfur must be +14, or +7 on each. It is a neutral molecule so the net charge must be zero.

6. Polarity is the result of a lack of symmetry in the molecule and the presence of atoms with electronegativity differences. CH_4 has symmetry so it will not be polar. Both CHI_3 and $CHBr_3$ lack symmetry. The difference in electronegativity between C and Br is 0.3 and C and H is 0.4 which leaves only a slight chance for polarity. The C and I have no difference in electronegativity.

Answer to Chapter 10 Crossword

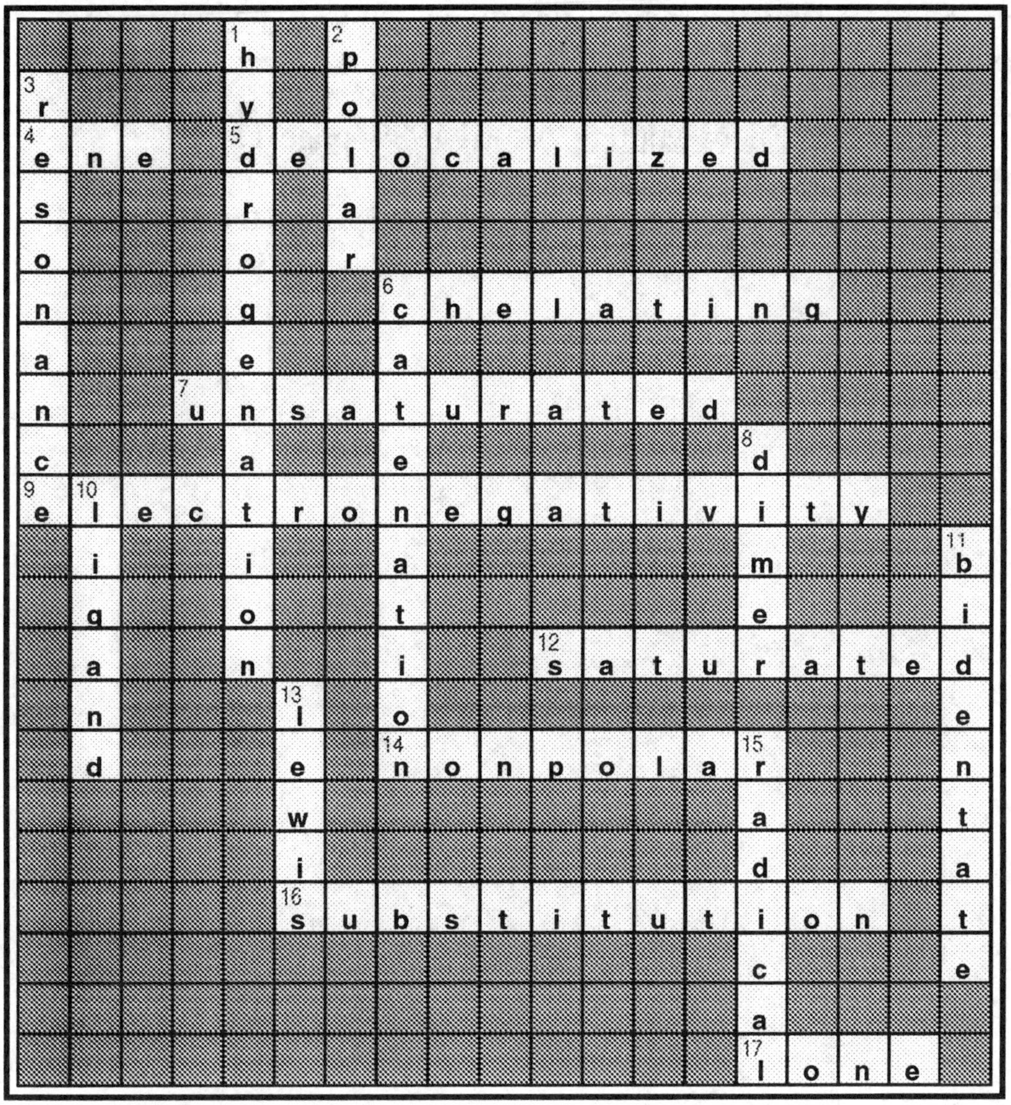

Some Supplementary Readings Relating to This Chapter:

1. Two Scientists Find a Way To Make Cheaper Methanol, The New York Times , January 17, 1993.

2. The Fruits of Ethylene, Chem Matters , April 1989, page 11

3. Alcohol in Your Tank, Chem Matters , December 1988, page 10

Chapter 11 Molecular Geometry

This chapter continues the development of ideas started in chapter 10. The VSEPR Theory of bonding, that allows for prediction of molecular geometry, is developed and its ties to Lewis electron dot structures are shown. The importance of bond polarity, the tetrahedral structure, and various types of isomers are clarified. This chapter also introduces some of the important modern instrumentation in use by chemists and outlines some of the data we obtain from their use.

Focus Your Study Efforts

You need to become familiar with the various notations used by chemists to represent bonds and electrons in atoms. Dots, lines, hashed lines, wedge models are all common. Some of these may overlap or look the same, for example, --- may look like either a bond below the page or a hydrogen bonded system.

VSEPR theory predicts geometry based on 5 basic structures.

Objectives:

1. You should be able to use VSEPR model to predict shapes of molecules and polyatomic ions.
2. You should be able to distinguish between the geometry of electrons pairs and geometry from the ions present.
3. You should be able to predict the effects of bond pair electrons and lone pair electrons on geometry.

The VSEPR theory enables prediction of geometry. This is in contrast to the Lewis dot system which has no geometry prediction ability. The VSEPR model also is useful only with the representative elements.

How do the O-S-O bond angles in SO_2 and SO_3 compare?

Drawing Lewis structures:

SO_2

S = 6e
O = 12e
18e or 9 pair

SO_3

S = 6e
O = 18e
24e or 12 pair

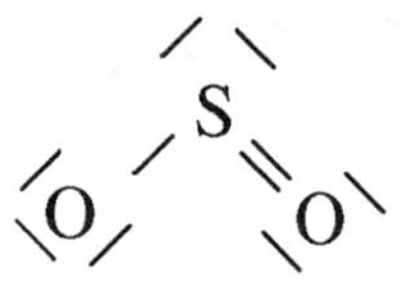

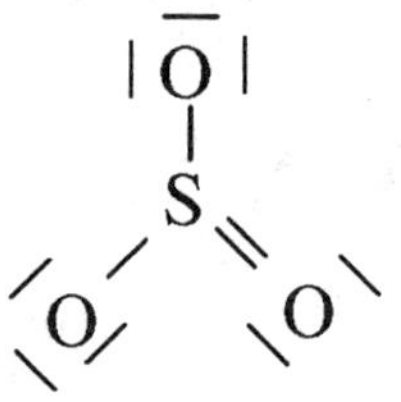

Both will have resonance structures.

The SO_2 has four sets of electron groups on the central atom which would be tetrahedral but two are lone pairs so the molecular geometry is vee shaped.

The SO_3 also has three sets of electron groups on the central atom which would give it a trigonal shape.

The VSEPR theory works with five basic structures that have definite bond angles. These geometries provide minimum interaction between electron pairs either bonding or lone pairs. See figure 11.5 on page 456.

A confusing aspect of the VSEPR theory is that the geometry it refers to can be either electron geometry or molecular geometry. The molecular geometry refers only to the atoms about the central atom and ignores any lone electron pairs. The electron geometry considers both in the geometric shape. Chemists are interested in the molecular aspects.

Bond-pair and lone-pair (non-bonding) electrons do not occupy the same amount of space. Since bond pairs are aligned between nuclei they tend to require less space. Lone pairs, being less restrained by only one nucleus, tend to spread out and occupy extra space. This results in a distortion of the geometry on the remaining atoms.

Draw the VSEPR structure for IF_3.

Some simple rules help us draw these structures.

1. Identify the central atom. This is the one there is a single atom of. It often is the atom present which is farthest to the right in the periodic table.
2. Use all the valence electrons for the central atom.
3. Identify the ligand atoms and determine the number of unpaired valence electrons from the Lewis structure.
4. Use only the unpaired valence electrons from the ligands.
5. Determine the total number of electrons available. Convert this to pairs.
6. This number of pairs sets the initial geometry.
7. Determine the number of bond-pairs and lone-pairs. Each ligand must have a bond pair.
8. If bond lone-pairs exist you can discuss either electronic or molecular geometry.

Central atom =I and it has 7 valence electrons.
The ligands are F and each has only one of the 7 valence electrons unpaired. Only one of these is used in the structure.

I = 1 x 7e = 7e
F = 3 x 1e = 3e

10 electrons total or 5 pairs
There are 3 bond pairs (because three ligand atoms) and two lone pairs.

The initial geometry is trigonal bipyramidal. However, that will not be the molecular geometry since the two lone pairs will occupy equatorial positions. The molecular geometry will be T shaped.

Lone pair electrons will occupy positions that present the minimum number of interactions. These would be the equatorial positions in several of the geometries.

Multiple bonds are no different that single bonds in the VSEPR geometries. Even though there are more electrons they still are aligned between the nuclei.

Electronegativity of the atoms affects molecule or polyatomic ion properties.

Objectives:

1. You should be able to use electronegativity values to predict polarity properties of molecules.
2. You should be able to explain the importance of the tetra-hedral shape to the chemistry of carbon.

A molecule containing atoms with significant differences in electronegativity values will have the bonding electrons shifted toward the atom of higher electronegativity. This makes the molecule slightly unbalanced in charge and a molecule with a dipole results.

Bond polarities can be effectively canceled if a molecule has symmetry. As a result, a molecule may have polar bonds, but not have a dipole moment.

Would $SiCl_4$ have polar bonds? Would the molecule be a dipole?

Polar bonds depend upon a significant difference in electronegativity values between the atoms. A common electronegativity range difference is 1.2 to 1.8. A dipole requires a molecule with unequal charge distribution.

The electronegativity of Si is 1.8 and Cl is 3.0 for a difference of 1.2. That is enough to make the range of polar bonding.

Since the molecule is symmetrical these polar bond characteristics are the same in all directions so there is no dipole.

The tetrahedral structure is one of the most important in molecular geometry. It shows up in several systems but especially with carbon, silicon and phosphorus.

Isomers are very important to the chemistry of carbon.

Objectives:

1. You should be able to identify isomeric structures of a molecule.
2. You should be able to identify the various classes of isomers.
3. You should be able to explain how isomers are named.

Isomers are the result of being able to put the same atoms together with different structures. Although the formulas have the same atoms the different arrangements lead to very different chemical and physical properties.

Isomers are classified into several general groupings. Constitutional isomers may result as a consequence of different arrangements about a given atom or it may be the result of branches in hydrocarbon chains rather than a single long chain. These branches are known as alkyl groups. Alkyl groups are just a hydrocarbon less one hydrogen. A multiple bond in the molecule could be located in different positions, and if so, each new location results in a new molecule.

Naming of isomers with multiple bonds and/or side chains has very specific rules. For our purposes, in introductory chemistry, only a few rules are needed. You should learn these rules and know how to apply them.

Draw structural formulas for:
(a) 1,3,4- trichloroheptane
(b) 3-ethyl-4- octyne

(a) First draw the hydrocarbon with C atoms only. Hept = 7. After that place Cl atoms on the 1st, 3rd and 4th carbons from either end. Then fill out each carbon to four bonds with hydrogen.

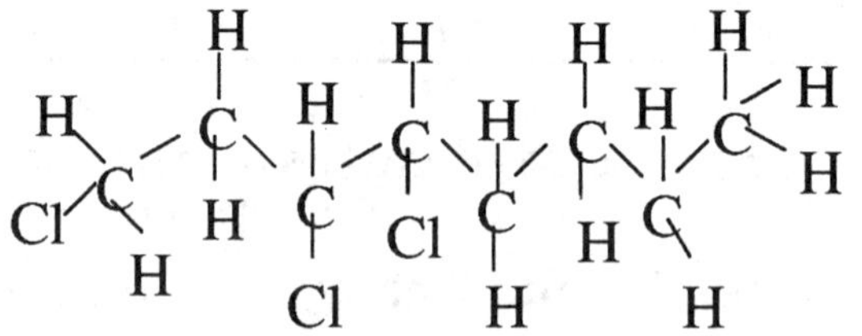

The position of the Cl atoms on the carbons does not matter. They could have been above, or one above and one below.

(b) Oct means eight, yne means a triple bond is present, ethyl is C_2H_5-, 3 means the ethyl is on the third carbon atom and the 4 means the triple bond is after the forth carbon.

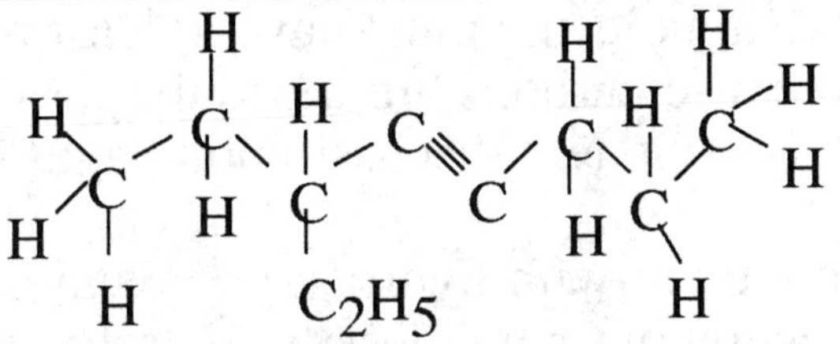

A second major grouping of isomers is stereoisomerism. This can be related to positions about a double bond known as cis (same side) or trans (opposite side). The other refers to optical isomers which require that a handedness exist in the isomers.

Draw the isomers possible for a molecule made of 2 carbons with a double bond, 2 hydrogens and 2 bromines.

Start with the carbon system and then place the bromines in all different possible configurations. With this type of situation is you can get the structure by just flipping the molecule it is not a different isomer.

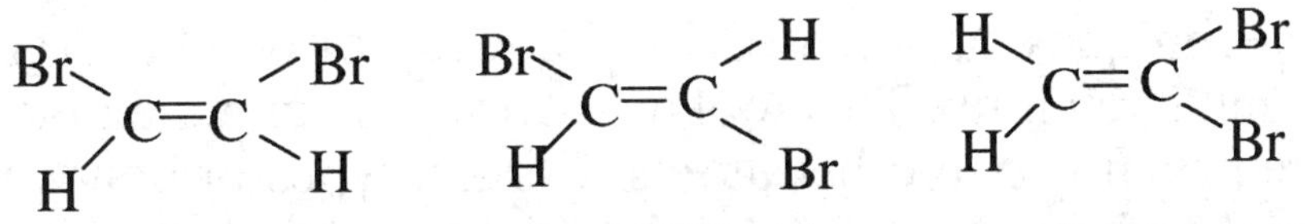

There are several terms associated with optical isomers. If you want to understand this concept you must know the terminology and you will probably need to spend some time working with models.

Non-superimposable mirror image molecules are known as enantiomers and they have the unique ability to rotate plane polarized light in opposite directions. A mixture of equal amounts of these isomers is known as a racemic mixture.

To be optically active a molecule must have at least one asymmetric carbon atom. Asymmetric carbons are those that have four different groups bonded to it. See the text discussion on pages 482-485.

Isomerism is very common with hydrocarbons and other organic molecules. It also occurs, but much less frequently, with the coordination compounds. Not all geometries involved in coordination compounds will allow isomers however.

Various techniques are used to study isomeric compounds.

Objectives:

1. You should be able to describe the spectroscopic techniques used to evaluate the structure of molecules.

Chemists use many sophisticated tools to study chemical compounds. There are two excellent techniques in common use that use the electromagnetic spectrum for studying molecular structure. These are infrared (IR) spectroscopy and nuclear magnetic resonance (NMR) spectroscopy.

Infrared techniques can be used for the study of bond energies related to bending and stretching. This tool also allows us to get an idea of what atoms are proximate to other atoms. Certain types of bonds will show up in detail and give us a fingerprint of the molecule.

NMR spectra use radio waves to gain information about atoms bonded to hydrogens as well as how many hydrogen atoms are present in a molecule.

Symbols found in Chapter 11

Page	Symbol	
454	◥	Represents a bond that is above the page plane.
454	- -	Represents a bond that is below the page plane.
454	—	Represents a bond that is on the page plane.
457	··	Represents a lone pair (non bonding) of electrons.
465	∂^+	Slightly positive charge area on a molecule.
465	∂^-	Slightly negative charge area on a molecule.
465	+→	Indicated the direction of molecular polarity.
466	**D**	Symbol for dipole moment in debye units.
472	ATP	Adenosine triphosphate - high energy source.
472	ADP	Adenosine diphosphate - lower energy value.
475	sec-	Means central C bonded to 2 other carbon atoms.
475	t-	Means central C bonded to 3 other carbon atoms.
479	cis-	Isomer formation with items across the double bond
479	trans-	Isomer formation with items on same side of bond.
484	D	Refers to molecule handedness. Means to right.
484	L	Refers to molecule handedness. Means to left.
487	C^*	Represents an asymmetric carbon atom.
494	IR	Infrared spectroscopy.
496	NMR	Nuclear magnetic resonance spectroscopy.

Terms in Chapter 11

The following terms are listed for your convenience. They should not become a list for you to memorize. You should have a working definition for each. If you have doubt about any particular term check its definition in the glossary and/or review it in the chapter. Some of these terms are included in the crossword puzzle later in the chapter.

VSEPR	triangular planar	tetrahedral
triangular bipyramidal	octahedral	bond angles
electron-pair geometry	molecular geometry	equatorial positions
axial positions	dipole moment	dipole-dipole force
Zeolites	ion-exchange	phosphate builder
isomers	constitutional isomers	alkyl groups
branched chain isomers	normal hydrocarbons	aromatic compounds
carcinogens	stereoisomerism	optical isomerism
chiral	achiral	enantiomers
asymmetrical	plane polarized light	optical isomers
optically active	polarimeter	racemic mixture
spectroscopy	photons	electronic spectroscopy
infrared spectroscopy	NMR spectroscopy	magnetic resonance

Concept Maps for Molecular Geometry

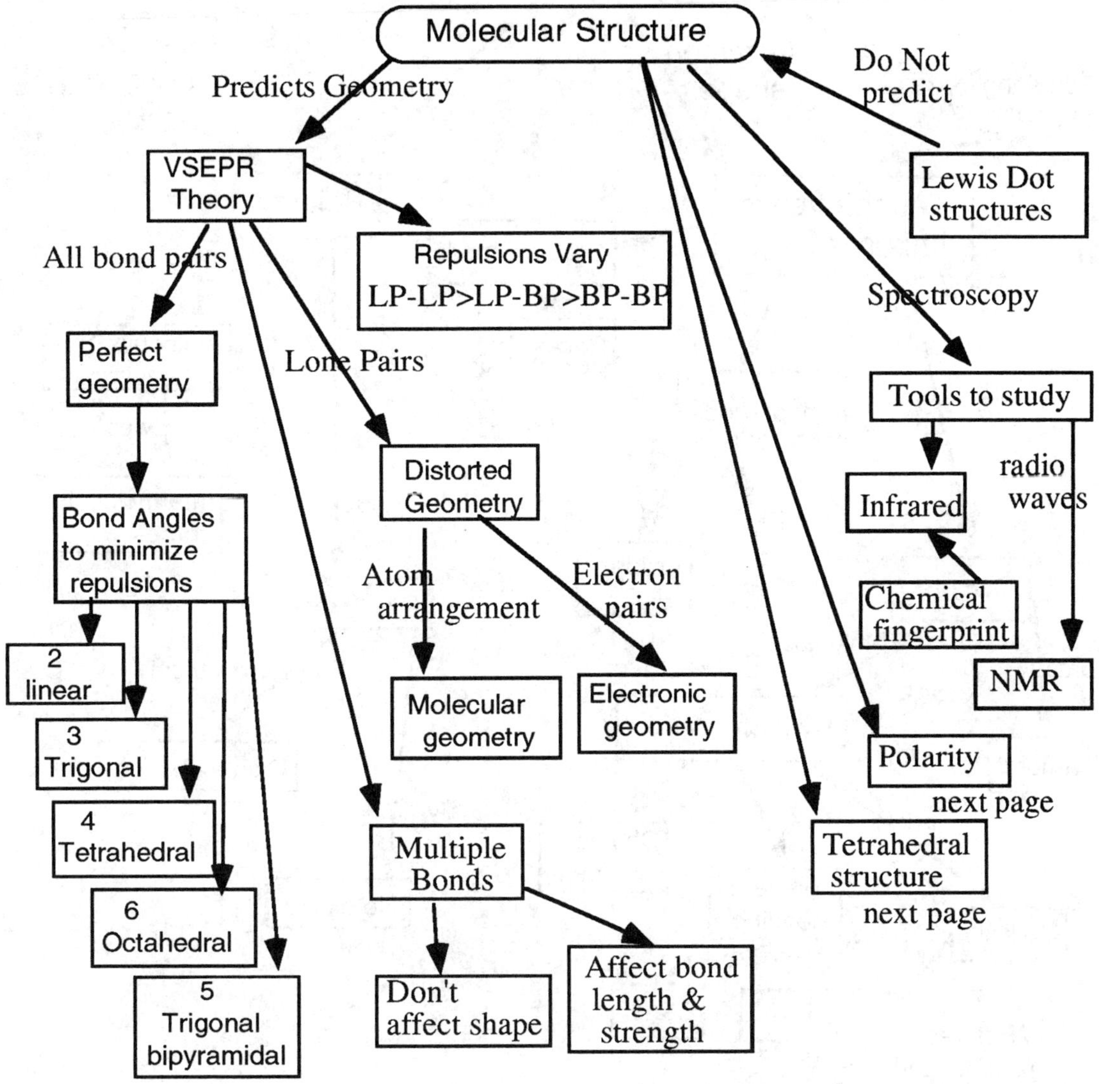

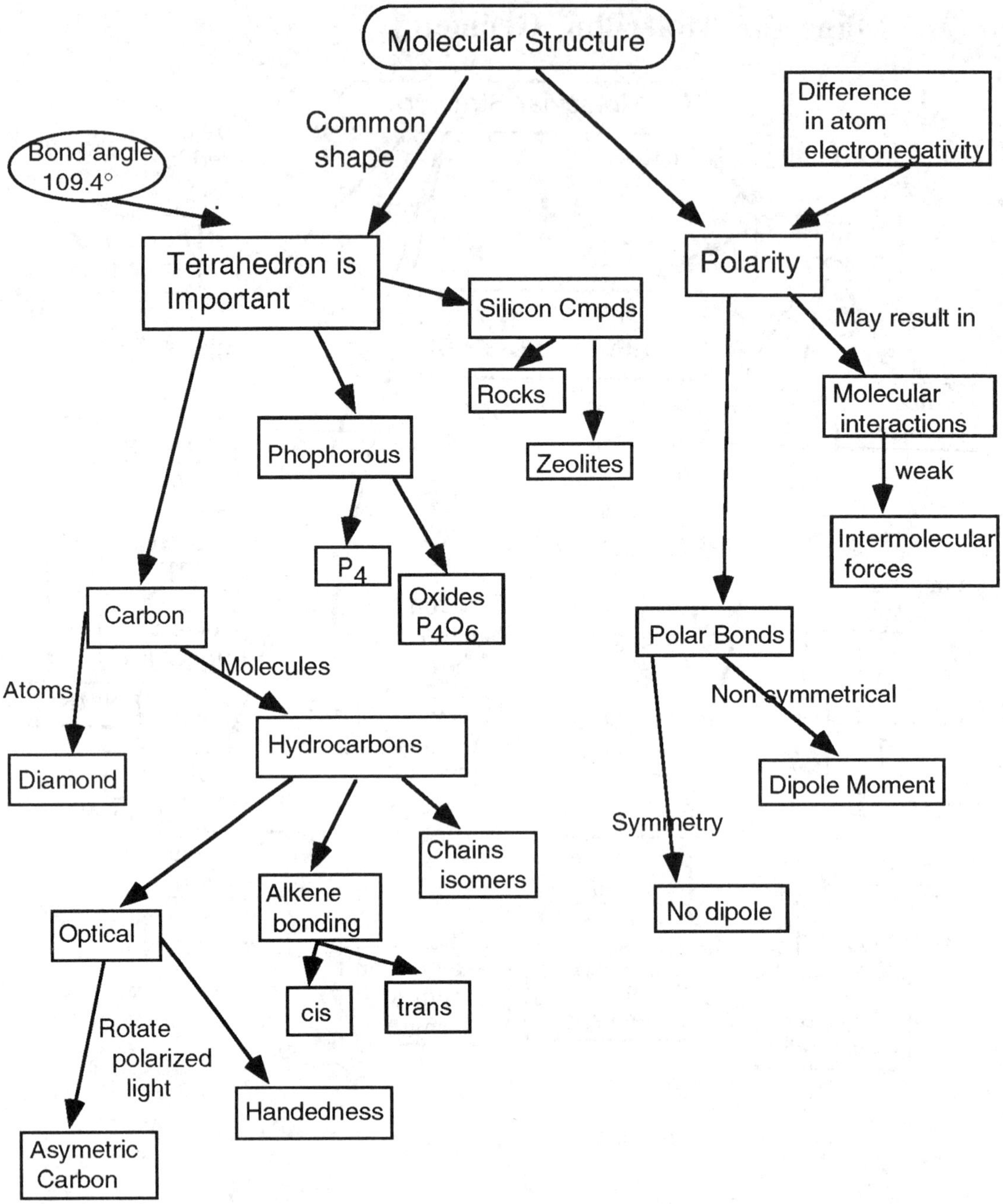

Molecular Structure
Common shape
Bond angle 109.4°
Tetrahedron is Important
Silicon Cmpds
Rocks
Zeolites
Phophorous
P4
Oxides P4O6
Carbon
Atoms
Diamond
Molecules
Hydrocarbons
Optical
Alkene bonding
Chains isomers
cis
trans
Rotate polarized light
Asymetric Carbon
Handedness
Difference in atom electronegativity
Polarity
May result in
Molecular interactions
weak
Intermolecular forces
Polar Bonds
Non symmetrical
Dipole Moment
Symmetry
No dipole

Molecular Geometry True/False Statements

Answers are at the end of the chapter.

1. The VSEPR theory has the capability of some molecular geometry predicting.

2. VSEPR theory works well for the transition elements.

3. Lone pairs will distort a "preferred" geometry because they interact differently than bonding electron pairs.

4. When four pairs of electrons are involved both square planar and tetrahedral structures are possible.

5. VSEPR theory predicts bond angles.

6. Lone pairs in axial positions will result in less distortion than those in equatorial positions.

7. Multiple bonded atoms are assumed to have an effect identical to single bonds in the VSEPR system.

8. Lewis dot structures and VSEPR theory give exactly the same information.

9. VSEPR theory is not capable of handling more than one kind of ligand atom.

10. The molecules NCl_3 and PBr_3 would have the same bond angle.

11. All polar bonds result in a dipole moment.

12. Water is a polar molecule with a dipole moment.

13. Dipole-dipole forces are stronger than hydrogen bond forces.

14. The tetrahedral structure is important only to the carbon atom.

15. It would be possible to have isomers for the molecule with the formula C_2H_6S.

16. Alkanes will allow for cis and trans type isomers.

17. Names of isomers may vary considerably from the names of the "normal" or straight chain compound with the same formula.

18. 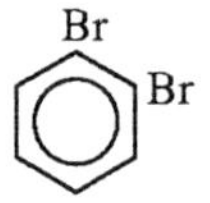  are the same material.

19. Your feet would constitute a set of chiral objects.

20. Any pair of molecules would be a set of enantiomers.

21. The symbols **D** and **L** are used to represent molecular handedness.

22. It is possible for a molecule with the formula $CH_3CBrICl$ to have an asymmetric carbon atom.

23. Racemic mixtures will always rotate polarized light to the right.

24. Square planar coordination compounds may have isomeric forms.

25. Infrared spectroscopy works because some energy in a chemical bond matches some energy in the IR radiation.

Testing Your Understanding of Concepts in Chapter 11.
Answers are at the end of the chapter.

1. What are the favored geometries in the VSEPR Theory? Why?

2. What is meant by the term electron pair geometry?

3. Why are lone pair to lone pair interactions greater than interactions of bond pairs next to bond pairs?

4. What affect do resonance structures have on the predicted molecular geometry?

5. Why is AlF_3 considered to be polar bonded but have no dipole moment?

6. Make a perspective drawing of the molecule CH_3I.

7. Outline a method to determine how the geometry of a molecule such as $SbCl_3$ should be drawn by VSEPR theory.

8. What is meant by a branched chain isomer?

9. When naming alkenes a number is used in the name. What does the number tell us?

10. Draw the structure of 1,4-dichlorbenzene. What other name might it have?

11. What does handedness mean?

12. Why doesn't a racemic mixture rotate plane polarized light?

13. Which instrument method, infrared spectroscopy or NMR spectroscopy, uses the higher energy electromagnetic radiation?

Testing You on Some Analytical Concepts.
Answers are at the end of the chapter.

1. Is there a net dipole in the molecule CH_2F_2?

2. Predict the shape of the IF_5 molecule.

3. Write the structural formula for 2-methylhexane

4. Describe the geometry(s) for ICl_4^{1-}.

5. Predict a reasonable bond angle for the F-O-F bond in OF_2.

6. If present, identify a chiral carbon atom in the molecule below:

```
      H
      |
CH3-C-CH2CH3
      |
      F
```

Crossword - Molecular Geometry

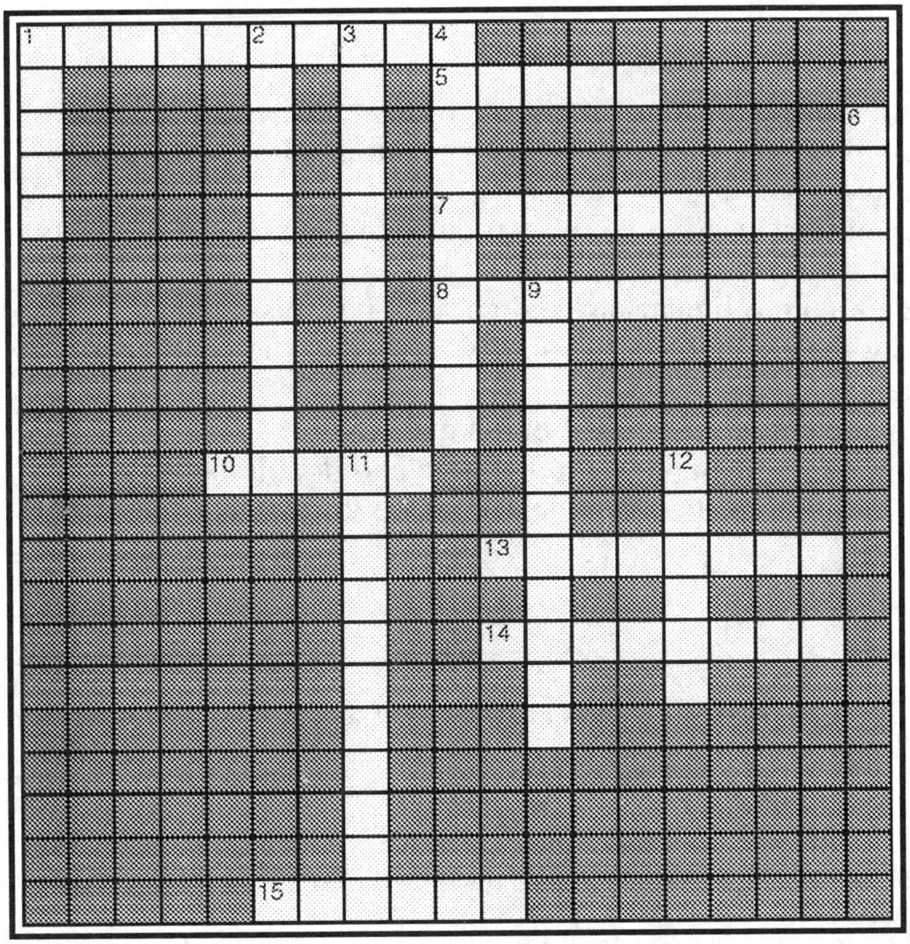

Crossword - Molecular Geometry Clues

ACROSS

1 A carbon atom with four different groups attached to it.
5 Geometry positions is VSEPR Theory that are above and below the equatorial band.
7 This technique is used to study structure by the way it interacts with bond bending and stretching.
8 A molecule with 6 ligands spaced about it has this geometry.
10 A bonding theory that has the ability to predict the geometry of a molecule.
13 Silicon oxide compounds used in ion exchange.
14 Ring compounds with alternating single and double bonds.
15 This is what unbranched hydrocarbons are known as. The prefix is often not used.

DOWN

1 Groups such as methyl or ethyl are known as _______ groups.
2 Mirror images of non-superimposable molecules are called this.
3 A mixture of equal amounts of enantiomers.
4 Materials capable of causing cancer.
6 This may be the result of two atoms with relatively large electronegativity value differences.
9 An extremely important shape in carbon and silicon chemistry.
11 A devise to measure the rotation of light by optically active molecules.
12 This type of molecule can not be superimposed on its mirror image.

Ties to Previous Chapters

An introduction to molecular compounds occurred in chapter 4. Here a very basic idea of organic compounds was made. The discussion of types of formulas may be a helpful review. Naming of organic structures was started in chapter 4 and expanded in the current chapter.

Chapter 9 presented an introduction to electromagnetic radiation which is applied to infrared and NMR spectroscopy in this chapter.

Bonding ideas were developed in some detail in chapter 10. Many of the concepts receiving extensive use in this chapter were discussed in chapter 10 and any questions you have may be answered by a review of material in chapter 10.

Answers to True/False Statements:

1. True. This is the advantage over Lewis structures.

2. False. The VSEPR theory is used for the representative elements not the transition elements.

3. True. The lone pair electrons are less restrained to space since only one nuclei affects them. They spread out and squeeze the bonded atoms together some.

4. True. Generally the tetrahedral structure will be favored.

5. True. The shapes are based on fixed geometries that minimize interactions and these are common mathematical shapes.

6. False. The equatorial positions have fewer close interactions than the axial positions. The equatorial would be have less interaction.

7. True. The extra sets of electrons are still aligned between the atomic nuclei so they are geometrically equivalent to a single bond.

8. False. Lewis dot structures do not predict geometry.

9. False. The number of kinds of ligands is immaterial.

10. True. Both N and P are group five elements so they have the same number of valence electrons and undergo similar reactions.

11. False. Often polar bonds are offset by molecular symmetry. When this occurs the polarity will cancel out.

12. True. The atoms involved have a sufficiently large electronegativity difference so polarity is present. The molecule is not symmetrical so the polarity is not canceled out.

13. False. The hydrogen bond forces are the greatest of the intermolecular forces. They may be up to 1/20 the strength of a regular bond.

14. False. Several other atoms utilize the tetrahedral geometry. Silicon and phosphorous are two common examples.

15. True. CH_3CH_2SH and CH_3SCH_3.

16. False. A double bond is required.

17. True. The molecule is named from the longest continuous chain present. With many side chains the number of carbons may be large but the longest continuous chain may be rather short.

18. True. To get the second you only need to rotate the first molecule by 60 degrees.

19. True. They are mirror images and they are not superimposable.

20. False. To be enantiomers they must be mirror images and non-superimposable.

21. True. At one time these were common notations with one being to the right and the other to the left.

22. True. This could be written so that a carbon atom could have four different groups attached to it.

23. False. Racemic mixtures have equal amounts of both left and right rotating materials. The net effect is cancellation.

24. True. Cis-platinate is a good example.

25. True. The bond bending and stretching energies are in the infrared region.

Answers to Understanding of Concepts Questions:

1. There are five regular geometries that fit the VSEPR theory's ability to predict angles and shapes. They are linear, trigonal, tetrahedral (sometimes square planar), trigonal bipyramidal and octahedral. These are all regular mathematical shapes and fit the electron requirements of the representative elements.

2. Geometry can be thought of in terms of the atoms only or the atoms with lone pair electrons also. Electron pair geometry takes into consideration the positions of the lone pair electrons.

3. Lone pair electrons are less constrained to space than bond pair electrons so they often take more space and thus can interact with nearby electrons more than bond pair electrons are able to.

4. Resonance structures have no effect. The geometry is the same regardless of which structure you work with.

5. The Al to F bonds are polar because the fluorine has a very high electronegativity value and pulls the electron pair toward itself. However the molecule is symmetrical so there is no net distortion in a single direction. This molecule is also probably ionic.

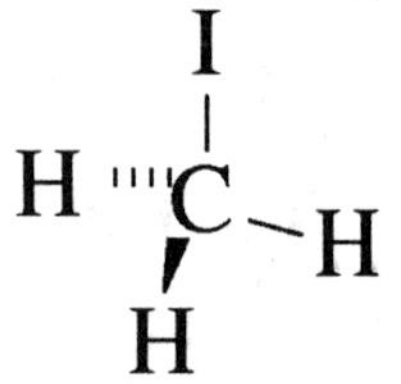

6.

7. This involves a series of steps:
   1. Identify the central atom   Sb the material with 1 atom
   2. Use all the electrons in the valence shell   5A thus 5
   3. Use the unpaired valence electrons from the ligands
      Cl has 7 but 6 are paired   thus 1 for each Cl or 3
   4. Determine total available electrons   5 + 3 or 8 for 4 pair
   5. Determine bond and lone pairs - bond pair for each ligand
      3 bond pair and 1 lone pair
   6. Determine initial geometry - 4 pair is tetrahedral
   7. The molecular geometry is slightly pyramidal

8. Branched chain isomers are hydrocarbons which have side chains of alkyl groups such as CH_3- or C_3H_7- coming off the longest continuous carbon chain.

9. The number indicates the position of the double bond in the molecule. The naming is done so that the smallest numbers possible are used.

10. Another name for 1,4-dichlorobenzene would be para dichlorobenzene.

11. Handedness refers to the arrangement of groups on a molecule which has optical activity. If the arrangement is clockwise or counterclockwise.

12. A racemic mixture has equal amounts of the optical isomer that rotates plane polarized light to the right and also the one that rotates the light to the left.

13. The infrared is higher than the NMR. Infrared is very close to the visible spectrum but NMR uses radiation in the radio wave region which is much lower energy.

Answers to Testing the Analytical Concepts

1. The molecule CH_2F_2 would have a tetrahedral shape. The two fluorine atoms would be on one side of the carbon atom and the two hydrogens on the other so the molecule would not be symmetrical. The electronegativity differences between C-F is much greater than that between C-H so a dipole would exist with the F end being partially negative and the hydrogen end partially positive.

2. The IF_5 molecule geometry involves both bond pair and lone pair electrons.

$I = 1 \times 7e = 7e$
$F = 5 \times 1e = 5e$

Total of 12 electrons for 6 pairs. Five bond pairs are needed. The molecule will have 5 bond pair and 1 lone pair.

The shape would be octahedral with no atom below the equatorial plane. It would look like a square planar system which is puffed up on top.

3. This molecule has a methyl group on the second carbon from the end and the longest chain has six carbons in it.

$$\begin{array}{l} CH_3\text{-}\underset{\displaystyle |}{C}H\text{-}CH_2\text{-}CH_2\text{-}CH_2\text{-}CH_3 \\ \quad\;\; CH_3 \end{array}$$

4. ICl_4^{1-} has a total of 12 electrons, 7 from the I , 4 from Cl and 1 from the charge. This would give 6 pair and there must be 4 bond pair since there are 4 Cl atoms attached to I. The electron geometry would be octahedral but the molecular geometry would be square planar.

5. OF_2 would be shaped like H_2O. The geometry would be V-shaped. It would also be polar and have a dipole with the fluorine end partially negative.

6. A chiral carbon atom needs four different groups attached to it. Looking at the given molecule we see that is present.

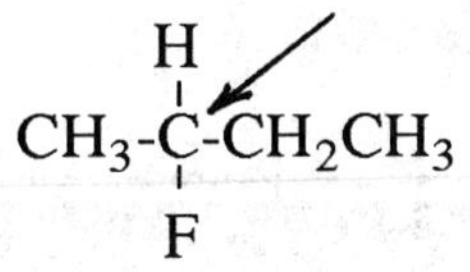

The H,F, CH_3 and CH_2CH_3 groups are all different.

Answer to Chapter 11 Crossword

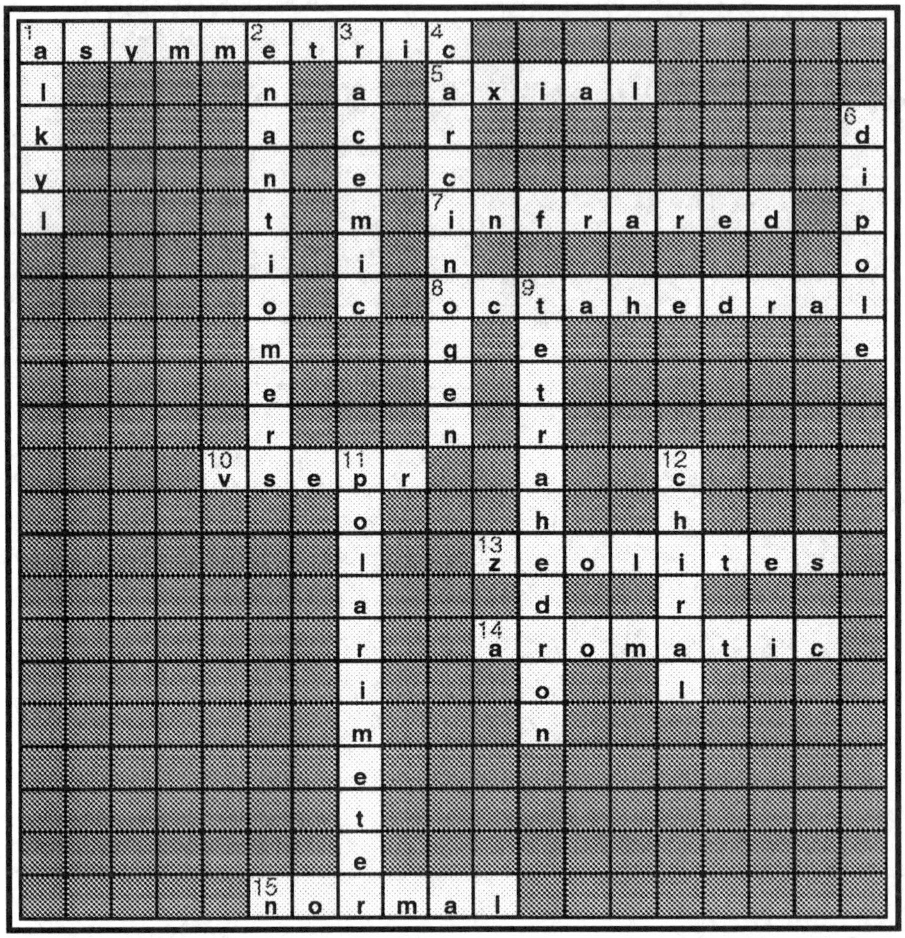

Some Supplementary Readings Relating to This Chapter:

1. Mystery Matters- Hit and Run, <u>Chem Matters</u> , page 14

2. The Color of Gems, <u>Chem Matters</u>, December 1988, page 7

3. Largest Hydrocarbon Molecules Have Rigid Dendritic Structures, <u>Chemical and Engineering News</u> , April 12, 1993, page 26

4. Separating 'Mirror' Molecules for Better Drugs, The New York Times, February 12, 1992.

5. Lead-Poisoning Harm Held to Be Partly Reversible, The New York Times , September 30, 1992.

Chapter 12 Gases, Forces and Liquids

The focus of this chapter is "states of matter". Two of the three common states, gases and liquids, are discussed. The solid state will be looked at in detail in chapter 14. The various forces that affect the properties of a gas such as temperature, pressure and composition will be studied. The intermolecular interactions that account for non-ideal behavior and the conversion from the gas to liquid state are also presented.

Focus Your Study Efforts

Several natural laws apply to gas relationships.

Objectives:

1. You should be able to state and explain the significance of standard conditions.
2. You should be able to identify what gas law relationships apply to gases under various changes in conditions.
3. You should be able to explain the properties of gases and the relationships expressed by the gas laws.

Chemists have two measurable reference points that are essential to the study of gases. They are standard pressure (1 atm.) and standard temperature (0°C or 273K). They are referred to as S.T.P.

A given gas had a pressure of 680 torr. Express this in mm of Hg and in atms

The most significant part of this question is being aware that pressures of gases can be expressed in many units. All you need are appropriate conversion units. For this question we need to know that 1mm Hg = 1 torr and that 1 atm = 760 torr.

$$\text{mm Hg} = 680 \text{ torr} \times \frac{1 \text{ mm Hg}}{1 \text{ torr}} = 680 \text{ mm Hg}$$

$$\text{atm} = 680 \text{ torr} \times \frac{1 \text{ atm}}{760 \text{ torr}} = 0.895 \text{ atm}$$

There are several laws associated with the study of gases: Boyle's law, pressure; Charles's law, temperature; Gay-Lussac's law, combining volumes; Avogadro's law, moles; Dalton's law of partial pressures; and Graham's law, diffusion. As you can see, each law addresses a specific property and its relationship to volume. These laws, however, are not independent and often are coupled together such as in the ideal gas law.

Sometimes we can get so involved with trying to determine what law to apply that we forget to use our common knowledge. We all know that if pressure is applied to a gas it will compress the volume, or that heating a gas will cause expansion. When working with problems make use of that knowledge then set up dimensional analysis solutions.

Remember the gas laws apply to a fixed amount of gas, which is a specific molar amount, not to a fixed volume.

Boyle's law is an expression relating the inverse relationship of pressure applied to a gas and the volume the gas occupies. This law depends upon temperature remaining constant.

A 300.0 mL flask with methane gas was at standard pressure and temperature. When transferred to a 500.0 mL flask what was the pressure if temperature is constant?

This is a pressure volume problem and Boyle's law applies. However we know that when the volume of a given gas sample increases the pressure must decrease. This tells us which way to set up our conversion factor.

$$P_{std} = P_{300} \times \frac{300.0 \text{ mL}}{500.0 \text{ mL}} = 760 \text{ torr} \times \frac{300.0 \text{ mL}}{500.0 \text{ mL}} = 456 \text{ torr}$$

The direct relationship that exists between a gas volume and the **absolute temperature** is expressed by Charles's law. When using this expression the temperature **must** be expressed in Kelvin temperature.

An 11.0 L sample of SO_2 was collected at 25.0°C and 745 torr. If the pressure is constant and the gas is subjected to a temperature change so it occupies 2.5 L what is the new temperature?

As we know, temperature affects volume, so Charles's law is applied. A decrease in volume has to be the result of temperature lowering. The volume change is directly proportional to absolute temperature.

$$T_f = 298\text{K} \times \frac{2.5 \text{ L}}{11.0 \text{ L}} = 67.7\text{K} \quad \text{or } -205.4°\text{C}$$

A gas sample in a 1.00 L container at 25.0°C and 720.0 mm Hg was heated to 100.0°C and the pressure was increased to 800.0 torr. What is the new volume?

This problem involves volume with pressure (Boyle's law) and temperature (Charles's law). The other factor to keep in mind is that temperature must be in Kelvin. °C + 273 = K; torr = mm Hg.

The temperature is increasing which will cause a volume increase. However, the pressure is also increasing which will result in a volume decrease.

$$\text{Vol} = 1.00\ \text{L} \times \text{Pressure factor} \times \text{Temperature factor}$$

$$\text{Vol} = 1.00\ \text{L} \times \frac{720\ \text{mm Hg}}{800\ \text{mm Hg}} \times \frac{373\ \text{K}}{298\ \text{K}} = 1.13\ \text{L}$$

A point sometimes not intuitive, but very important to remember, is that when pressure and temperature are constant, any gases of equal volume will have the same number of particles. This is known as Avogadro's law. Any volume change is directly proportional to the change in moles of gas involved.

Two identical containers were at the same temperature and pressure. One container has 25.0 g of propane and the other has 25.0 g of xenon tetrafluoride. Which container has the greatest number of molecules present?

According to the law of Avogadro the same number of particles would have the same volume. Or the volume is related to the number of particles present. In this case the constant mass does not mean the number of particles is the same because propane has a molar mass of 44 and that of xenon tetrafluoride is 207.3. The number of moles of propane is much greater so it has the greater number of particles.

Gay-Lussac established the relationship between gas volumes as they react. He provided evidence that the changes were always in the ratio of whole numbers.

A mole of any gas at STP (273K and 760 torr) will always occupy 22.4 liters of volume. This volume is known as the **molar volume**.

Given: $2\ C_4H_{10} + 9\ O_2$ ===> $8\ CO_2 + 10\ H_2O$
If the butane burns completely in a 22.4 L container what pressure would result from the CO_2 produced if 1.0 grams of fuel is burned? (Assume the system goes to standard conditions.)

We need to convert the mass of fuel to moles so we can determine the moles of CO_2 and then that can be translated to pressure.

$$\text{mol } CO_2 = 1.0 \text{ g } C_4H_{10} \times \frac{1 \text{ mol } C_4H_{10}}{58 \text{ g } C_4H_{10}} \times \frac{8 \text{ mole } CO_2}{2 \text{ mole } C_4H_{10}}$$

mol CO_2 = 0.069

A mole of gas in 22.4 liters would have a pressure of 1 atm. We have 0.069 moles in 22.4 liters.

$$\text{pressure} = \frac{760 \text{ torr}}{\text{mole}} \times 0.069 \text{ mole} = 52.4 \text{ torr}$$

When the relationships present in Boyle's, Charles's, and Avogadro's laws are combined we get the ideal gas law. Volume is directly proportional to temperature and moles of particles but inverse to pressure. This gives the relationship:

$$V \propto \frac{n\ T}{P} \quad \text{converted to an equation} \quad V = \frac{n\ R\ T}{P}$$

This proportionality can be made into an equation by using a proportionality constant. This constant is symbolized with **R** and is the ideal gas constant.

When you use the ideal gas constant, R, care is needed because this constant is applied in many situations and units are not always the same.

A 1000. L sample of propane gas is at 1100 mm Hg and 35.0°C. How many moles of gas are present at these conditions?

We can apply the ideal gas equation to solve this problem.

VP = nRT

$$n = \frac{RT}{PV} = \frac{0.082 \text{ L atmK}^{1-}\text{mol}^{1-} \times 308 \text{ K}}{1100 \text{ mm}/760 \text{ mm/atm} \times 1000. \text{ L}} = 0.0174 \text{ mol}$$

Expression of gas densities need be done with care because changes in temperature and pressure have major effects on the value. For this reason, it is best to compare all density values at a standard set of conditions, STP.

A sample of freon gas, CCl_2F_2, occupies 800.0 mL at 25.0°C and 300.0 torr. What volume will the gas occupy at 0.0°C and 700.0 torr? What happens to the gas density?

To solve these questions we need the answer to the first problem before we can solve the second.

Both Boyle's and Charles's laws apply. Since temperature decreases the volume will decrease and an increase in pressure will bring about a volume decrease.

$$\text{Vol} = 800.0 \text{ mL} \times \frac{273\text{K}}{298 \text{ K}} \times \frac{700.0 \text{ torr}}{760.0 \text{ torr}} = 675.0 \text{ mL}$$

The gas density would increase. We have the same amount of gas but it is in a smaller volume.

When more than one gas is present in a sample, each gas will exert a pressure. The total pressure measured is a sum of the individual gas pressures. This is Dalton's Law of Partial Pressures.

A gas sample was found to contain the following gases and the pressure exerted by each was measured as noted. An unidentified gas was also present.

SO_2	120 torr
CH_4	85 torr
CO	135 torr

The total pressure was 735 torr. What is the partial pressure of the unknown gas? Which of the identified gases has the greatest average molecular speed?

$P_t = P_{\text{sulfur dioxide}} + P_{\text{methane}} + P_{\text{carbon monoxide}} + P_{\text{unknown}}$
$P_t = 120 \text{ torr} + 85 \text{ torr} + 135 \text{ torr} + P_{unk} = 735 \text{ torr}$
$P_{unk} = 395 \text{ torr}$

The gas with the lowest molecular mass would have the greatest average molecular speed. Of the known gases that would be methane, CH_4.

A mole fraction is the ratio of the moles of a given gas to the total number of moles of gas present in the container. (See text page 526.)

The kinetic-molecular theory is used to explain gas behavior.

Objectives:

1. You should be able to state the fundamental concepts of the kinetic-molecular theory.
2. You should be able to use the kinetic-molecular theory to explain gas effusion and diffusion.
3. You should be able to relate the physical process of vapor-

ization, evaporation and vapor pressure to the kinetic-molecular theory.

The kinetic-molecular theory is the basis for our understanding of gases. It has five basic concepts, some of which may not seem rational, but are all based on extensive experimental evidence.

All particles are in constant random motion in any material, but very constrained in solids. In gases the particles speed depends on the gas temperature and its molar mass. Higher temperatures and lower mass result in more particles being at higher speeds.

When gases move they can escape through a hole by a process known as effusion. When different gases come in contact, and if they do not react, they will mix with each other by diffusion.

Without calculating, arrange the following gases in <u>decreasing</u> order of rate of effusion from a container.

He UF_6 CO NO_2

Gases with lower mole weight will effuse faster. He has a mass of 4, UF_6 is 352, CO is 28 and NO_2 is 46.

He > CO > NO_2 > UF_6

This concept was used to make enriched uranium in the Manhatten Project during World War II.

The terms volatility, vapor pressure, vaporization, and evaporation all refer to the liquid-gas interchange. Be sure you know which term to apply in specific situations.

Various forces affect ideal gas behavior.

Objectives:

1. You should be able to define and explain the various intermolecular forces and explain the importance of hydrogen bonding.
2. You should be able to explain how intermolecular forces influence the liquid properties of surface tension, vapor pressure, and condensation / boiling point.

Ideal behavior is lost by gases at some increased pressure and/or reduced temperature. Several common gases will lose this ideal property even when small changes have occurred. The loss of ideal behavior is caused by the intermolecular forces molecules exert on each other.

Intermolecular forces are the cause of several properties of liquids and gases. They are what cause gases (even the noble gases) to condense and behave as if they are not independent particles. Intermolecular forces also raise boiling temperatures to unexpected values and give liquids surface properties.

All of the intermolecular forces are weak. However, the effect they have is often significant because of the large numbers of particles interacting. Many small interactions will result in gases having non-ideal behavior. Of all the intermolecular forces the strongest, by far, is the hydrogen bond. The hydrogen bond however is limited to molecules containing hydrogen and an element of high electronegativity.

A gas will condense to the liquid state when the intermolecular forces overcome the kinetic energy of the particles.

A liquid will become a gas when the particles receive enough energy to overcome the attractive forces between the particles. The vaporization process requires energy and is endothermic.

Some particles of every liquid, at a given temperature, have sufficient energy to escape. Those particles that make it into the vapor phase will exert a partial pressure which is called the vapor pressure.

When sufficient energy is added so the vapor pressure is equal to the atmospheric pressure we have reached a temperature known as the boiling point.

The boiling temperature varies with the applied pressure. When the applied pressure is one atmosphere the boiling occurs at what is called the "normal" boiling point.

Once a particle escapes from the liquid it may again come in contact with the surface and recombine with the liquid state. There is a dynamic exchange constantly occurring between the liquid and vapor. Given time the rates of exchange will be equal and a condition known as equilibrium has been established.

Symbols found in Chapter 12

Page	Symbol	
508	mm Hg	Pressure unit in millimeters of mercury.
508	torr	Pressure unit of torr, equivalent to mm Hg.
508	kPa	Pressure unit of kiloPascal.
508	bar	Pressure unit of bar, equivalent to 0.01 kPa.
508	lb/sq in	Common English pressure unit.
508	atm	Pressure unit of 1 atmosphere at sea level. 760 torr.
511	N	Newton, a unit of force.
518	n	Represents the number of moles of material present.

519	R	The ideal gas constant. This can have many different sets of units. Here it is 0.0821 L·atm/K mol.
526	X_a	Symbol for mole fraction. The subscript varies.
528	···	Represents a hydrogen bond between molecules.
545	VOCs	Volatile organic compounds.

Terms in Chapter 12

The following terms are listed for your convenience. They should not become a list for you to memorize. You should have a working definition for each. If you have doubt about any particular term check its definition in the glossary and/or review it in the chapter. Some of these terms are included in the crossword puzzle later in the chapter. There are also 7 Laws discussed.

Pressure	Barometer	Standard atmosphere
Newton	Pascal	Bar
Compressibility	<u>Boyle's law</u>	<u>Charles's law</u>
<u>Gay-Lussac's law</u>	<u>Avogadro's law</u>	<u>Ideal gas law</u>
Ideal gas constant	Standard temperature	Standard pressure
Standard molar volume	General gas equation	Partial pressure
<u>Dalton's law of partial pressure</u>		Mole fraction
Kinetic-molecular theory		Volatile
Gaseous diffusion	Effusion	<u>Graham's law</u>
Intermolecular forces	Condensation	Dispersion forces
Induced dipole	Polarization	Dipole-dipole force
Hydrogen bonding	Surfactants	Surface tension
Capillary action	Vaporization	Evaporation
Latent heat of vaporization		Heat of vaporization
Vapor pressure	Equilibrium vapor pressure	
Boiling	Boiling point	Normal boiling point

Concept Maps for Gases, Forces & Liquids

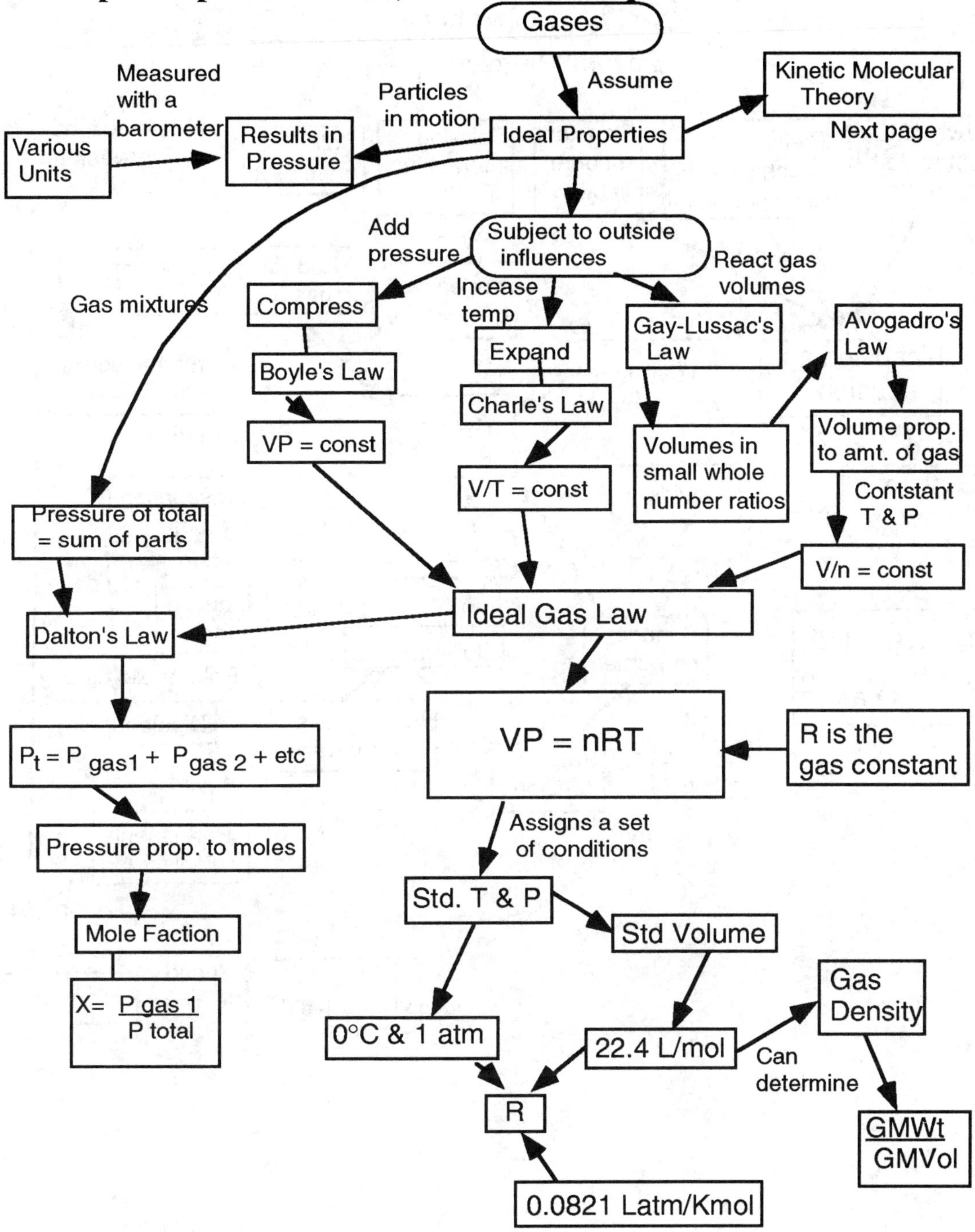

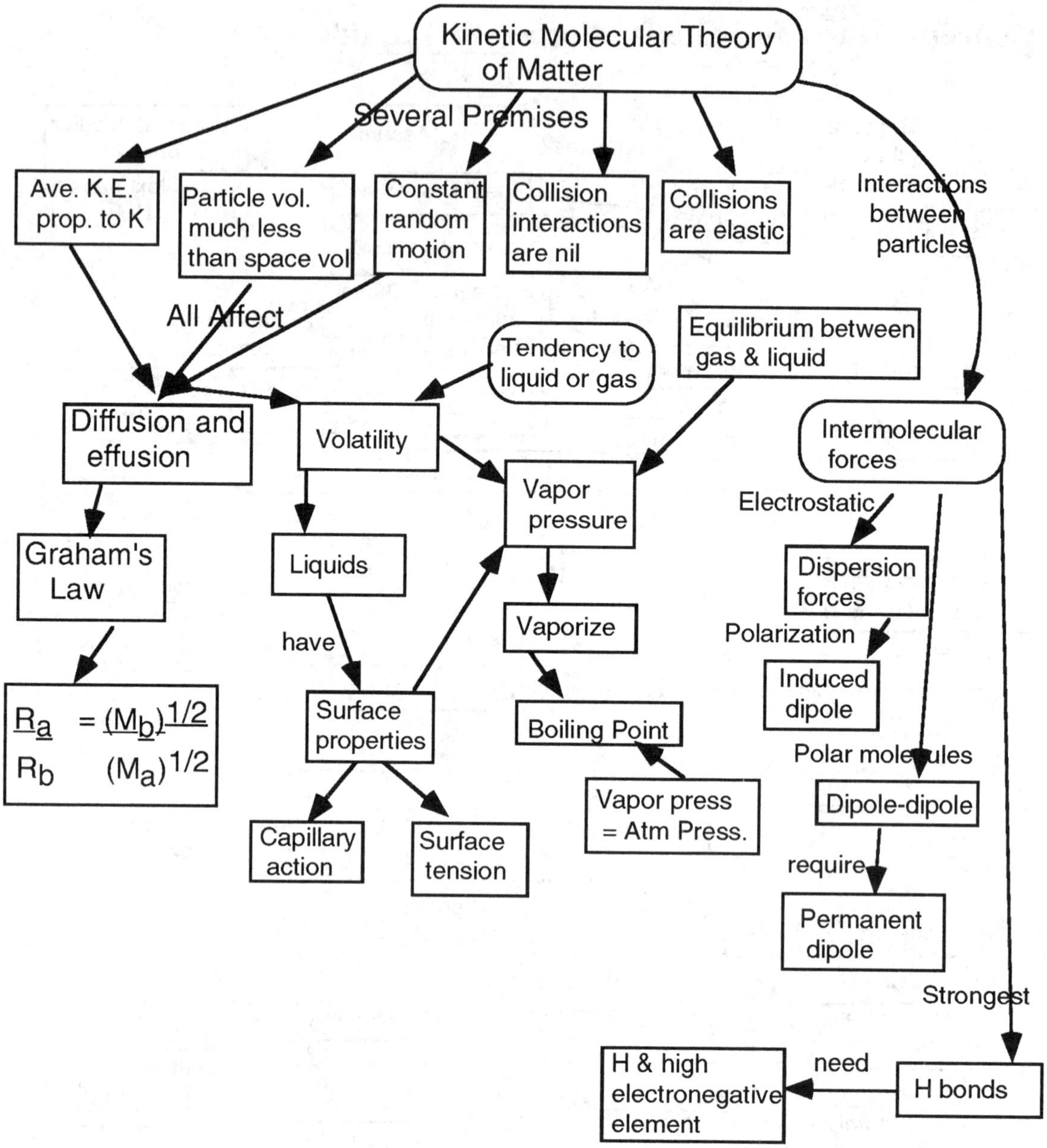

Kinetic Molecular Theory of Matter
Several Premises
Ave. K.E. prop. to K
Particle vol. much less than space vol
Constant random motion
Collision interactions are nil
Collisions are elastic
Interactions between particles
All Affect
Tendency to liquid or gas
Equilibrium between gas & liquid
Diffusion and effusion
Volatility
Intermolecular forces
Vapor pressure
Electrostatic
Graham's Law
Liquids
Dispersion forces
Vaporize
Polarization
have
Induced dipole
$\frac{R_a}{R_b} = \frac{(M_b)^{1/2}}{(M_a)^{1/2}}$
Surface properties
Boiling Point
Polar molecules
Vapor press = Atm Press.
Dipole-dipole
require
Capillary action
Surface tension
Permanent dipole
Strongest
need
H & high electronegative element
H bonds

Gases, Forces & Liquids True/False Statements

Answers are at the end of the chapter.

1. Particles in the liquid state contain less energy than those in the gaseous state.

2. A gas will exert a pressure in proportion to the number of particles present.

3. Standard pressure is 100 mm Hg.

4. According to Boyle's law the volume of a gas is directly proportional to the pressure applied to it.

5. Since heating a gas causes it to expand, a mole of gas varies in mass as the temperature changes.

6. If a gas is cooled and has added pressure applied to it, these two factors work together to decrease the gas volume.

7. If you know the volume of a given amount of a gas at a specific temperature and pressure you would know the volume of twice that amount of gas at the same temperature and pressure.

8. Five liters of any gas at STP will have the same number of moles as five liters of any other gas at STP.

9. The ideal gas law is valid only at standard temperature and pressure.

10. As the pressure applied to a gas sample, for example a balloon, is increased the density will also increase.

11. Every gas present in a mixture exerts a pressure.

12. If water vapor were removed from the atmosphere there would be no change in the air pressure.

13. SO_2 gas will effuse from a container faster than CO_2 gas.

14. Hydrogen bonding has a significant effect on the boiling point of some molecules.

15. When a gas condenses, energy is released to the environment.

16. One function of shampoo is to break forces between water molecules.

17. The boiling point of a given liquid is always constant.

Testing Your Understanding of Concepts in Chapter 12. Answers are at the end of the chapter.

1. Why is pressure of a the gas dependent on temperature?

2. Why are standard conditions important?

3. When gases are compressed they eventually lose their ideal behavior. Why does pressure cause this loss of ideal behavior?

4. What is the significance of -273°C or 0 kelvin?

5. Why does an increase in both temperature and pressure make it difficult to predict the net effect on volume without doing a calculation?

6. How does gas density relate to the molar mass of the gas?

7. If all of the oxygen were suddenly removed from an atmospheric gas sample what would happen to the pressure?

8. If 50.0 grams of NO_2 gas and 50.0 grams of CH_2F_2 gas were introduced into a cylinder which would present the greater mole fraction?

9. Give two examples of the advantages of high volatility in a sample and also give an example of a disadvantage.

10. What is the difference between an induced dipole and a dipole-dipole interaction?

11. Why does surface tension exist in a liquid?

12. What is the significance of the term "equilibrium" in the expression equilibrium vapor pressure?

Testing You on Some Analytical Concepts.

Answers are at the end of the chapter.

1. A certain engine has a cylinder volume of 246.0 mL that decreases to 24.0 mL. at full compression. If the initial gas pressure was 0.981 atm. what is the pressure at the point of maximum compression?

2. If liquid nitrogen at -195.8°C was poured on a balloon of 15.0 liters volume at 25°C, what would the final volume be?

3. What would 50.0 mL of carbon dioxide gas weigh at STP?

4. Laboratory gases are often stored in compressed gas cylinders. What weight of nitrogen gas is stored at 21.0°C and 170 atm in a cylinder whose volume is 60.0 liters?

5. A 5.00 liter container at 25.0°C has 1.00 gram of He and 1.00 gram of H_2 added to it. What is the pressure in the container?

6. If the average velocity of a molecule of hydrogen is 2000 m/sec how fast would a molecule of oxygen travel at the same set of conditions?

Crossword - Gases, Forces & Liquids

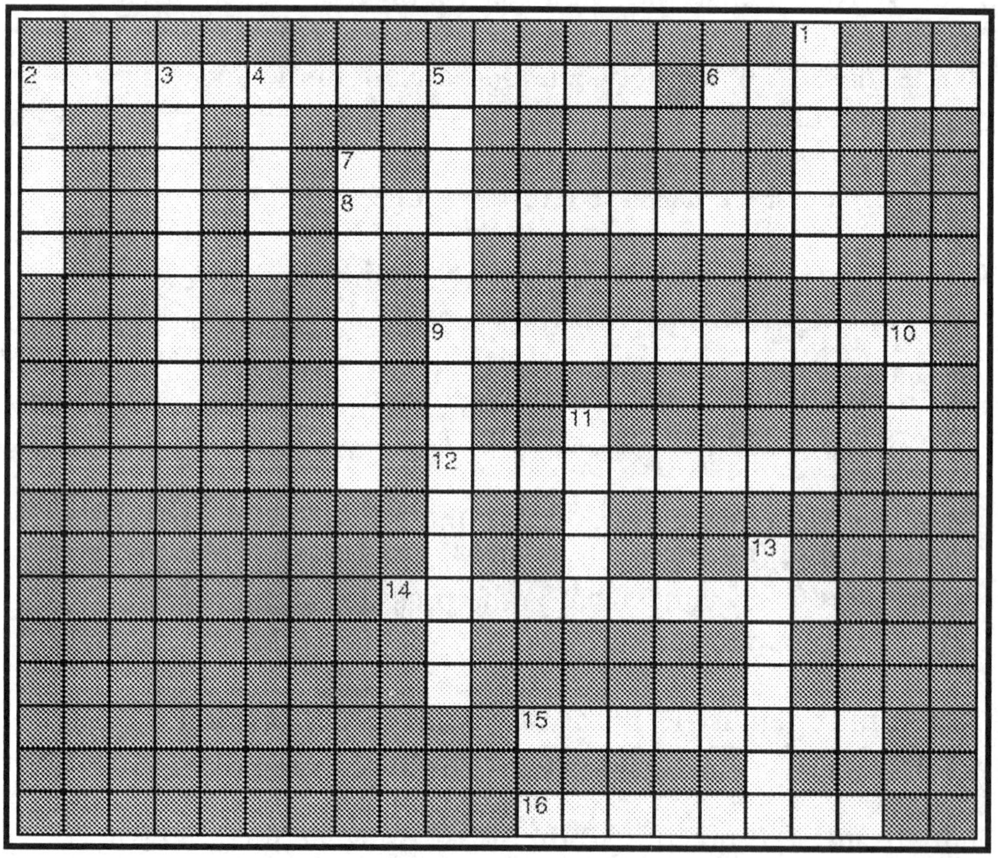

Crossword - Gases, Forces & Liquids Clues

ACROSS

2 These are forces between molecules.
6 The law referring to gas diffusion is named after this person.
8 The process of becoming a gas but not necessarily boiling is called ________.
9 These materials will weaken surface forces such as surface tension.
12 This devise is used to measure gas pressure, such as the pressure of the atmosphere.
14 The group of weak forces that exist between molecules are called ____ forces.
15 A material that readily becomes a gas is said to be _____.
16 This is the strongest of the weak forces between molecules and requires the presence of an atom with high electronegativity.

DOWN

1 The law named for this person refers to pressures in a gas mixture.
2 The laws relating to gases assume the gases have this behavior.
3 The process of escaping through a small hole is called _______.
4 22.4 liters is known as the ________ volume.
5 This is a characteristic of gases when pressure is applied.
7 The law relating the number of particles of gas to the gas volume is named after this person.
10 Abbreviation for standard conditions of temperature and pressure.
11 The law relating the effect of pressure on the volume of gas is named after _______.
13 A liquid will start this when the vapor pressure equals the atmospheric pressure.

Ties to Previous Chapters

Mixtures were introduced in chapter 2. Since gases are often mixtures the material discussed in chapter 2 will also apply to aspects of chapter 12.

Gases react with other gases as well as solids and liquids. Whenever chemical reactions occur stoichiometry becomes an important issue to the chemist. Chapter 6 was devoted to a discussion of chemical stoichiometry. This chapter continues work with amounts in chemical reactions and a working knowledge of chapter 6 is assumed.

When liquids and gases states are in contact, exchanges take place and this involves energy. Many of the ideas and terms from chapter 7 are important in this chapter also.

There is a significant amount of bonding that is present between gases and liquids.This is not intramolecular bonding but bonding of the intermolecular nature. This is a different kind of bonding than was discussed in chapters 4 and 10. You need to know what these differences are and when the various types of bonds are important.

Answers to True/False Statements:

1. True. Energy is required to convert a liquid to a gas. This is known as the heat of vaporization.

2. True. This is a statement of Avogadro's law.

3. False. Standard pressure is 760 mm of Hg.

4. False. The volume is inversely proportional to the pressure. As the pressure increases the volume decreases.

5. False. A mole of gas is a mole of gas. The mass does not change. The density of the gas would vary.

6. True. Both of these factors tend to decrease volume. This is a combination of Boyle's and Charles's laws.

7. True. The volume is directly proportional to the amount of gas present.

8. True. This is just a variation of the fact that a mole of any gas at STP occupies 22.4 liters.

9. False. The ideal gas law will apply until the gas loses ideal behavior due to particle interactions.

10. True. Since the balloon can be compressed the same amount of gas would be confined to a smaller volume so the density would increase.

11. True. Every gas would exert a partial pressure. This is Dalton's Law of Partial Pressure.

12. False. This would violate Dalton's Law of Partial Pressure.

13. False. The greater the molar mass, the slower the rate of effusion. SO_2 has a molar mass of 64 and CO_2 has a molar mass of 44.

14. True. This is very evident in the case of water.

15. True. The energy that was consumed in converting it to a gas would be released when it is condensed.

16. True. Shampoo acts as a surfactant which lets the water wet a surface so it can interact.

17. False. The boiling point depends upon the pressure applied to the liquid. The higher the pressure the greater the boiling point.

Answers to Understanding of Concepts Questions:

1. Pressure is the result of the particle striking the container. As the temperature is increased the average kinetic energy of the gas increases so the frequency and intensity of the particle collision is increased. Consequently raising or lowering the temperature will change the pressure.

2. It is important to report information regarding gases. To make this more efficient and uniform the information is reported at a given set of conditions so that comparisons can be made by anyone at any place with the same results.

3. As the gas is compressed the volume between the particles decreases. This gives the particles a better chance to interact with each other. Once this occurs they lose their individual characteristics and take on group properties. This interaction results in such things as gas condensation.

4. This temperature is the coldest possible. There is a total lack of molecular motion. At this temperature all particles would interact with each other and every material would be non-ideal.

5. These two effects work against each other. Without knowing the magnitude of each effect there is know way to determine which has the greater influence.

6. Density is mass per unit of volume. For gases, the density can be calculated from the molar mass and molar volume. Since the molar volume is constant, as the mass increases so does the density.

7. Since the atmosphere is 21% oxygen if all of the oxygen were removed the pressure would decrease by 21%.

8. The molar mass of CH_2F_2 is slightly greater than that of NO_2 so the sample of nitrogen dioxide would have a slightly greater number of moles for the same mass. The NO_2 would have the greater mole fraction of gas present.

9. High volatility in a vapor is needed when the sample is useful or needed in the vapor state. For example, to get a liquid to burn, the vapors are necessary. Materials which give a scent such as flowers do so because of the volatility of the components. A disadvantage would be when the liquid may be hazardous in the vapor form.

10. An induced dipole is the result of particle collisions in which a temporary polarization of the molecule occurs. This results in some very weak interactions. A dipole-dipole interaction is much more permanent and much stronger.

11. Surface tension is the result of liquid particles interacting with each other. These interactions are much stronger than the interactions between the particles in the atmosphere above the liquid and the liquid itself. The net result is a skin effect on the liquid.

12. A liquid and the vapor above it are constantly exchanging molecules. This is a dynamic condition. At some point the vapor can accept no more particles from the liquid because it is

saturated. Each time a particle enters the vapor state a particle from the vapor condenses to the liquid. This rate of exchange is constant and an equilibrium is said to be present.

Answers to Testing the Analytical Concepts

1. This is a Boyle's law problem so we know that at smaller volumes the pressure will be greater with the temperature constant.

$$P_{comp} = 0.981 \text{ atm} \times \frac{246.0 \text{ mL.}}{24.0 \text{ mL.}} = 10.1 \text{ atm.}$$

2. This problem requires the application of Charles's law. As a gas is cooled the volume will decrease.

$$V_{cool} = 15.0 \text{ L} \times \frac{77.3 \text{ K}}{298.1 \text{ K}} = 3.90 \text{ liters}$$

3. We know a mole of any gas at STP occupies 22.4 liters.

$$\text{Moles of } CO_2 = \frac{0.0500 \text{ L}}{22.4 \text{ L mol}^{1-}} = 0.00223 \text{ mol}$$

$$\text{g } CO_2 = 44.0 \text{ g mol}^{1-} \times 0.00223 \text{ mol} = 0.098 \text{ g}$$

4. We have enough information with this problem to apply the ideal gas equation. P = 170 atm, T = 21.0°C or 294.1K, vol = 60.0 L

$$PV = nRT \quad \text{or} \quad n = \frac{PV}{RT} \quad \text{and} \quad n = \frac{\text{Wt used}}{\text{Mol Wt}}$$

$$\frac{\text{Wt}}{\text{MW}} = \frac{PV}{RT}$$

$$\text{Wt.} = \frac{\text{Mol Wt x PV}}{\text{RT}} = \frac{28.0\text{ g mol}^{1-}\text{ x 170 atm x 60.0 L}}{0.082\text{ L atm K}^{1-}\text{ mol}^{1-}\text{ x 294.1 K}}$$

Wt. = 1.18×10^4 grams

5. This is a mixture of gases so each will contribute to the total pressure. This will involve Dalton's Law of Partial Pressures but before that is applied we need to determine the pressure of each of the gases.

$$\text{mol He} = 1.00\text{g x } \frac{1\text{ mol He}}{4.00\text{ g}} = 0.250\text{ mol}$$

$$\text{mol } H_2 = 1.00\text{g x } \frac{1\text{ mol } H_2}{2.00\text{ g}} = 0.498\text{ mol}$$

$$P_{He} = \frac{0.250\text{ mol x 0.0821 L atm K}^{1-}\text{mol}^{1-}\text{ x 298.1 K}}{5.00\text{ L}} = 1.22\text{ atm}$$

$$P_H = \frac{0.498\text{ mol x 0.0821 L atm K}^{1-}\text{mol}^{1-}\text{ x 298.1 K}}{5.00\text{ L}} = 2.43\text{ atm}$$

$$P_{total} = P_{He} + P_H = 1.22\text{ atm} + 2.43\text{ atm} = 3.65\text{ atm}$$

6. This problem involves the diffusion of gases which is expressed by a Graham's law expression. Since oxygen is more massive than hydrogen we realize it should diffuse at a slower speed.

$$\frac{V_{Hydrogen}}{V_{Oxygen}} = \frac{[MW_{oxygen}]^{1/2}}{[MW_{Hydrogen}]^{1/2}}$$

$$V_{oxygen} = V_{hydrogen} \text{ x } \frac{[MW_{hydrogen}]^{1/2}}{[MW_{oxygen}]^{1/2}}$$

$$V_{oxygen} = 2000 \text{ m/s} \times \frac{[2.01 \text{ mass units}]^{1/2}}{[32.0 \text{ mass units}]^{1/2}} = 501 \text{ m/s}$$

Answer to Chapter 12 Crossword

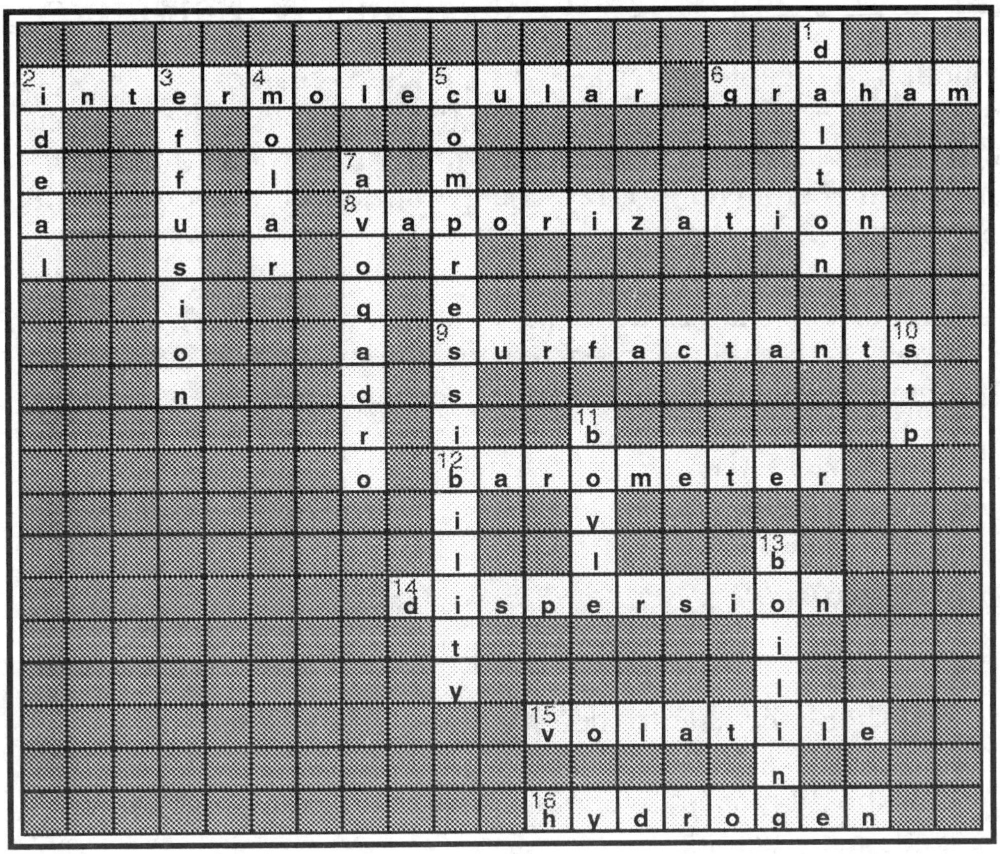

Chapter 13 Energy, Organics & Polymers

An introduction to organic materials was started in chapter 10, continued in chapter 11 and now extended in this chapter. The previous chapters were almost entirely concerned with hydrocarbons (molecules containing C and H atoms only). Previously introduced rules for naming apply here as well.

The major part of this chapter deals with various kinds of organic materials known as the hydrocarbon substitution products. These compounds are the organic materials of major interest to chemists and industry for chemical production.

Focus Your Study Efforts

Petroleum Chemistry

Objectives:

1. You should be able to identify the fossil fuels and their origin.
2. You should be able to explain how crude petroleum can be made more useful.

Coal, petroleum, and natural gas are known as fossil fuels because of how they were formed.

Fossil fuels are the major source of organic chemicals, especially petroleum. However,of all the petroleum being recovered, only about 6% is used for production of organic chemicals.

Petroleum is a complex mixture which is separated into other, less complex, mixtures.

Burning of petroleum offers little chemistry of interest to the chemist. There is, however, a great deal of chemistry involved in the modification and formulation of petroleum fuels. Since the major use of petroleum is for fuels, a very technological industry has developed to meet the ever changing needs for types, quantities, and special characteristics.

Refining of crude petroleum involves several steps: fractional distillation to separate the material into more similar components; catalytic reforming to convert straight chain hydrocarbons into more useful branch chain molecules; catalytic cracking to convert large less useful molecules into smaller useful ones; and enhancement or reformation to make mixtures of components to meet special demands.

A fuel blend is rated on its ability to burn smoothly under specific conditions. The rating scale is open ended and gives each material a number value known as the octane number. (See table 13.2 of the text.)

Other fossil fuels

Objectives:

1. You should be able to explain the differences in supplies of natural gas and coal.

Natural gas appears to be of two distinct types. That found in North America is a mixture of three hydrocarbons and several other gases. Natural gas from Europe and Japan is nearly all methane, CH_4.

Coal also is a complex mixture. When coal undergoes pyrolysis we get several components: coal gas, coal tar, and carbon known as coke. Coal tar yields many chemicals upon chemical workup. Coke can be converted to synthesis gas.

To some extent coal can be converted to hydrocarbons by hydrogenation. This may enhance the use of a major resource since hydrocarbons can be made into other useful compounds.

Hydrocarbons & Functional Groups

Objectives:

1. You should be able to identify and/or write the reactions that hydrocarbons undergo.
2. You should be able to explain why saturated and unsaturated hydrocarbons undergo different reactions.
3. You should be able to identify various classes of organic compounds by the functional group present in the formula.

When hydrocarbons are converted to other compounds their value is much greater. Saturated hydrocarbons undergo substitution reactions and the unsaturated hydrocarbons will more readily undergo addition reactions.

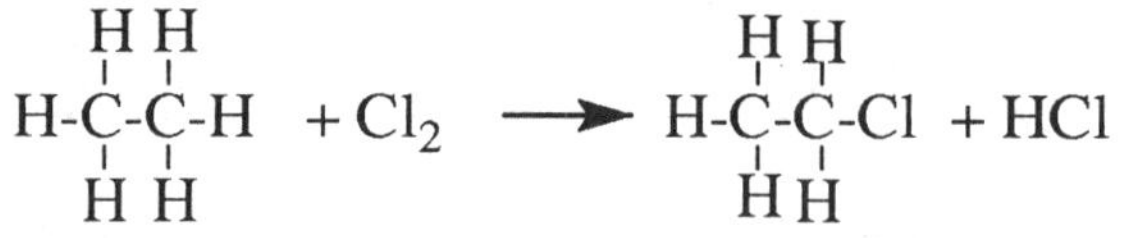

Substitution

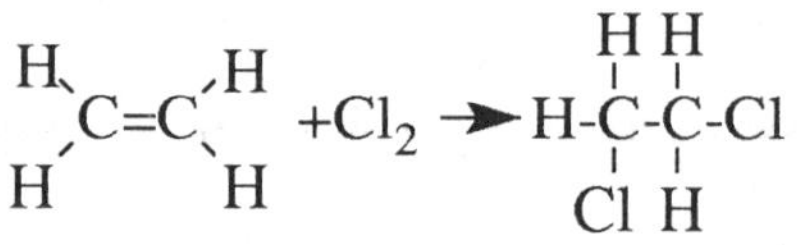

Addition

Hydrocarbon substitution products are compounds of major interest to chemists. Some common examples are: alcohols, aldehydes, acids, ketones, esters, amines, halides, and amides. These classes are each identified by a distinctive functional group.

When referencing organic materials we tend to write a generic system for multiple compounds. To do this we use R or R' etc. to represent an alkyl group such as CH_3-.

Alcohols, R-OH. These materials have an -OH group attached to a carbon with no other oxygens attached.

Alcohols are classified in two common ways. The first uses primary, secondary and tertiary as names referring to the number of carbons bonded to the carbon with the alcohol functional group. The second refers to the number of OH groups in the molecule. For example, if there are two, it would be called a diol.

There are many alcohols but a few are very common and have major commercial value. Methanol, wood alcohol; ethanol, grain alcohol; 2-propanol, rubbing alcohol; ethylene glycol, antifreeze solution; and glycerol in metabolism and fats.

Aldehydes, $R\text{-}C(=O)\text{-}H$,are a common product as a result of primary alcohol oxidation. Ketones, $R{-}C(=O){-}R$, result when secondary alcohols are oxidized.

The $R\text{-}C(=O)\text{-}O\text{-}H$ group is known as the carboxylic acid functional group. Several of these acids are common to us all. The common material vinegar is 4-5% ethanoic acid (acetic acid). One we try to avoid is butanoic acid, found in rancid butter (butter not margarine).

Reactions of alcohols with acids will produce a class of compounds

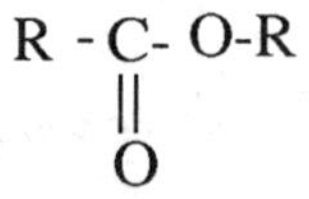

known as esters, . These are commonly found in flavorings and also make excellent solvents.

By now you should have noticed that many of these functional groups are similar in appearance and you need to look at the formulas very carefully to avoid mistakes. Acids, aldehydes, and esters are especially similar.

Draw structural formulas for: (a) propanoic acid, (b) a secondary alcohol, (c) an amine with two carbons.

Remember that a structural formula shows how the various atoms are bonded to each other. This formula gives us the most information.

(a) The acid functional group is 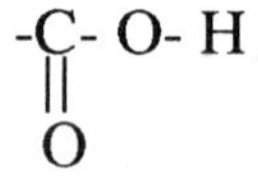 and the prefix prop- means three carbons.

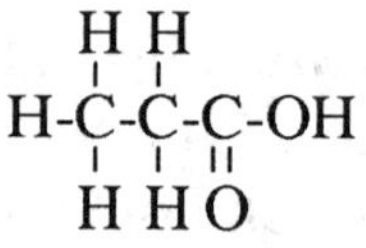

(b) The alcohol functional group is -OH. Secondary means 2 carbons are attached to the carbon atom with the functional group. The example given is only one of many possible answers.

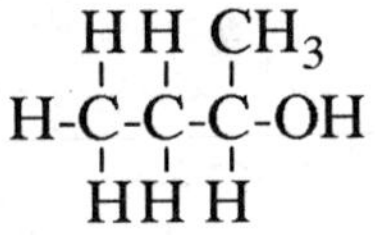

(c) The amine functional group is $-NH_2$.

CH_3-CH_2-NH_2

Place the following functional groups on the carbon atom indicated. Fill the other sites with hydrogens as needed.

Ketone on #3 Alcohol on #5 Acid on #1 Amine on #4

1 2 3 4 5 6 7 8
C- C - C - C- C- C - C- C

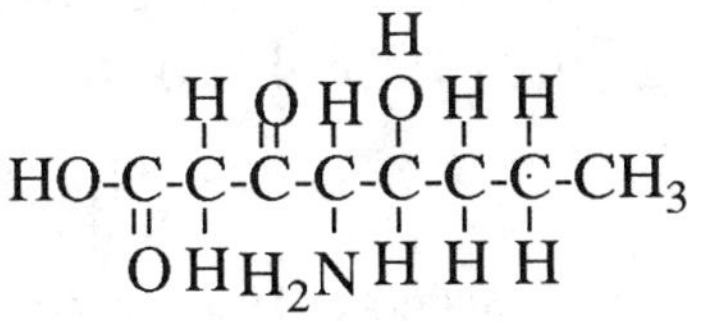

The groups could be above or below.

Draw the structural formula for 3,4-dimethyl-2-hexanol.

The ol ending tells us we have an alcohol. The hex means 6 carbons. Methyl is the CH_3 group and di means two of them. The numbers are the locations of the groups.

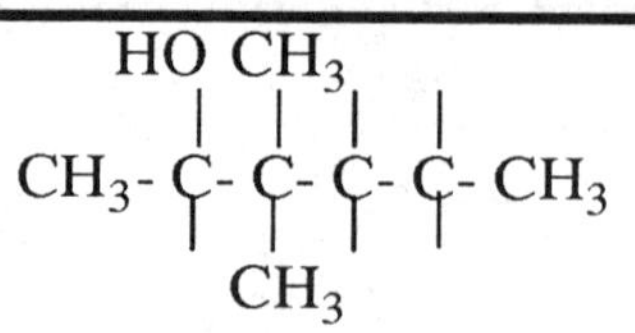

hydrogens fill the blank bonds.

Organic Polymers

Objectives:

1. You should be able to define what a polymer is and the two major reactions used to form them.
2. You should be able to identify and give examples of natural and synthetic polymers.

Organic polymers make up the largest segment of modern organic chemistry and probably will continue to grow in importance.

Plastics are a subgroup of polymers. There are two broad subdivisions for plastics based on what happens to them if heated. Thermoplastics are capable of being reformed but thermosetting plastics, which are cross-linked, once formed can not be redone.

Reaction of monomers occurs by two major processes. If monomers have double bonding they will form polymers by addition to the double bond. Monomers with functional groups such as acids or amines react by a condensation reaction.

Addition and condensation polymers are produced by two very different reactions, with very different conditions, and you need to know the difference.

Ethylene, $H_2C{=}CH_2$, is the starting material for many different kinds of polymers with many varied uses. Table 13.10, on page 585 of your text, shows several examples.

We often think about polyethylene as being a single material when really it can be several very different materials. Three common forms are low density polyethylene, high density polyethylene, and cross-linked polyethylene.

One natural material that has been reproduced in the laboratory, and then modified many ways, to meet very special needs is rubber. In the natural state the monomer is isoprene.

Polymers need not be a single monomer. There often are two separate monomers which combine and a good example is nylon. Some natural polymers, like proteins, have several different monomers present.

Many synthetic fabrics such as nylon and polyester are condensation type polymers. Also several biochemical polymers such as proteins and carbohydrates are condensation polymers.

A very common combination of materials used to make condensation polymers are amines and carboxylic acids. When these two react they will form an amide linkage between the monomers.

How could the molecule below form a polymer? What polymer material would it be like?

$$H_2N\text{-}CH_2\text{-}CH_2\text{-}CH_2\text{-}CH_2\text{-}CH_2\text{-}\underset{\underset{O}{\parallel}}{C}\text{-}OH$$

The amine end of one molecule could react with the acid end of another while the acid end of the molecule could react with the amine end in endless sequence. This would form a long chain. This would be a polymer like nylon.

Many new materials are a polymer unit with materials embedded within the structure. These materials are known as composites. Glass and graphite are two common embedded materials.

Symbols found in Chapter 13

Page	Symbol	
560	MTBE	Methyl-tertiary-butyl ether. A gasoline additive.
567	-OH	Functional group that identifies alcohols.
567	R R’	These represent alkyl side groups in molecules.
569	-C=O \\ H	Aldehyde functional group.
569	-C- O- H ‖ O	Carboxylic acid functional group.
569	- C- ‖ O	Ketone functional group.
571	M100	100% methanol fuel
571	M85	Blend of 85% methanol and 15% gasoline as a fuel.
571	FFV	Flexible-fueled vehicles
572	ZSM-5	A zeolite catalyst used to make fuels.
573	E85	Ethanol blend fuels.
575	- C- O-R ‖ O	Ester functional group.
582	HDPE	High density polyethylene
582	LDPE	Low density polyethylene
582	CLPE	Cross-linked polyethylene
586	PVC	Polyvinyl chloride
590	PET	Polyethylene terephthalate polymer.

591 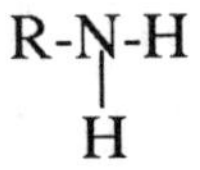 Amine functional group.

591 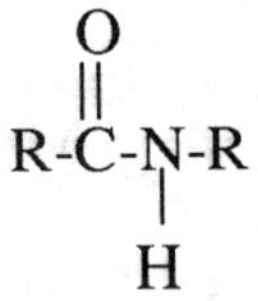 Amide functional group.

Terms in Chapter 13

The following terms are listed for your convenience. They should not become a list for you to memorize. You should have a working definition for each. If you have doubt about any particular term check its definition in the glossary and/or review it in the chapter. Some of these terms are included in the crossword puzzle later in the chapter. This chapter also makes use of several terms introduced in chapters 10 and 11.

Fractional distillation	Octane number	Catalytic reforming
Naptha fraction	Adsorb	Octane enhancer
Reformulated gasoline	Catalytic cracking	Coal tar
Coal gas	Synthesis gas	Substitution reaction
Addition reaction	Primary alcohol	Secondary alcohol
Tertiary alcohol	Aldehyde	Carboxylic acid
Ketone	Wood alcohol	Grain alcohol
Proof	Denatured alcohol	Gasohol
Polymer	Plastic	Thermoplastics
Thermosetting plastic	Monomer	Addition polymer
Condensation polymer	Initiation step	Free radical
Vulcanized rubber	Elastomers	Stereoregulation
Copolymer	Polyester	Amine
Amide	Polyamide	Reinforced plastics
Composites		

Concept Maps for Energy, Organics & Polymers

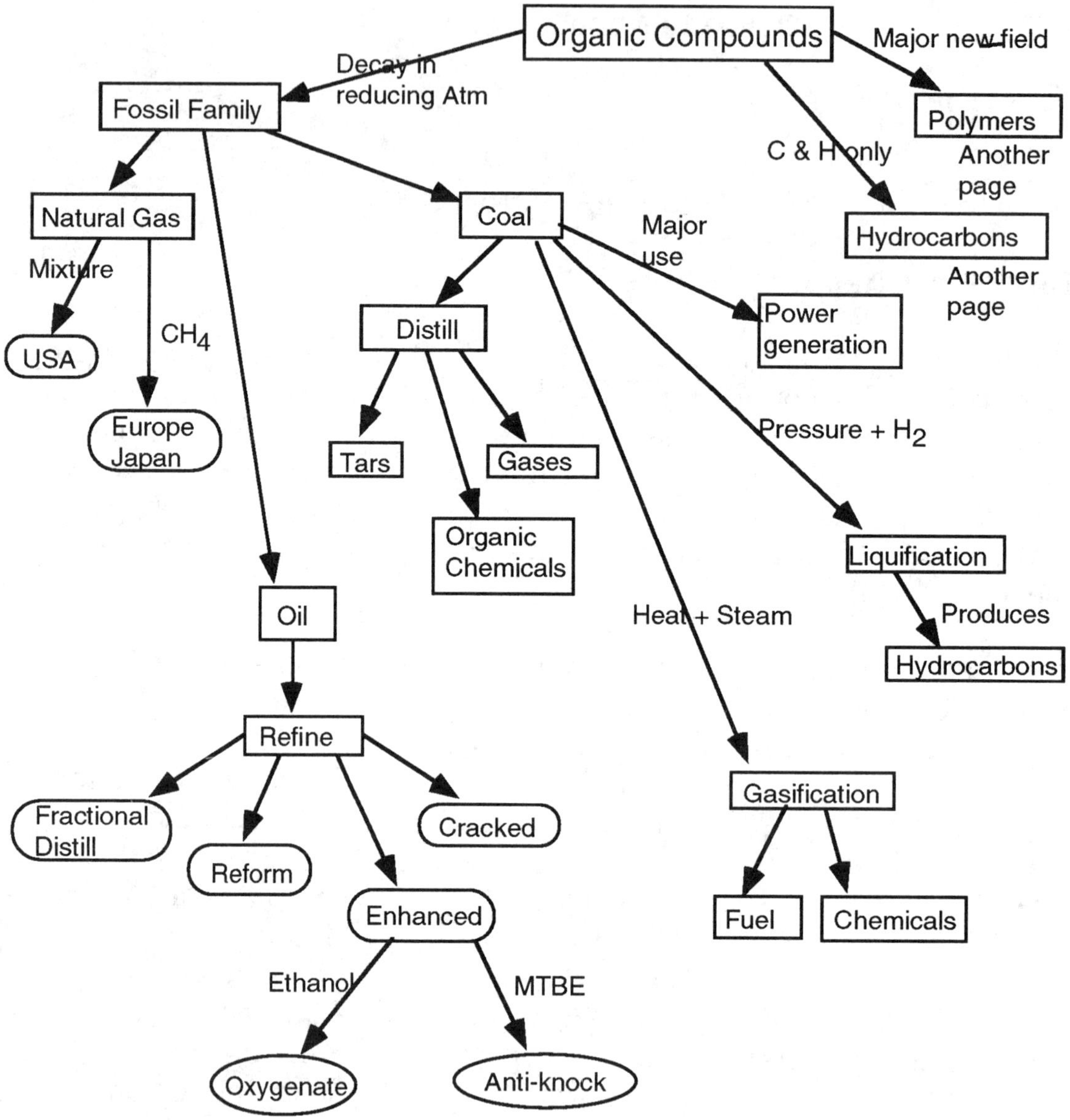

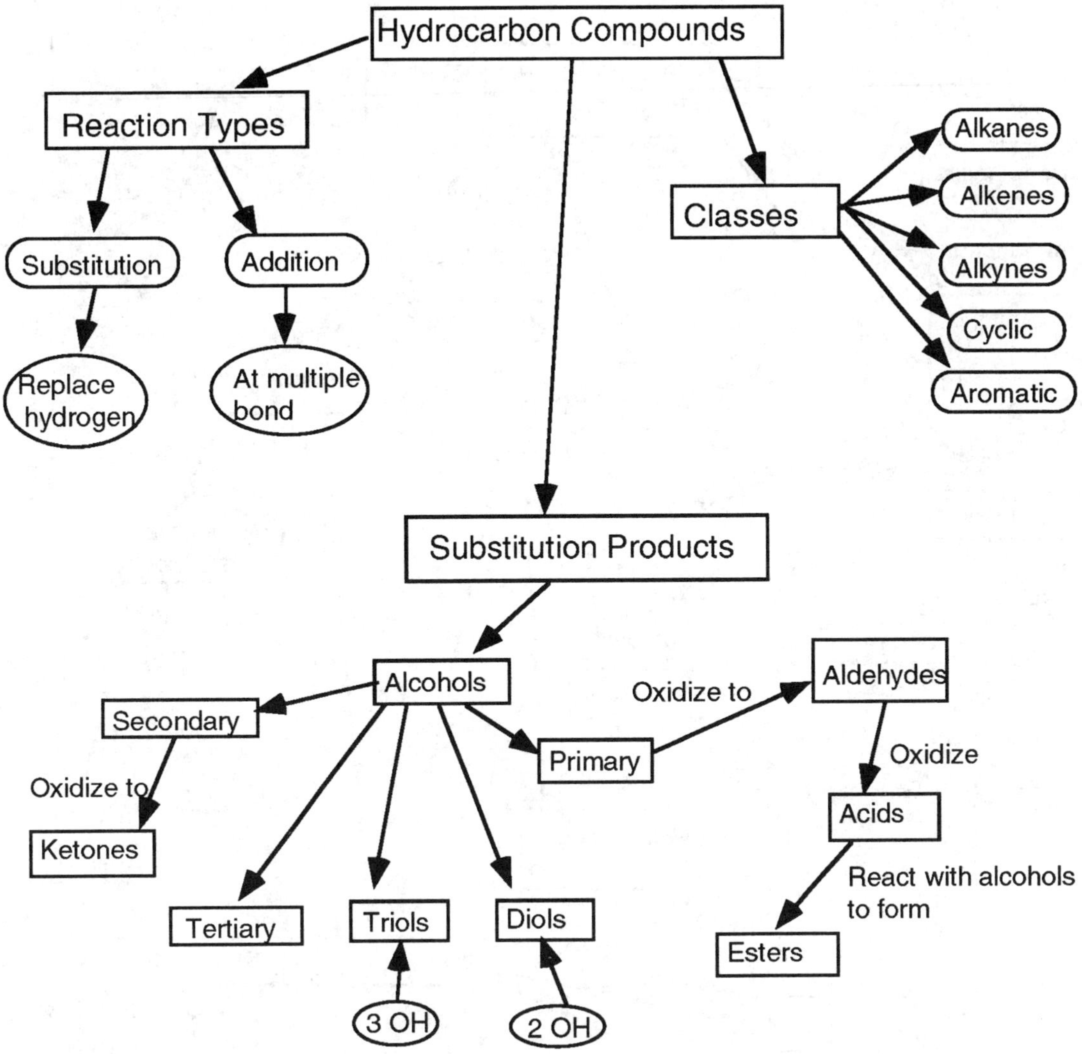
Hydrocarbon Compounds
Reaction Types
Substitution
Addition
Replace hydrogen
At multiple bond
Classes
Alkanes
Alkenes
Alkynes
Cyclic
Aromatic
Substitution Products
Alcohols
Secondary
Oxidize to
Ketones
Tertiary
Triols
3 OH
Diols
2 OH
Primary
Oxidize to
Aldehydes
Oxidize
Acids
React with alcohols to form
Esters

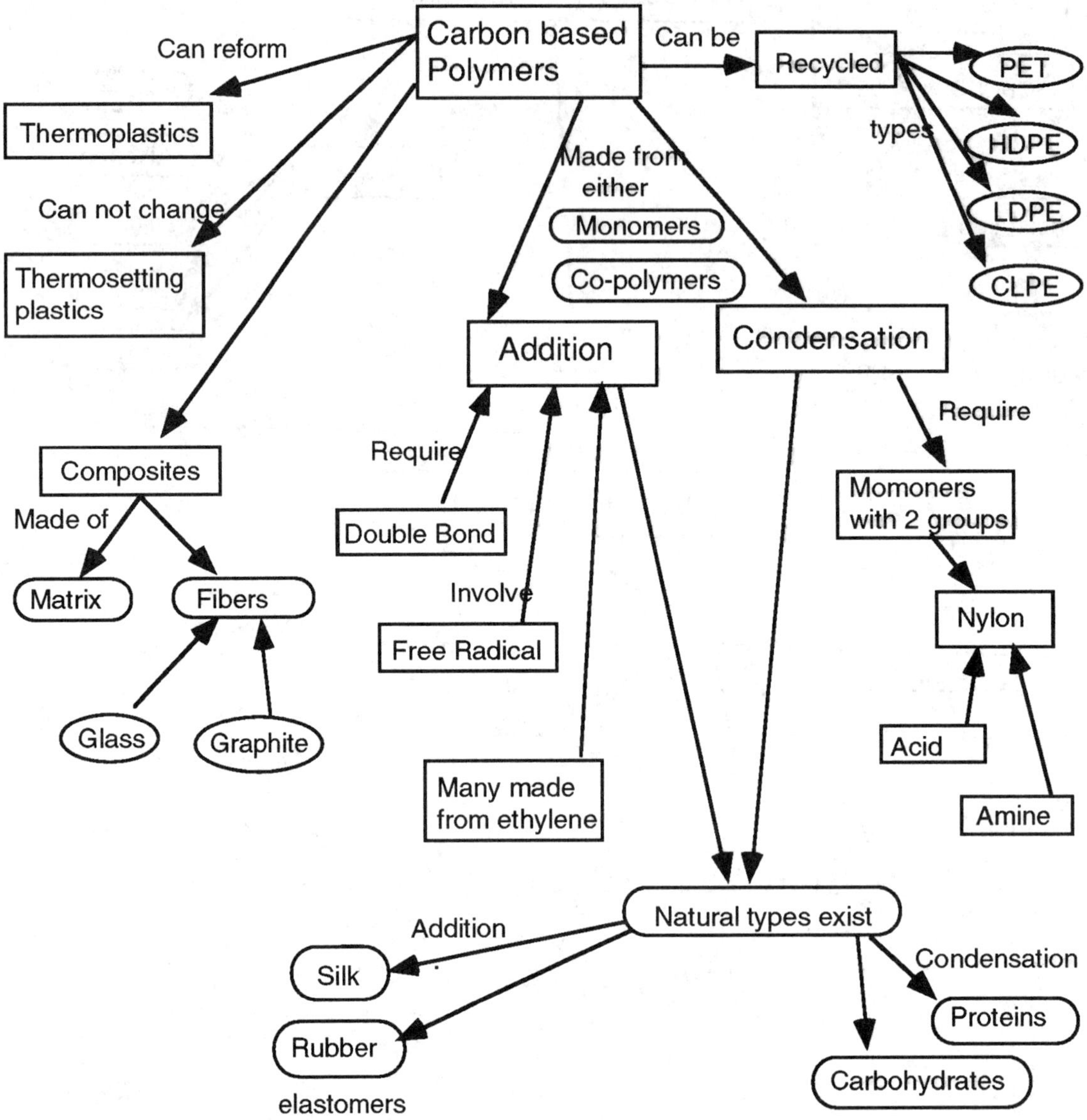
Carbon based Polymers
Can reform
Thermoplastics
Can not change
Thermosetting plastics
Can be
Recycled
types
PET
HDPE
LDPE
CLPE
Made from either
Monomers
Co-polymers
Addition
Condensation
Composites
Made of
Matrix
Fibers
Glass
Graphite
Require
Double Bond
Involve
Free Radical
Many made from ethylene
Require
Momoners with 2 groups
Nylon
Acid
Amine
Natural types exist
Addition
Silk
Rubber
elastomers
Condensation
Proteins
Carbohydrates

Energy, Organics & Polymers True/False Statements

Answers are at the end of the chapter.

1. Coal has some potential to replace petroleum as a source of organic compounds.

2. Fractional distillation separates each component of petroleum.

3. Octane numbers have a range limit from 0 to 100.

4. Noble metals are the catalysts often used in catalytic reforming.

5. The function of catalytic reforming is to convert large hydrocarbon chains to shorter hydrocarbon chains.

6. Octane enhancers allow the use of materials in fuels that have lower octane number ratings.

7. Oxygenated gasolines are used in an effort to reduce CO emissions from the exhaust.

8. Evaporation rates of gasoline blends can be reduced by lowering the amount of aromatic hydrocarbons present in the fuel.

9. Catalytic cracking allows the petroleum industry to make better use of crude oil.

10. Natural gas and coal gas are the same material from different sources.

11. Synthesis gas is a mixture of hydrocarbons.

12. Coal liquification is a hydrogenation process.

13. The primary reaction for alkenes is substitution.

14. CH_3-CH_2-CH_2-OH is a tertiary alcohol.

15. Glycerol is an alcohol known as a triol.

16. All alcohols when oxidized produce an aldehyde.

17. Methanol is called "wood alcohol" because that is the primary source of the material.

18. An 80% pure ethyl alcohol solution is also 80 proof.

19. Gasohol is an ethanol gasoline blend.

20. Extensive oxidation of alcohols will lead to formation of carboxylic acids.

21. The only use for esters is food flavorings.

22. A certain plastic material after production was found to be flawed. The material was then reheated and recast. This indicates that the plastic was of the thermosetting type.

23. A polymer made from monomers of acid and alcohol is a condensation type polymer.

24. Packing in low density polyethylene is less efficient because of the side chains in the product.

25. Condensation polymers involve a free radical in the process of formation.

26. Composite polymers have an advantage in that they add strength to the material.

Testing Your Understanding of Concepts in Chapter 13.
Answers are at the end of the chapter.

1. What do the fossil fuels have in common?

2. What is the difference between distillation and fractional distillation?

3. What processes are used to increase the amount of gasoline that is obtained from a barrel of crude oil?

4. What difference is their in oxygenated gasoline and octane enhanced gasoline?

5. What is the difference between coal gas and synthesis gas?

6. When the following functional groups react what is produced in each case? (a) acid and alcohol. (b) acid and amine.

7. How could two chemists, with a compound having the formula $C_5H_{12}O$, get two different classes of compounds upon initial oxidation?

8. Why can you not get a material with over 12% ethanol content by the process of fermentation?

9. If you wanted the opportunity to reuse a plastic product in another shape what type of plastic should you start with?

10. Why is ethylene so important in the production of addition polymers?

11. Why is stereoregulation important in the polymer industry?

12. Which type of polymerization process always splits out a small molecule at each monomer reaction site?

13. What type of material is fiberglass?

Testing You on Some Analytical Concepts.
Answers are at the end of the chapter.

This chapter does not cover quantitative data. Numerical calculations are not emphasized.

Crossword - Energy, Organics & Polymers

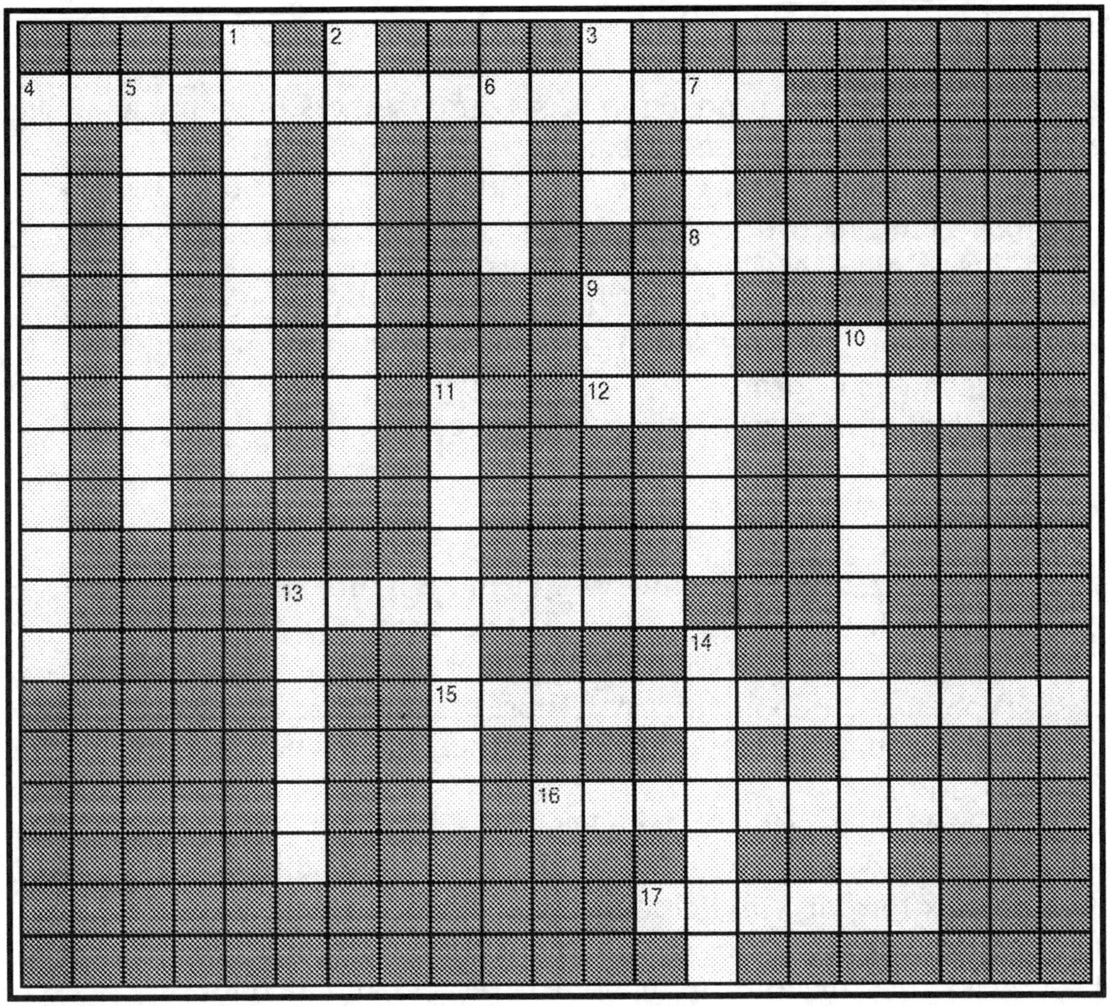

Crossword - Energy, Organics & Polymers Clues

ACROSS

4 Important in the production of specific isomers of isoprene.
8 A blend of ethanol and gasoline.
12 Process used to break large molecules into smaller ones in the petroleum industry.
13 This results when primary alcohols are oxidized.
15 A polymer that can be reformed after it has once been prepared.
16 A catalytic process used to convert straight chain molecules to branch chain forms.
17 Product of oxidation of secondary alcohols.

DOWN

1 Polymerization by mixing two or more monomers.
2 Ethanol that has been made unfit to consume.
3 Stands for methyl-tertiary-butyl ether.
4 Type of reaction alkanes will most often undergo.
5 Material that will return to its original shape when stress is released.
6 Means low density polyethylene.
7 Fuel blends that contain alcohol are said to be this.
9 Stands for polyvinyl chloride type of polymer.
10 Method of separation used in petroleum refining.
11 A polymer made from monomers of acid and alcohol.
13 Means to bind to the surface.
14 A starting material in the production of polymers.

Ties to Previous Chapters

An introduction of organic nomenclature was started in chapter 4. It would be worth your time to review that part of chapter 4. The bonding and isomerism in chapter 11 deals mainly with organic molecules. That information is of value in this chapter. Functional groups were also introduced in chapter 11.

Answers to True/False Statements:

1. True. It may not replace but could supplement. There are processes that extract liquids from the coal and these can be refined into various chemicals.

2. False. This process only separates into smaller size fractions not individual materials.

3. False. The octane scale is open ended.

4. True. Since they do not react they have an advantage.

5. False. Reforming converts straight chain materials to branch chain compounds.

6. True. The function they provide is to raise the overall octane value.

7. True. They supposedly allow the fuel to burn more completely to form CO_2 instead of CO.

8. True. The aromatic materials are the more volatile.

9. True. This process allows larger molecules which have less demand to be converted into smaller molecules, such as those used in gasoline, that are in greater demand.

10. False. The coal gas is a mixture of many more materials and has much less hydrocarbon content.

11. False. This gas is a mixture of hydrogen and carbon monoxide.

12. True. Under pressure the hydrogen can be added.

13. False. The double bond more readily undergoes addition.

14. False. This is a primary alcohol. A tertiary alcohol would have three carbon atoms attached to the carbon with the OH.

15. True. Glycerol has three OH groups. The OH makes it an alcohol and the three make it a triol.

16. False. Secondary alcohols would produce a ketone.

17. False. This was the original source. Now it is made synthetically.

18. False. The proof number is twice the % of ethanol. 160 proof.

19. True. The blend is 90% gasoline and 10% alcohol.

20. True. This compound is the oxidation product after the aldehyde. The final product would be CO_2 and water.

21. False. This is but one use. Esters are often used as solvents.

22. False. Since it can be reworked it must be a thermoplastic.

23. True. The acid and alcohol react to form as ester linkage and split out a water molecule.

24. True. The side chains prevent efficient packing since they get in the way of the next monomer.

25. False. The addition reaction process involves free radicals.

26. True. The glass or graphite material added gives strength and flexibility.

Answers to Understanding of Concepts Questions:

1. They are all prepared from living material that was held under high pressure and temperature for millions of years. These materials where held in a reducing environment so they were not oxidized. They all have very high content of carbon and hydrogen.

2. The main difference is in the degree of separation. The fractional distillation separates into ranges which are narrow but still a mixture. Distillation separates the materials into single component parts.

3. Two methods will provide more gasoline. The process of catalytic reforming makes higher octane components. The catalytic cracking makes large molecules smaller so they can be burned in a gasoline engine.

4. Oxygenated fuel has additives, such as ethanol (up to 10%), added. Other materials that contain oxygen are also used. Octane enhanced gasoline has materials such as MTBE or toluene added.

5. Coal gas is the material obtained by destructive distillation of coal. That is, the heating of coal in an oxygen free atmosphere. This process drives the volatile materials out. Synthesis gas is the result of heating carbon with steam to produce a H_2 and CO mixture.

6. (a) This would produce an ester.
(b) This would produce an amide.

7. If the material were a primary alcohol the product would be an aldehyde. However, if the starting material was a secondary alcohol, the product would be a ketone.

8. Fermentation is process brought about by certain organisms. When the alcohol content gets to the 12% range it becomes toxic to the organisms and they die off.

9. Plastic materials that are reusable are of the thermoplastic type. You could not do this with thermosetting plastics.

10. Ethylene can be the starting material to form many other monomers. Table 13.10, on page 585 of the text, lists many but not all examples.

11. Stereoregulation enables the chemist to produce a specific isomer. It could be the cis form rather than a mixture of cis and trans.

12. Condensation reactions split out a small molecule such as water. Addition polymers do not do this because they add to a double bond and there is no need to eliminate any material.

13. Fiberglass is a composite material. The polymer resin has glass particle embedded in it.

Answers to Testing the Analytical Concepts

No quantitative problems.

Answer to Chapter 13 Crossword

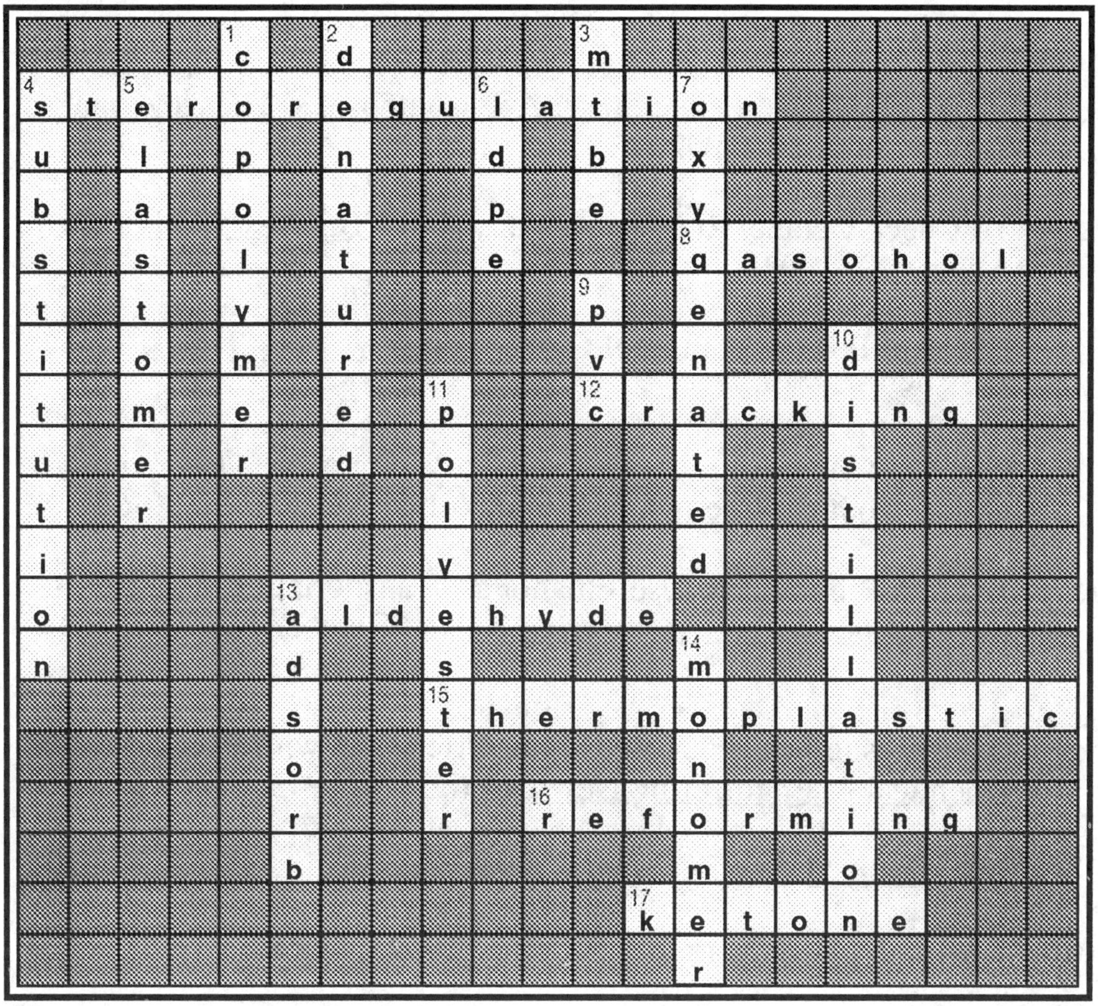

Some Supplementary Readings Relating to This Chapter:

1. Making Coal More Useful By Cooking Out the Water, The New York Times, September 30, 1992.

2. Two Scientists Find a Way To Make Cheaper Methanol, The New York Times, January 17, 1993.

3. Needed by Space Plane: Space-Age Composites, The New York Times, August 1, 1992.

4. To Check Octane, Refiner Sees the Light, The New York Times , August 12, 1992.

5. Efficient Power From Fuel Gases, The New York Times, June 10, 1992.

6. Crackers: How Refiners Learned To Wring More Gas From Crude, Today's Chemist at Work, Vol. 2, NO.9, Page 59-60.

7. Brazil's Program To Use Ethanol as Transportation Fuel Loses Steam, Chemical & Engineering News , October 18, 1993, page 13-15.

8. Advances in Feedstock Recycling Offer Help with Plastic Waste, Chemical & Engineering News, October 4, 1993, page 11-14.

9. Ink, Chem Matters, February 1993, page8.

10. Perfume, Chem Matters , February 1992, page 8.

11. Something's Rotten, Chem Matters, April 1989, page14.

12. Natural Dyes, Chem Matters , December 1986, page4.

13. Dissolving Plastic, Chem Matters, October 1987, page12.

14. Biodegradable Bags, Chem Matters, October 1991, page 4.

15. Nylon, Chem Matters, December 1990, page 4.
16. Denatured Alcohol, Chem Matters, December 1990, page 14.

17. Oxygenated fuel cost may outweigh effectiveness, Chemical & Engineering News, April 12, 1993, page 28-30.

18. Conductive Polymers Recharged, Science, Vol 246, October 13, 1989, page 208.

19. Alcohol in Your Tank, Chem Matters, December 1988, page 10.

20. The Fruits of Ethylene, Chem Matters, April 1989, page 11.

Chapter 14 Solids

This chapter discusses the properties of the third common state of matter- solids. Various physical properties are discussed as are the relationships that exist between it and the liquid state.

Various bonding types and forces, along with their relationship to physical properties, are identified. The structure of solids is discussed in detail. Several special types of solids are noted.

Focus Your Study Efforts

Changes of state in solids.

Objectives:

1. You should be able to explain the effect of adding, or removing, energy from matter.

The study of solids will be greatly facilitated by using three dimensional models. If models are available make use of them.

The physical state of matter depends on the motion of the particles in the material. The solid state is characterized by a minimal amount of free motion.

Particle interactions in solids are greater than in any other state.

State changes occur when energy is added or removed. Consequently the temperature is significant in determination of the physical state.

MP BP

Solid <========> Liquid <=========> Gas

FP CP

Low energy High energy

The MP (melting point) and FP (freezing point) are the same temperature just like BP (boiling point) and CP (condensation point) are the same. What we name a particular temperature depends on what is happening in the system.

> A 0.250 kg piece of lead is heated from 20.0°C to its melting point at 327.0°C and then melted. If the specific heat for this range is 0.135 $Jg^{-1}°C^{-1}$ and it has a heat of fusion of 4.77 kJ per mole what total energy is consumed.
>
> Total energy = Energy to warm + energy to melt
> = mass x ΔH x sp ht + mass x heat of fusion
>
> $= 250.0 \text{ g} \times 307°C \times 0.135 \text{ J/g°C} + 250\text{g} \times 4770\text{J/mol} \times \frac{1 \text{ mol}}{207.2 \text{ g}}$
>
> = 10361 J + 5755 J = 16116 J = 16.1 kJ

Forces between molecules are important to structure

> **Objectives:**
> 1. You should be able to explain the importance of various intermolecular forces on properties of materials in the solid state.

Even in the solid state particle motion is present and a considerable amount of vacant space exists.

Arrangement of particles has a significant impact on properties. Both diamond and graphite are solids made from the element carbon. The atom arrangements are very different and that results in very different properties shown by the two solid forms.

Hydrogen bonding may be very important in determining properties of some solids. Water, as ice, is an excellent example since ice is less dense than the liquid form.

Several solids, including ice, are capable of going directly from the solid to the gaseous state by a process known as sublimation.

Solids may be made from ions, polar molecules or non polar molecules. The amount of ionic or polar character present in the intermolecular forces has an influence on physical properties such as melting point.

Solids may be amorphous or crystalline

Objectives:

1. You should be able to use characteristic properties to distinguish between amorphous and crystalline solids.

Solids are subdivided into two main groups depending on their structure. Crystalline or amorphous.

Amorphous solids have little predictable structure. They do not have real melting points, but soften over a wide temperature range until they are liquid.

Some amorphous materials can have properties altered in the process of cooling. This is known as annealing. Glass is a good example.

Crystalline solids have a lattice structure

Objectives:

1. You should be able to define the unit cell for a crystalline lattice.
2. You should be able to explain what a crystal lattice is and how the lattice affects the properties of the solid.

3. You should be able to distinguish the difference between categories of cubic crystal lattices.
4. You should be able to identify differences between lattices of materials that contain metal ions and those with nonmetal atoms only.

Crystalline solids have a repeating pattern of particles. They exhibit what is known as a crystal lattice.

The smallest repeating structure of a crystal lattice, which when joined to other like units that gives the macro structure, is known as the Unit Cell.

Crystal lattice types fall into seven major categories. Some of these, such as cubic, may have sub categories. These are outlined in the text on page 611.

When you compare unit cells it becomes apparent that they are not packed with equal efficiency. This results in significant differences in densities of the solids. Models will be helpful in seeing this.

Data has established the length of a unit cell for the element nickel at 350 pm. It is also known that this element is in a face centered cubic structure. What radius does the nickel atom have?

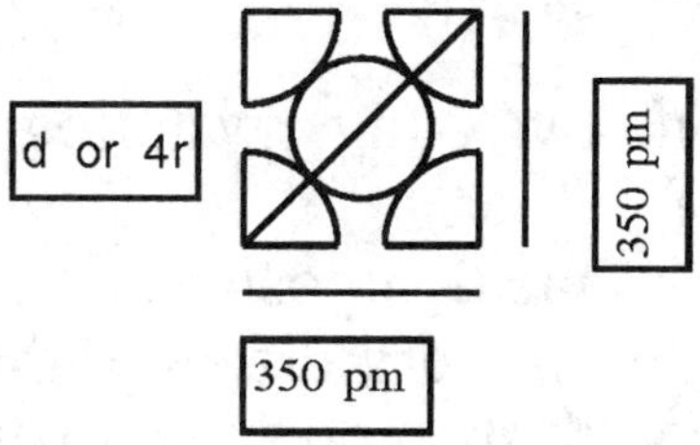

The diagonal is 4r (all units are pm)

$(4r)^2 = (350)^2 + (350)^2$
$16r^2 = 122500 + 122500 = 245000$
$r^2 = 15313$

r = 124 pm

Atom or ion sizes have a major impact on packing. When packed the particles will leave open space, called holes, of varying sizes. It is possible other particles may fit into those holes.

> Data has confirmed that one form of iron is a body centered cubic structure. The unit cell has an edge length of 287 pm. A bcc unit cell contains the equivalent of two atoms. What is the density of Fe?
>
> $$D = \frac{\text{mass}}{\text{volume}}$$
>
> $$\text{mass} = 2 \text{ atoms Fe} \times \frac{55.8 \text{ g Fe}}{6.02 \times 10^{23} \text{ atoms Fe}} = 1.85 \times 10^{-22} \text{ g}$$
>
> $$\text{Volume} = (\text{length})^3 = (287 \text{ pm})^3 = (2.87 \times 10^{-10}\text{m})^3$$
> $$= (2.87 \times 10^{-8}\text{cm})^3 = 2.36 \times 10^{-23} \text{ cm}^3$$
>
> $$D = \frac{1.85 \times 10^{-22} \text{ g}}{2.36 \times 10^{-23} \text{ cm}^3} = 7.83 \text{ g cm}^{-3}$$

Ionic crystals have a lattice, most often made of anions, with the smaller cations filling various holes of the anion lattice.

A majority of solid structure studies are done by X-ray crystallography.

Metals and nonmetals can be thought of as chemical opposites. Chemical and physical properties of metals are, as a group, very different than those of nonmetals. You should be familiar with the general properties associated with each group.

There are a group of solid elements that are not clearly defined as being either metallic or nonmetallic in all their properties. These solid elements we call metalloids. They have major uses in semiconductors and solar cells.

Solids have different types of bonding

Objectives:
1. You should be able to define and give examples of various types of bonding found in solids.
2. You should be able to explain the significance of energy bands and gaps on electrical conductivity.

The bonding types in various solids may be very different. Molecules may be held together by intermolecular forces; by ions electrostatically interacting to produce ionic solids; or by metals that seem to behave as if the valence electrons are free to move throughout the entire lattice.

In metals, the valence electrons interact and energy adjustments occur in the energy levels. This results in energy bands that may overlap, have a large gap, or just a narrow gap. This would result in conductor, insulator, or semiconductor capability depending on the gap energy. Impurities in the metal can have a significant effect on the gap that exists between energy bands.

Extensive purification followed by controlled doping of the metal will result in semiconducting devises such as transistors.

Some solids are very large network structures

Objectives:
1. You should be able to explain what a macromolecule, or network, structure is and identify examples of materials with this structure.

Some materials set up extensive interactions that involve large amounts of the material and there is the appearance of a macromolecular structure. These substances are also known as network solids. The diamond and graphite allotropes of carbon are examples. The molecule silica, SiO_2, also does this very effectively.

Mixtures of solids can be bound together by high temperature firing resulting in a set of materials we call ceramics. These are also network solids.

Other mixtures will form large solid matrix structures when allowed to hydrate. Cement is an example of this type of network solid.

Conductivity of an electric current by a metal decreases as the metal is heated. This can be explained in terms of energy bands and gaps. Heating results in increased particle motion so bands vary in energy and overlay is not consistent. This means electrons are less free to move.

When sufficiently cooled a material may lose all electrical resistance and an electric current reaches infinite conductance. This type of material is known as a superconductor.

Currently certain mixed metal oxide ceramic materials become superconducting at relatively high temperatures. Superconductivity will be used much more when chemists can raise the threshold temperature and expensive coolants will not be required.

Symbols found in Chapter 14

Page	Symbol	
610	sc	Simple cubic.
610	bcc	Body-centered cubic crystal.

611	fcc	Face-centered cubic crystal.
621	ΔH_{fusion}	Energy involved in solid to liquid state change.
625	n-p-n p-n-p	Notation for doped silicon used in transistors.
625	ICs	Integrated circuits.

Terms in Chapter 14

The following terms are listed for your convenience. They should not become a list for you to memorize. You should have a working definition for each. If you have doubt about any particular term check its definition in the glossary and/or review it in the chapter. Some of these terms are included in the crossword puzzle later in the chapter. Terms from the chapters on liquids and gases may also find application in this chapter.

Melting point	Crystallization	Sublimation
Heat of sublimation	Crystal lattice	Unit cell
X-ray crystallography	Ductility	Malleability
Closest packed systems	Metallic bonding	Energy band
Insulators	Band gap	Semiconductor
Zone refining	Doping	Transistor
Solar cell	Network solid	Pyroxenes
Asbestos	Crocidolite	Feldspar
Corundum	Ruby	Ceramics
Cement	Tensile strength	Ceramic composites
Superconductor	Amorphous	Mole %
Annealed	Pyroceram	

Concept Maps for Solids

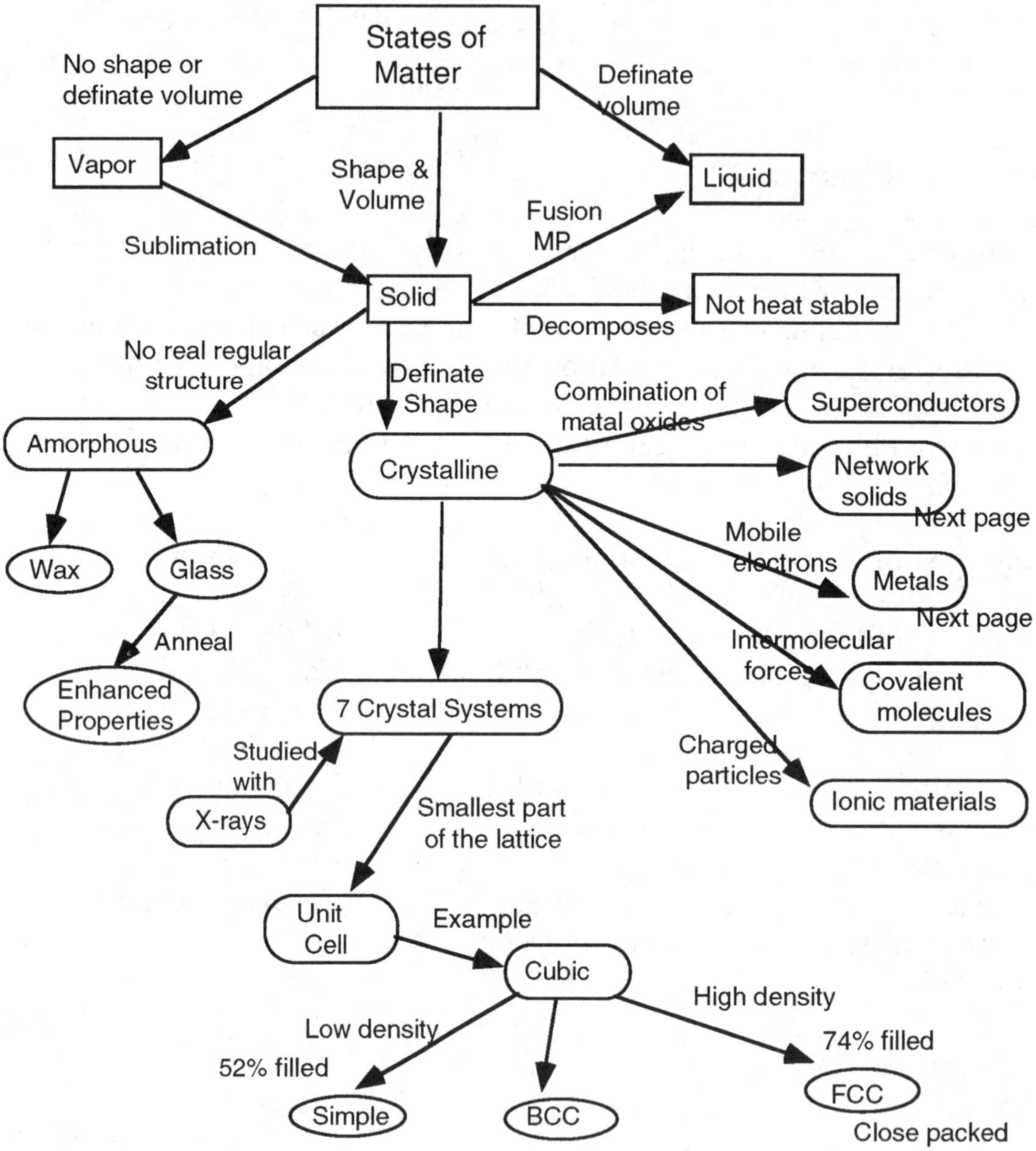

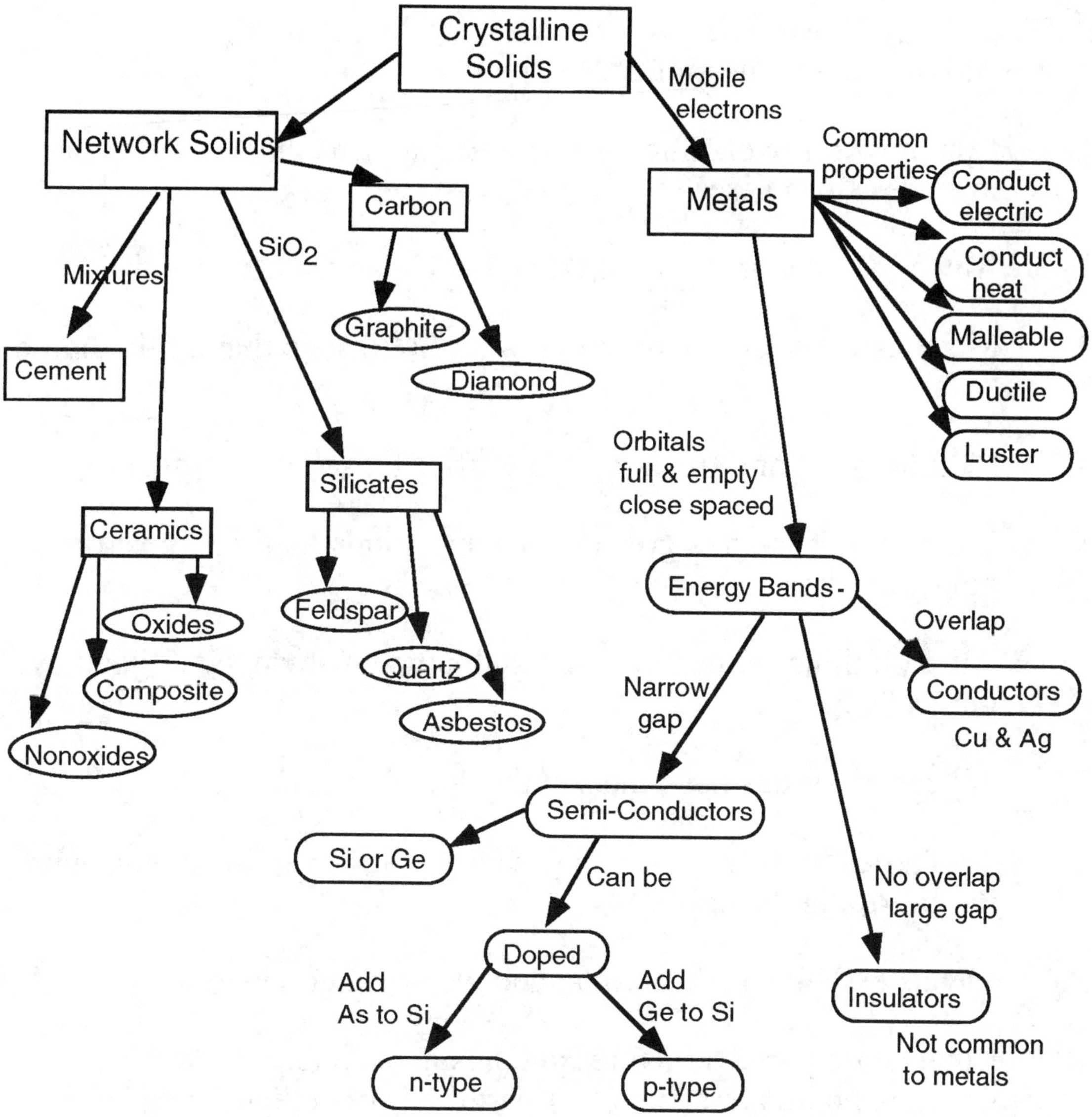

Crystalline Solids
Mobile electrons
Network Solids
Carbon
Metals
Common properties
Conduct electric
Conduct heat
Malleable
Ductile
Luster
Mixtures
Cement
SiO2
Graphite
Diamond
Silicates
Ceramics
Oxides
Composite
Nonoxides
Feldspar
Quartz
Asbestos
Orbitals full & empty close spaced
Energy Bands -
Overlap
Conductors
Cu & Ag
Narrow gap
Semi-Conductors
Si or Ge
Can be
Doped
Add As to Si
Add Ge to Si
n-type
p-type
No overlap large gap
Insulators
Not common to metals

Solids True/False Statements

Answers are at the end of the chapter.

1. Compressibility of solids is about the same as that of liquids but much less than gases.

2. Any crystalline material must be hard.

3. Most materials are more dense in the solid state than in the liquid state.

4. Melting point and freezing point differ by several degrees.

5. The heat of fusion is greater for ionic solids than for most other kinds of solids.

6. All crystalline solids have a crystal lattice that can be defined by a unit cell.

7. All cubic crystals have the same unit cell.

8. The lattice in ionic crystals is usually made from the anions with the cations in the holes.

9. Metals exhibit a high thermal and electrical conductivity.

10. Cubic close paced structures and hexagonal close packed structures both are efficient at particle packing with 74% of the space filled.

11. Insulator materials have a narrow band gap between energy bands.

12. n-p-n transistors could use gallium as a doping agent.

13. Asbestos is a network solid.

14. Composite ceramics are all superconductors.

15. Amorphous solids have no real melting point.

Testing Your Understanding of Concepts in Chapter 14.
Answers are at the end of the chapter.

1. Explain the changes that occur when a solid, such as ice, melts.

2. Why do some materials, such as water, have properties so much different than expected?

3. Draw a graph showing temperature on the vertical axis vs energy added on the horizontal axis to illustrate what happens when ice is heated from -20°C to 101°C. Label the various parts of the graph.

4. Of what value is the concept of crystal lattice to the chemist?

5. What is meant by holes in a crystal lattice? Why are they important to ionic materials?

6. What are energy bands and band gaps?

7. Identify several properties that are used to characterize metals?

8. Why are some transistors called p-n-p and others n-p-n ?

9. How can an element, like carbon, and a complex material, such as cement or ceramic, both form a network solid?

10. Why is there so much interest in superconductivity?

11. Identify several characteristics of amorphous solids.

Testing You on Some Analytical Concepts.

Answers are at the end of the chapter.

1. Silver metal forms face centered cubic crystals. The silver atom has a radius of 1.44×10^{-8} cm. What is the density of the metal?

2. Sodium fluoride has an enthalpy of fusion of 29.3 kJ/mol. What energy must be removed to convert 100.0 grams of liquid NaF to a solid?

3. Sodium metal melts at 98°C and boils at 881°C. A chemist converted 10.0 g of the metal at 25.0°C to a liquid at its boiling point. The heat capacities are 28.1 J/mol-K for the solid and 32.6 J/mol-K for the liquid. The ΔH_{vap} is 97.4 kJ/mol and the heat of fusion is 2.60kJ/mol. How much energy was added to the metal?

Crossword - Solids

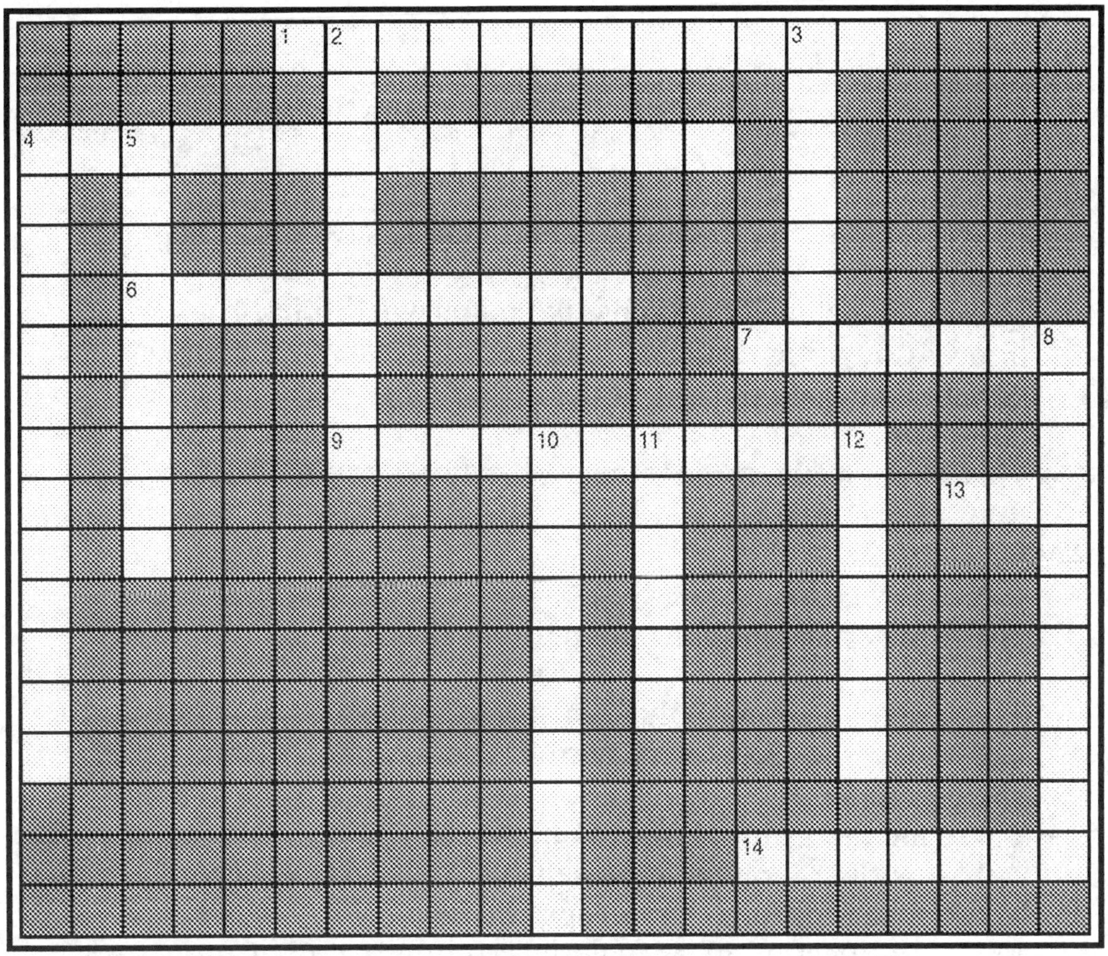

Crossword - Solids Clues

ACROSS

1 A metal property that allows the material to be beaten into thin sheets.
4 A material that has nearly zero electrical resistance.
6 This is a large type hole in a lattice that is surrounded by six anions.
7 This class includes materials like pottery and bricks.
9 Process of going directly from the solid state to the vapor state.
13 Symbolism used for a very dense and close packed lattice structure.
14 This refers to the arrangement of materials in a structure.

DOWN

2 Solid materials that have no definite structure are said to be _________.
3 This refers to strength that resists stretching.
4 These materials have very narrow energy gaps between full energy bands.
5 These are the simplest silicate networks.
8 Asbestos fiber thought to be capable of causing cancer.
10 These materials have large gaps between energy bands.
11 Process of heating glass materials to improve their properties.
12 These types of solids are very large, like one big molecule.

Ties to Previous Chapters

Solids are one state of matter, and as such, are related to the liquids and gases of chapter 12. Forces in all states have common characteristics. Some of the same information applies to all states. Some solids such as molecular solids require an understanding of bonding found in chapters

10 and 11. Chapter 7 discussed thermodynamics, or energy changes, and a review of parts of that chapter may be useful.

Answers to True/False Statements:

1. True. Compressibility is a function of particle packing. In both solids and liquids the particles are closely packed. Gases however have large distances between the particles.

2. False. There is hardness scale, known as the Moh's scale, which has a range of 1 to10. Talc is a crystalline material but is very soft, being 1 on the scale.

3. True. A few exceptions exist. Water is probably the most significant.

4. False. These two values are the same.

5. True. The strong interactions between the ions cause this.

6. True. Both parts are correct.

7. False. There are three types of crystal lattices for the cubic system and each is different.

8. True. Anions are the larger. The smaller cations are more likely to fit in the hole sites created by the lattice building anions.

9. True. These are two classic characteristics of metals.

10. True. These two lattice types are nearly identical.

11. False. A large energy gap is required to make a material an insulator.

12. False. The first n indicates negative, or an excess of electrons. The doping material would have to be able to provide electrons so it would need to be from the 5A group of the periodic table.

13. True. This material is a silicate type material.

14. False. Only very special materials exhibit the superconductivity property at temperatures other than close to absolute zero.

15. True. These materials gradually soften until they are liquid.

Answers to Understanding of Concepts Questions:

1. When a material melts the added energy affects the bonds between the atoms, ions, or molecules. When the energy supplied provides enough kinetic energy to overcome the forces between the particles the crystal lattice falls apart and the liquid form results.

2. A material such as water has hydrogen bonded to a highly electronegative atom such as oxygen. This results in a weak bond forming between the molecules - from the hydrogen end of one molecule to the oxygen end of another. This is called a hydrogen bond and although weak individually, collectively they are strong. When these bonds interact the result is an unexpected force which must be overcome before a change, such as melting can occur. This raises the melting point.

3.

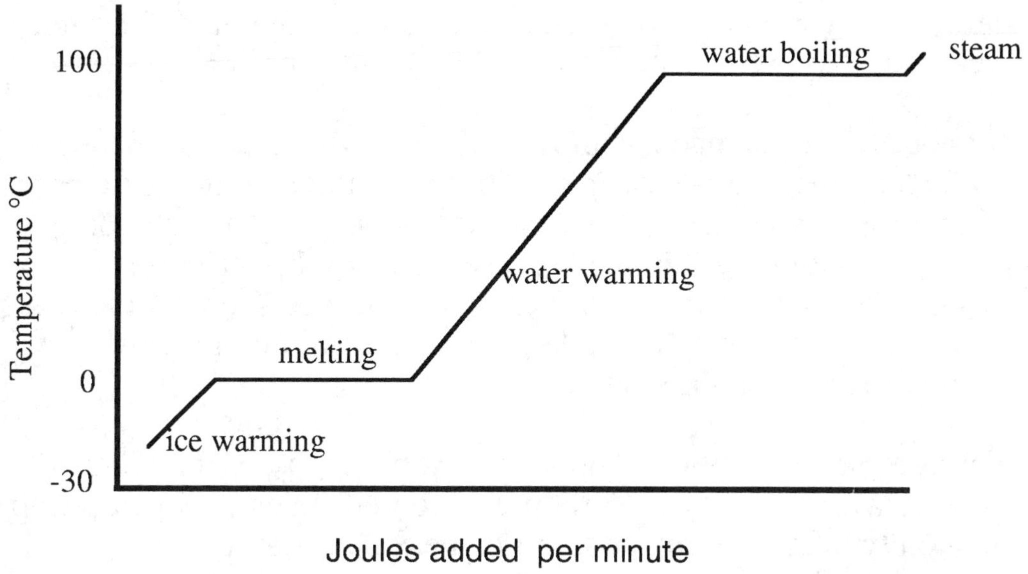

4. The concept of crystal lattice allows the chemist to explain the shape of materials, how materials may break, why certain ionic structures are favored over others, variations in melting points, and hardness in solids. These are some of the factors.

5. We assume atoms or ions are spheres. When these pack they do not fill all the available space. In some cases, the packing is such that various size voids are present, and they can be formed by various numbers of particles. When six spheres come together they form a void known as an octahedral hole.

6. An energy band is composed of a large group of orbitals that have nearly the same energy content. These bands are separated by energy spaces called gaps. The size of the gap determines some properties, such as the ability to conduct.

7. Metals have a luster, they can be drawn into wire, they can be deformed by pounding on them, they conduct heat and electricity well, and they have almost no solubility in water.

8. If the transistor is made of mostly silicon ,which has 4 valence shell electrons, it can be doped with small amounts of an element from either group 3A or 5A. Those elements in 3A are short electrons, with only 3 in the valence shell, so they take on a positive character relative to the silicon - p type. The 5A elements have 5 valence electrons, so they are in excess, and give the negative character - n type.

9. The complexity is not as important as how many of the units will interact with each other. Carbon can bond 4 times so it can set up large networks. The materials in cement or ceramics are molecules which can interact in a similar way.

10. Although superconductors have little commercial value now their potential for use is great. The potential can not be realized until we have a better understanding of the materials and how they work.

11. Amorphous solids have no melting point, they soften over a wide range. They will not break along predictable patterns such as crystal do.

Answers to Testing the Analytical Concepts

1. The face of the silver unit cell has the following appearance:

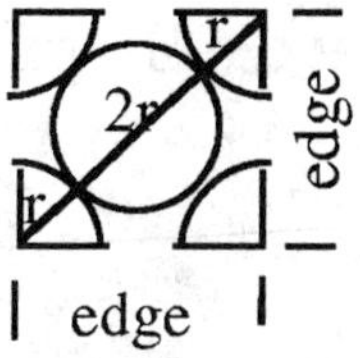

We know the unit cell contains 4 atoms and we know the edge value is 1.44×10^{-8} cm. We also know the atomic weight of silver is 108 g/mol.

We can calculate the length of the edge using the Pythagorean theorem.

$e^2 + e^2 = (4r)^2$

$2e^2 = 16r^2$ or $e^2 = 8r^2$ and $e = r(8)^{1/2} = 2.878r$

$e = (1.44 \times 10^{-8}\ \text{cm}) \times (2.8780) = 4.073 \times 10^{-8}\ \text{cm}$

$$D = \frac{\text{Mass}}{\text{Volume}} = \frac{4\ \text{atoms} \times 108\ \text{g/mol} \times 1\text{mol}/6.023 \times 10^{-23}\text{atoms}}{(4.073 \times 10^{-8}\ \text{cm})^3}$$

$$D = \frac{7.173 \times 10^{-22}\ \text{grams}}{6.756 \times 10^{-23}\ \text{cm}^3} = 10.62\ \text{g / cm}^3$$

2. Energy must be removed since we are starting with molten material. The freezing temperature is 992°C so at the end we will still have a solid at 992°C.

$$\text{kJ} = 100.\ \text{gNaF} \times \frac{1\text{mol NaF}}{42.0\text{g NaF}} \times \frac{29.3\ \text{kJ}}{\text{mol}} = 69.8\ \text{kJ}$$

3. A flow chart of the system would look like:

	Sp Ht_{solid}		Ht fusion		Sp Ht_{liq}	
Rm $Temp_{solid}$	-------->	MP_{solid}	--------->	MP_{liq}	----------->	BP_{liq}
25°C		98°C		98°C		881°C

Note we do not need the ΔH_{vap} data given since we only go to the boiling point but do not boil.

$$\text{Mol Na} = \frac{10.0 \text{ g Na}}{23.0 \text{ g/mol}} = 0.435 \text{ mol}$$

Note: we need all energy units the same- convert to joules.

Energy = 0.435 mol x 28.1J/molK x 73K + 0.435 mol x 2600 J/mol + 0.435 mol x 32.6J/mol K x 783K

= 892 J + 1131 J + 11104 J = 13127 J = 13.1 kJ

Answer to Chapter 14 Crossword

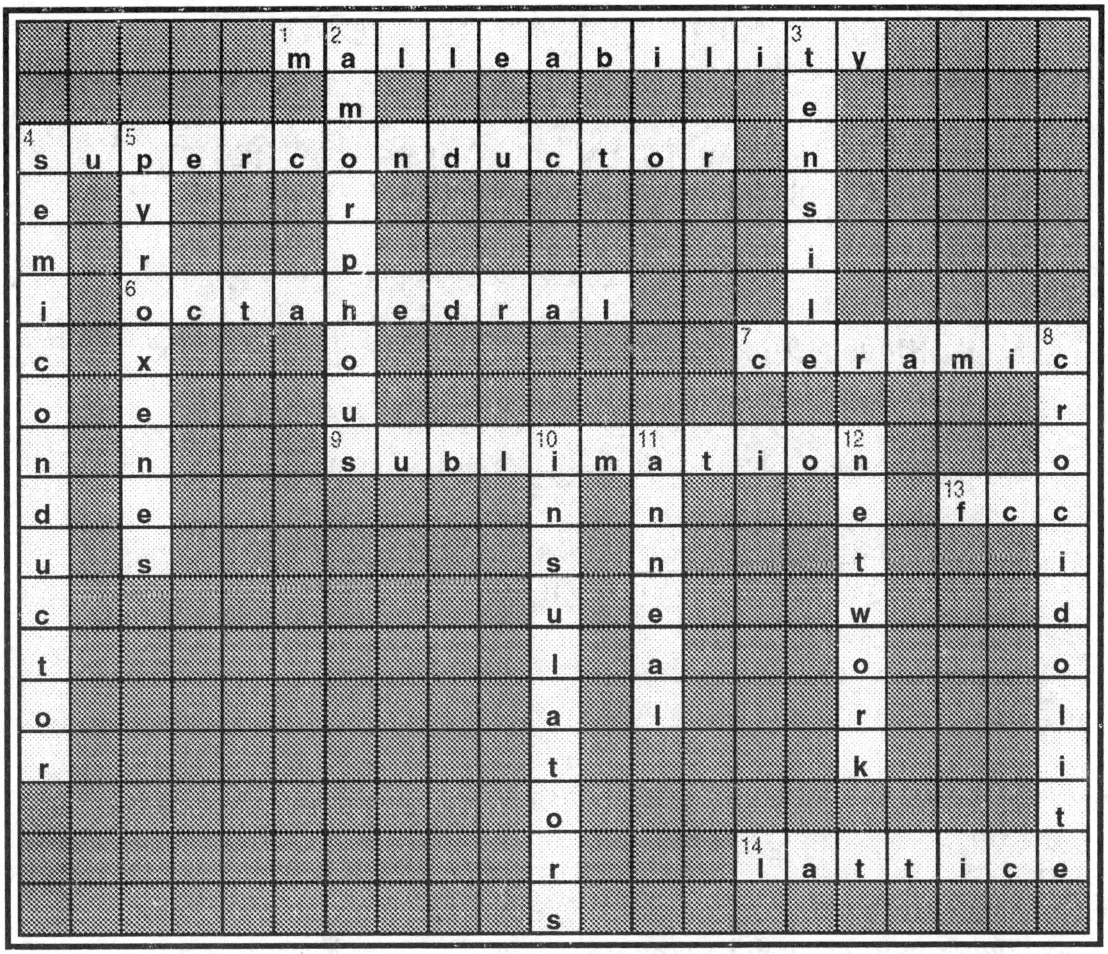

Some Supplementary Readings Relating to This Chapter:

1. Between a Rock and a Liquid Place, Science , Vol 260, 4 June 1993, page 1424.

2. Diamond: Glittering Prize for Materials Science, Science, Vol 250, 21 December 1990, pages 1640-1641.

3. Superconductivity, Chem Matters , October 1987, page 18.

4. Super-Student Superconductivity, Chem Matters , October 1987, page 22.

5. Memory Metal , Chem Matters , October 1993, page 4.

6. A Diamond-Coated Tool Etches the Tortuous Path of New Technology, The New York Times , September 23, 1992.

7. Superconductivity: Is the Party Over? , Science , Vol 244, 26 May 1989, page 914.

8. One MOre Family of Superconductors , Science , Vol 242, 25 Nov. 1988, page 1125.

9. The New Superconductors , Chemical & Engineering News , December 21, 1992, page 24.

10. Finally, a Hotter Superconductor , Science Vol. 260, 7 May 1993, page 755.

Chapter 15 Solution Chemistry

Much of the chemistry that affects us occurs in solution. Solutions are an effective way to bring chemical particles together and transport them from one place to another. Solution concentrations are expressed in many different ways, often depending upon what is being used and what system is involved. When a solute is added to a solvent the solution has significant changes in properties from those of the pure solvent.

Focus Your Study Efforts

Solution chemistry involves many terms.

Objectives:

1. You should be able to explain what the terms concentrated, dilute, saturated, and miscible mean.
2. You should be able to explain what dynamic equilibrium means when applied to the solutions.
3. You should be able to explain why water is a good solvent for many ionic materials.

Discussion of solutions involves a rather large vocabulary. Some terms are not as well defined as you may like. Terms like concentrated and dilute may involve various amounts depending on the solute.

Saturated is a temperature dependent situation. If the temperature of a saturated solution is raised or lowered the solution will become either unsaturated or supersaturated.

It may be confusing when it is stated that a solution is saturated but very dilute. This would mean, that at a given temperature, all that can dissolve has done so, but only a very small amount will go into solution.

Once a solution is saturated the dissolving process does not stop. However, the rate at which solute goes into solution is offset by the rate at which solute precipitates from the solution. This is an equilibrium process, and since it is on-going it is called a dynamic equilibrium.

Liquids that mix are not referred to as dissolved, instead they are labeled as miscible. Liquids which chemically resemble other liquids are more likely to be miscible than liquids that are chemically dissimilar. Some liquids will mix with water very well and are known as hydrophilic materials, water loving. Others are hydrophobic, or, do not mix with water.

A true solution is homogeneous, uniform throughout. This is an entropy driven situation.

The process of making a solution of solid in water is termed dissolving.

Solids will dissolve in a solvent that is similar to itself. A polar solvent is much better for dissolving ionic solutes than is a molecular solvent.

When water is the solvent, dissolving often involves the solvent particles surrounding the ionic solute particles and establishing forces that are stronger than the lattice forces in the solid. This is known as solvation or hydration.

Gas solubility follows different patterns than solids.

Objectives:

1. You should be able to explain gas solubility and pressure relationship as stated in Henry's law.

Gases will dissolve in liquids; however, the solubility relationship is the reverse of what we expect for solids. As the solvent temperature increases the gas solubility decreases. This is a very important factor with respect to oxygen solubility and its availability for fish in a body of water. This is a major concern with respect to thermal pollution.

Henry's law gives the relationship of gas solubility with respect to the pressure of the gas involved. Pressure and solubility are directly related.

Some gases are extremely soluble. Ammonia, for example, has a solubility of nearly 1700 liters of gas in a single liter of water at room temperature.

Solution concentrations are expressed in many ways.

Objectives:

1. You should be able to define the various terms used to express solubility.
2. You should be able to convert between the various methods used to express solubility.
3. You should be able to explain why all solutions of the same percentage composition do not have the same number of particles of solute.

When concentrations are discussed, various terms are possible, but the term **molarity**, M, best meets the needs of the chemist. This term has as its basis, particles. Particles are what react in chemical systems. Molarity is a temperature dependent system.

What is the molarity of a solution made by dissolving 35.0 grams of $AgNO_3$ in enough water to make 200.0 mL of solution?

$$\text{Molarity} = \frac{\text{moles involved}}{\text{Volume (L)}} = \frac{\text{grams / mol mass}}{\text{Volume (L)}}$$

Mol mass for silver nitrate is 169.9 g/mol

$$M = \frac{35.0\ g / 169.9\ g/mol}{0.2000\ L} \quad 1.03\ M$$

Concentrated and dilute are relative terms. Actually, any concentration less than maximum concentration could be considered dilute.

Molality, m, (don't confuse with molarity) moles per Kg of solvent, has limited application. It is used when working with what are known as colligative properties. These includes such factors as; freezing point lowering, boiling point elevation, and vapor pressure depression.

What is the molality of a solution made by dissolving 3.56 g of iodine in 55.0 mL of CCl_4 ? The density of the solvent is 1.54 g/mL.

$$\text{Molality} = \frac{\text{moles involved}}{\text{Kg of solvent}} = \frac{\text{grams / mol mass}}{\text{Kg of solvent}}$$

Mol mass for $I_2 = \frac{253.8\ g}{mol}$ & Mass of $CCl_4 = \frac{55.0\ mL}{1.54\ g/mL} = 35.7\ g$

$$m = \frac{3.56\ g / 253.8\ g/mol}{35.7\ g / 1\ kg / 1000g} = \frac{0.0140\ mol}{0.0357\ Kg} = 0.393\ m$$

Percentage, mass fraction, is very common, but the use for chemists is very limited. A 5% solution of sodium chloride has many more particles than a 5% solution of sucrose because the mole weights are very different. Percentage concentration does find use in medical solutions.

A solution was made by adding 35.0 grams of salt and 25.0 grams of sucrose to 75.0 grams of water. What is the weight percent of sucrose in this system?

$$\% \text{ sucrose} = \frac{\text{grams sucrose}}{\text{grams total}} \times 100\% = \frac{25.0}{135.0} \times 100\% = 18.5\%$$

The parts per method of concentration expression; ppm, ppb,ppt etc, finds use in environmental or pollution work.

Some properties depend only on the number of particles.

Objectives:

1. You should be able to identify various colligative properties and explain the cause of each.
2. You should be able to explain why solutions of ionic and molecular solutes of equal molality do not produce the same effect in the solution.

Properties that depend upon the number of solute particles but are independent of the kind of particle are known as colligative properties. It makes no difference if these particles are molecules or ions, or if they are large or small in size.

Ionic materials have a more significant effect than one may expect. When in solution they dissociate, and each individual ion has an effect. For example, a 1.0 molal NaCl solution, will have an effect nearly twice as great as a 1.0 molal sucrose solution. The effect often is not a whole number multiple since some ions do interact with each other and are not totally independent.

All solvents are not equally affected by solute particles. Some solvents may interact differently than others. As a result of this, each solvent has a proportionality constant, known as K_b or K_f, that accounts for this effect.

Camphor has a freezing point of 179.8°C and a K_f value of 40.0 °C/m. If 40.0 grams of CH_2Cl_2 were added to 200 grams of camphor at what temperature would the system melt?

$\Delta T = K_f\, m$ mole weight CH_2Cl_2 = 85.0 g/mol

$$m = \frac{40.0\ g / 85.0\ g/mol}{0.200\ Kg} = 2.35\ m$$

$\Delta T = 40.0\ °C\ m^{-1} \times 2.35\ m = 94.1°C$

New melting point = 179.8 - 94.1 = 85.7 °C

All colligative properties effects are entropy driven. Solutions are more disorganized than pure materials.

Osmosis is a biologically important colligative property.

Objectives:

1. You should be able to explain the significance of osmosis in biological systems.
2. You should be able to explain reverse osmosis.

When solutions of different concentration are separated from each other by a membrane, so that some particles can go through but not others, there will be a diffusion through the membrane to lower the concentration of the higher molar solution. This is known as osmosis. When this occurs it can be stopped by applying pressure. The amount of pressure required to stop the process is the osmotic pressure.

Osmotic processes are important in biological systems. Movement of materials in and out of cells depends on this principle. When materials are added to living systems it is important that they be isotonic with the cell concentration otherwise cell damage is likely to occur.

Sea water (assume it is NaCl) has a molarity approaching 0.55 M. What osmotic pressure would it develop if the temperature was 5.0°C ?

We need to remember that NaCl which is the solute will dissociate into 2 ions.

$\Pi = RTM = (0.0821 \text{ L atm K}^{-1} \text{ mol}^{-1})(278\text{K})(2\text{x } 0.55\text{M})$

$= 25.1 \text{ atm.}$

Removal of the solvent, such as water, from foods by brine solutions is the result of osmosis.

If pressures are applied to solutions it sometimes is possible to force small solvent molecules from the solution through a semipermeable membrane. This process is very energy intensive but can be used to purify water. The process is known as reverse osmosis since we are pushing the solvent from a dilute solution causing the solution to become more concentrated.

Water has properties that make in different.

Objectives:

1. You should be able to describe the effect the shape of the water molecule and the electronegativity of oxygen have on such properties of water as the boiling and freezing points.

Water has properties different than related molecules like H_2S. Much of this is do to the highly electronegative oxygen atom which sets up hydrogen bonding with other water molecules. This has an effect on the boiling and freezing point of water. The shape of the water molecule gives it polar character and this makes it an excellent solvent for ionic materials.

It is not uncommon for a temperature change to accompany the solution process. This is known as enthalpy of solution. It may be either positive or negative. The common hot packs and cold packs in use every day make use of this characteristic.

Solubility products are used for very insoluble materials.

Objectives:

1. You should be able to write K_{sp} expressions for slightly soluble ionic compounds and use K_{sp} values to estimate solubilities.
2. You should be able to explain the effect a common ion has on the solubility of a compound.

When materials are very insoluble we express their solubility in terms of solubility product. This takes each ion into account so if a formula, such as MgI_2, is involved the product is $[Mg^{+2}][I^{-1}][I^{-1}]$ or $[Mg^{+2}][I^{-1}]^2$. The easy way to keep this straight is to write the equation and use the coefficients as exponents in the solubility equation.

$$Ag_3PO_4 \Longleftrightarrow 3\ Ag^{+1} + PO_4^{-3}$$

$$K_{sp} = [Ag^{+1}]^3[PO_4^{-3}]$$

You need to exercise care when using K_{sp} values as a means to measure solubility. Several factors can affect actual solubility making the K_{sp} value erroneous.

Calculate the solubility of $BaCrO_4$ whose K_{sp} value is 1.2×10^{-10}. Would 1.0g saturate the system if it was placed in one liter of solution?

The equilibrium reaction is $BaCrO_4$ <===> $Ba^{+2} + CrO_4^{-2}$
Generally if the K_{sp} value is less than 10^{-5} we can assume the solubility is negligible compared to the amount present. We can let the ion amounts be represented by x.

$K_{sp} = [Ba^{+2}][CrO_4^{-2}] = 1.2 \times 10^{-10} = x^2$
$x = 1.1 \times 10^{-5}$ M

sol = 1.1×10^{-5} M x 253.3 g $BaCrO_4$/ mol = 2.8×10^{-3} g/ liter

One gram is more than enough to saturate the solution.

When solids are not very soluble there is an equilibrium established between solid and particles in solution. As noted in previous chapters, equilibrium conditions are addressed by LeChatelier's principle. If you are trying to dissolve a solid, such as zinc iodide, in water that already contains some iodide ion, the amount of the compound that dissolves is less than expected. The iodide ion in solution has the effect of shifting the equilibrium back toward the solid ZnI_2, or actually causing less to dissolve. This is referred to as the common ion effect.

If the solution, in the problem just completed above, already had barium ion from barium chloride present in the concentration of 0.1M how much of the barium chromate would dissolve?

This is a common ion type problem. Since an ion is present in the solution that is present in the solute we would expect the solubility to be decreased. The concentration of the barium ion will now be (x + 0.1M).

$K_{sp} = [Ba^{+2}][CrO_4^{-2}] = 1.2 \times 10^{-10} = (x + 0.1M)(x)$

The value of x is much less than 0.1 so the value of Ba^{+2} can be assumed to be 0.1.

$0.1x = 1.2 \times 10^{-10}$ or $x = 1.2 \times 10^{-9}$ mols will dissolve.

Colloids are not solutions.

Objectives:

1. You should be able to describe the difference between colloids and solutions.
2. You should be able to define surfactant, hydrophobic, hydrophillic and explain how soap functions as a cleaning agent.
3. You should be able to define and explain the term "biological oxygen demand" and tell why it is important to aquatic life.

If very small particles of solute are produced they may not dissolve but also may not separate from the solvent, at least for a very long time. Compared to ions these are large particles and they are not uniformly dispersed in the continuous phase. These systems are colloids and may involve all three states of matter. Some of these may interact well with water and are called hydrophilic colloids in contrast to those that are not attracted to water. The latter are known as hydrophobic colloids.

If another substance is present with liquid-liquid colloids it may act as a "glue" between the phases and produce an emulsion.

If a material is capable of interacting both with water and a nonpolar system we have a surfactant material. We find these very useful in cleaning (soaps and detergents) and also in industry, for example to extract oil from nearly depleted wells.

Surfactants often act in groups to surround a material and keep it suspended in the continuous phase. This is the way soap lifts dirt from a material.

Since water is such an excellent solvent for so many materials it also has the potential to dissolve and collect materials that are not desirable. When this happens we end up with polluted water.

Very often pollutants are organic or biological materials that can interact with oxygen in the water. The quantities of these materials can be measured by how much oxygen they consume under specific conditions. This is known as BOD, or biochemical oxygen demand. This demand for oxygen in a lake or stream is often the cause of fish kills. Those materials that can be oxidized are called biodegradable.

Symbols found in Chapter 15

Page	Symbol	
656	S_g	Solubility of a gas
656	P_g	Partial pressure of a gas above the solution
655	K_H	Henry's law constant
660	M	Molarity, concentration in mol/L of solution
661	m	molality, moles per Kg of solvent
662	ppm	Parts per million
662	ppb	Parts per billion
662	ppt	Parts per trillion
662	μg	Microgram or 10^{-6} grams
664	-SH	Sulfhydryl group, common in some enzymes
665	dL	Deciliter or 100 mL
665	EDTA	Ethylenediaminetetraacetate, used for lead poisoning
666	ΔT_b	Boiling point elevation
667	K_b	Boiling point elevation constant, solvent dependent
668	ΔT_f	Freezing point depression
668	K_f	Freezing point depression constant, solvent dependent
671	Π	Symbol for osmotic pressure
671	c	Symbol for solution molarity in osmotic pressure
676	ΔH_{soln}	Heat involved in solution formation

676	K_{sp}	Solubility product constant
686	BOD	Biological Oxygen Demand, measure of pollution

Terms in Chapter 15

The following terms are listed for your convenience. They should not become a list for you to memorize. You should have a working definition for each. If you have doubt about any particular term check its definition in the glossary and/or review it in the chapter. Some of these terms are included in the crossword puzzle later in the chapter. This chapter also makes use of several terms introduced in chapters 4, 7, and 10.

Solution	Solvent	Solute
Solubility	Insoluble	Saturated
Unsaturated	Supersaturated	Miscible
Hydrophilic	Hydrophobic	Hydration
Hydrated	Solvation	Henry's law
Thermal pollution	Concentrated	Dilute
Molarity	Molality	Mass fraction
Weight percent	Pica	Colligative property
Boiling point elevation	Semipermeable membrane	
Osmosis	Osmotic pressure	Hypotonic
Isotonic	Reverse osmosis	Enthalpy of solution
Solubility product constant		Common ion effect
Dispersed phase	Continuous phase	Emulsion
Tyndall effect	Hydrophilic colloid	Hydrophobic colloid
Coagulate	Surfactant	Detergent
Soap	Micelles	Biodegradable
Biochemical oxygen demand		Non-biodegradable

Concept Maps for Solution Chemistry

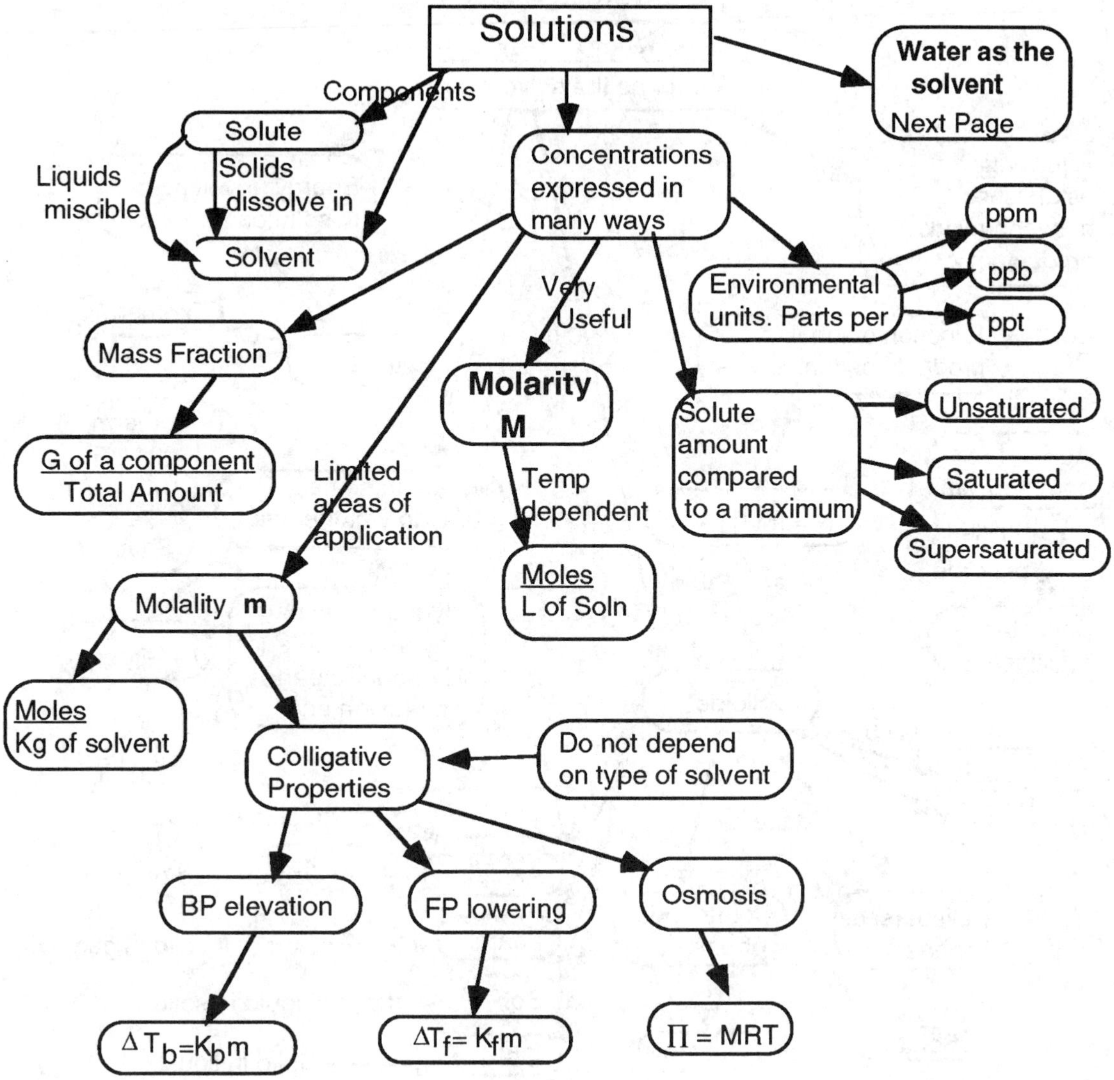

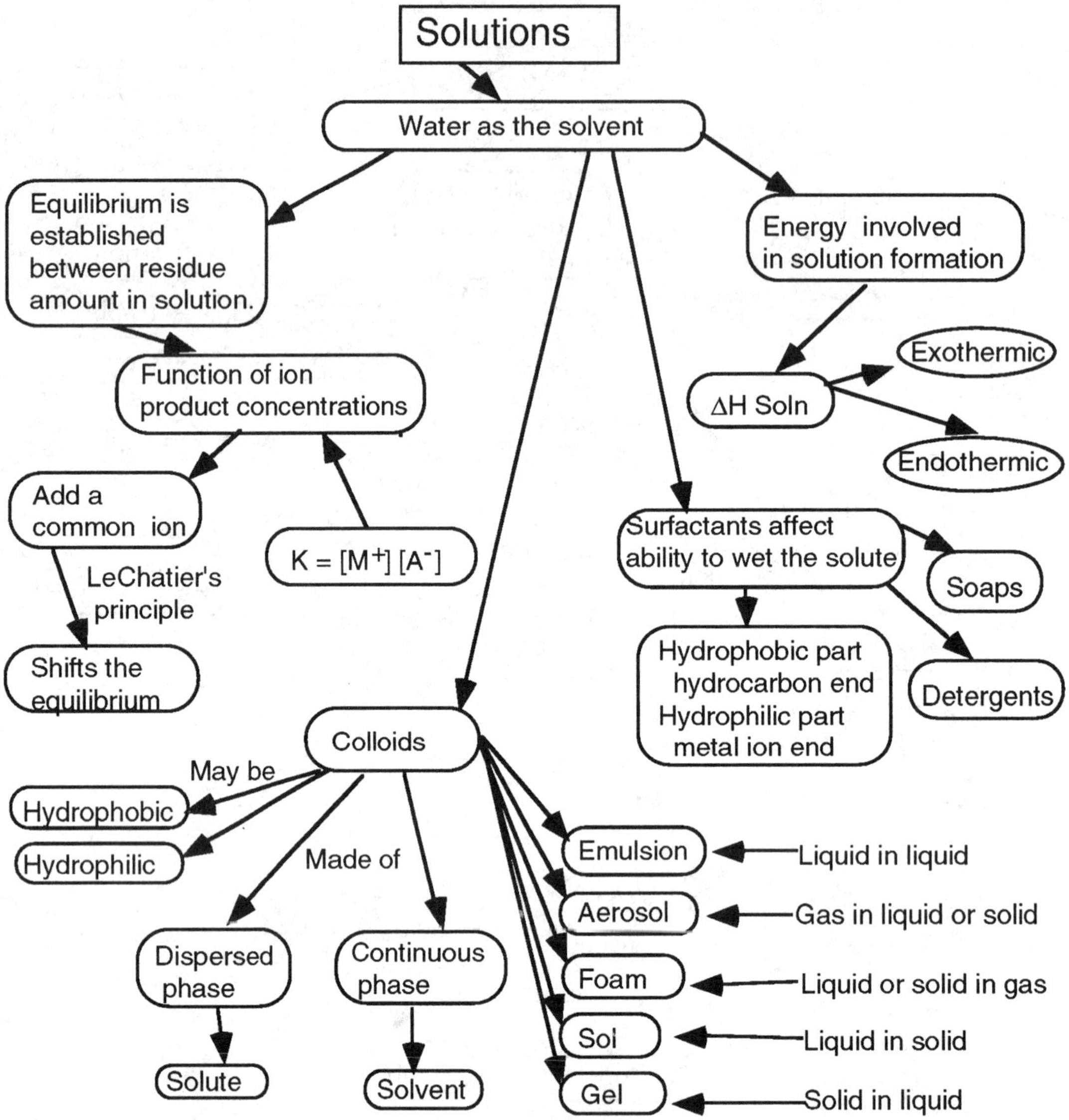
Solutions
Water as the solvent
Equilibrium is established between residue amount in solution.
Function of ion product concentrations
Add a common ion
LeChatier's principle
Shifts the equilibrium
K = [M+] [A-]
Energy involved in solution formation
ΔH Soln
Exothermic
Endothermic
Surfactants affect ability to wet the solute
Soaps
Detergents
Hydrophobic part hydrocarbon end
Hydrophilic part metal ion end
Colloids
May be
Hydrophobic
Hydrophilic
Made of
Dispersed phase
Continuous phase
Solute
Solvent
Emulsion
Liquid in liquid
Aerosol
Gas in liquid or solid
Foam
Liquid or solid in gas
Sol
Liquid in solid
Gel
Solid in liquid

Solution Chemistry True/False Statements

Answers are at the end of the chapter.

1. A true solution is a homogeneous mixture.

2. Solubility is temperature dependent.

3. Supersaturated solutions are stable.

4. A systems entropy is increased when a solution is made.

5. The hydrocarbon, toluene, would be a good solvent for a material like magnesium chloride.

6. All solvents that contain oxygen exhibit a good deal of hydrogen bonding.

7. Hydration is an important part of the dissolving of ionic materials in water.

8. According to Henry's law, if the pressure of a gas over a solvent is doubled the solubility of the gas will become half as great.

9. If radon gas was dissolved in water, one way to remove it would be to heat the water.

10. Molality, m, is a temperature dependent concentration expression.

11. Any 15% solution contains the same number of solute particles as any other 15% solution.

12. EDTA is one of the chief materials used in the treatment of lead poisoning.

13. The solute ethanol will have a different effect on the freezing point of water than the solute propanol would assuming all other factors remained constant.

14. The K_b value for the solvent ethanol is probably different than the K_b value for the solvent cyclohexane.

15. The movement of a solute through a membrane is called osmosis.

16. The mole weight of large biological molecules is often determined by measuring the osmotic pressure effect.

17. If you are being treated by intravenous methods it is important that the solution be hypertonic with respect to your body fluids.

18. If the enthalpy of solution for a particular material is negative the solution will get cold.

19. The more negative the exponent in a solubility product expression the more soluble the material is.

20. Adding an ion to a solution that already contains that ion will increase the solubility of the solute material.

21. Milk is a type of colloid known as an emulsion.

22. Detergent is just another word for soap.

23. A high value for BOD is important for a body of water to be of good quality.

24. Household waste materials seldom are classified as hazardous materials.

Testing Your Understanding of Concepts in Chapter 15.
Answers are at the end of the chapter.

1. How could you convert a saturated solution to one that is supersaturated without adding any more solute?

2. If one gram of solute is added to a solution that is saturated what effect will the extra gram of solute have?

3. When is the term miscible preferred to the term soluble?

4. Would a solute with several oxygen atoms in the material be considered hydrophilic or hydrophobic?

5. What is the difference between hydration and solvation?

6. Ammonia gas reacts chemically with water. How, if at all, does this affect the quantitative aspects of Henry's law for this gas?

7. What is the difference between a liter of solution used in molarity and a kilogram of solvent used in molality?

8. Why don't we express pollution quantities in terms of molarity rather than terms like ppm?

9. Would one mole of the two materials, methanol and calcium chloride, have the same effect on the boiling point of water?

10. If a system has an osmotic pressure of 560 torr could you stop the process of osmosis if you applied 400 torr of pressure?

11. Which material; AgCl, $K_{sp} = 1.8 \times 10^{-10}$, or AgBr, $K_{sp} = 3.3 \times 10^{-13}$ is more soluble?

12. What is the function of an emulsifier?

13. How can a single molecule have both a hydrophilic and hydrophobic end?

14. What effect, if any, does decaying plant material have on the BOD of a lake?

Testing You on Some Analytical Concepts.
Answers are at the end of the chapter.

1. At STP the water solubility of $N_{2(g)}$ is 23.55 mL per liter. If we wished to have a solution with 100.0 mL of nitrogen gas per liter what pressure of N_2 must be applied?

2. If you dissolved 0.25 mol of NaCl in water to make 500.0 grams of solution what is the weight percent of NaCl?

3. If 100.0 g of $BaCl_2$ was dissolved in enough water to make 1.50 liter of solution what would the molarity be? If the same amount was dissolved in 1.50 L of water whose density was 0.993 g/mL what is the molality?

4. An unknown solute was found to dissolve in the liquid $C_2H_4Br_2$. This solvent has a K_f value of 11.8°C/m and freezes at 10.0°C. When 0.700 g of an unknown was dissolved in 10.84 g of the solvent and cooled the solution froze at 6.51°C. What is the mol weight of the unknown?

5. A particular biological molecule was present in a solution. Analysis found 0.789 g in 30.0 mL of solution and the solution developed an osmotic pressure of 7.50 torr at 25.0°C. Calculate the mole mass of the molecule.

6. What amount of $La(OH)_3$ is present in 1.00 L at 25°C? The K_{sp} value for the solid is 1.0×10^{-19}.

7. What is the solubility of KBO_3 in a 0.50 M KNO_3 solution? The K_{sp} value for KBO_3 is 1.3×10^{-2} .

Crossword - Solution Chemistry

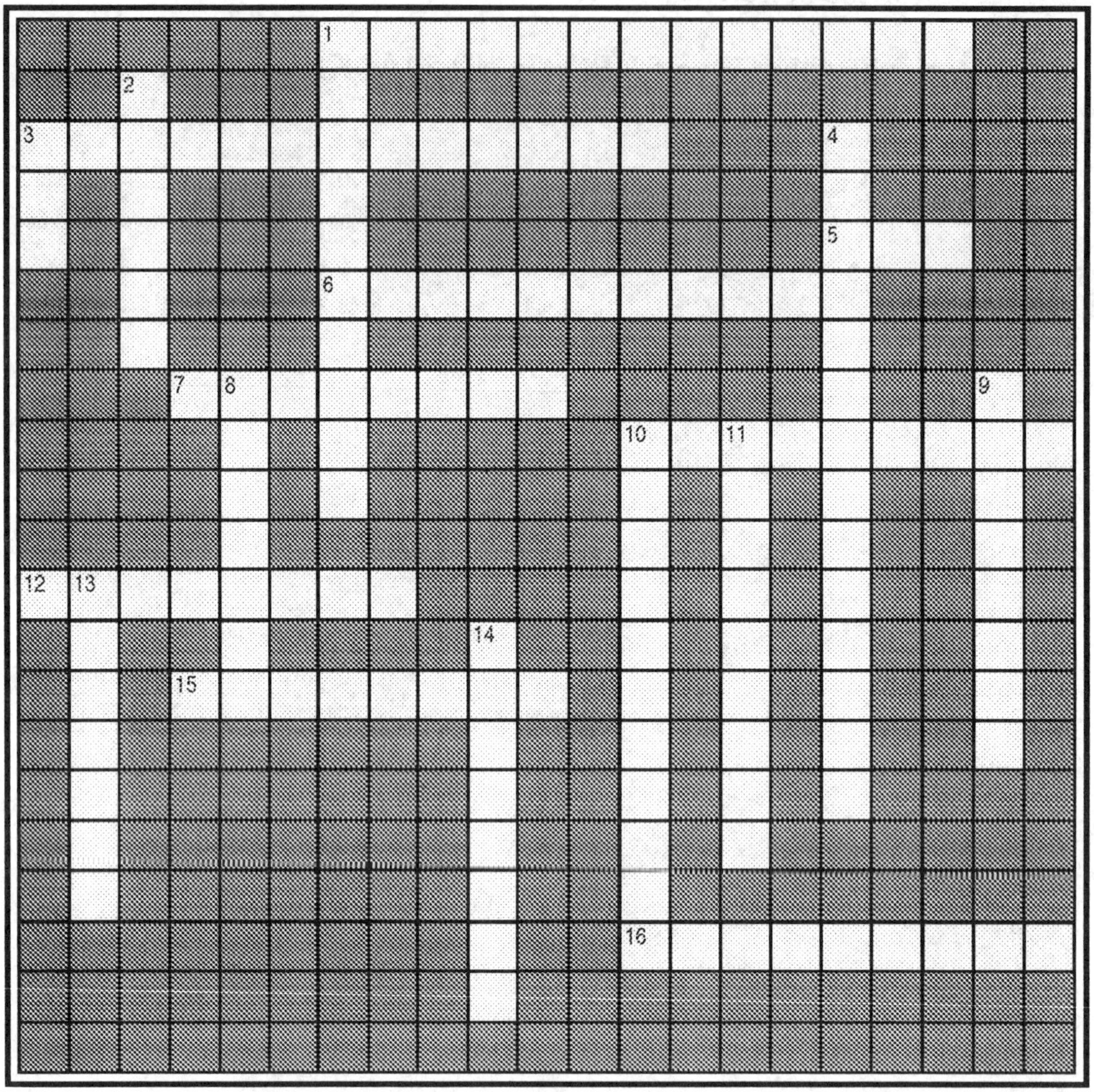

Crossword - Solution Chemistry Clues

ACROSS

1 A membrane that allows some materials but not others to move through it.
3 Materials that will undergo degradation in the environment.
5 Small quantity involving millions. An abbreviation.
6 Properties that depend on the number of particles but not the type.
7 Concentration term that is used for colligative systems.
10 Interaction of solvent (water) molecules with the solute. Often aids dissolving.
12 This material has a substance that coats and stabilizes the dispersed phase particles.
15 Solution of the same concentration as the fluids in body cells.
16 To cause the dispersed phase particles to join together resulting in large particles.

DOWN

1 Material with both hydrophobic and hydrophilic parts.
2 Material being added to the solvent.
3 Refers to the demand for oxygen in water systems.
4 Solution with more dissolved material than expected at the given conditions.
8 Movement of solvent through a membrane.
9 Most useful method of expressing solution concentrations.
10 Material that does not interact well with water.
11 The phase of the colloid that is distributed in the medium.
13 These are aggregates of colloidal sized particles.
14 Liquids that mix in all proportions are said to be this.

Ties to Previous Chapters

As you have noticed, the more we advance in the textbook the more we refer to subjects that received some discussion in previous chapters. The topic of this chapter is involved with many previous topics. Sections 4.2, 4.3, 4.4 and 4.8 involve types of compounds. In chapter 5, parts 5.3 and 5.6 have relationships to this chapter. Some ideas of energy from chapter 7; importance of bond polarity from chapter 10; intermolecular forces from section 12.5, and crystalline solids discussed in section 14.2, all have some relationship to topics in this chapter.

Answers to True/False Statements:

1. True. Solutions are mixtures. To be a true solution the mixture must be homogeneous.

2. True. For solids, solubility normally increases with temperature but the reverse is true for gases.

3. False. This is a system that has solute dissolved at a higher temperature where more is soluble than at the lower temperature. When the solution is cooled down. The extra solute will readily come out of solution.

4. True. This is a mixing process that leaves the system less organized. Less organization is characteristic of higher entropy.

5. False. Hydrocarbons are nonpolar materials. They will dissolve nonpolar solutes but not ionic ones such as salts.

6. False. Very few exhibit any noticeable hydrogen bonding.

7. True. The polar water molecules surround the solute with interactions that exceed those within the crystal.

8. False. As the pressure of the gas increases the solubility will increase. The relationship is direct not inverse.

9. True. Solubility of gases are lowered as the solvent temperature is increased.

10. False. Molarity is temperature dependent.

11. False. As the mole weight changes, the number of particles in a 15% solution will also change.

12. True. This material forms a chelate cage about the lead.

13. False. Both are molecules. The change in freezing point is dependent upon the number of particles, not the kind.

14. True. These constants are solvent specific. Each solvent has its own value. The same is true for the K_f constants.

15. True. The movement is always to dilute the more concentrated solution.

16. True. This method is very effective for large molecules.

17. False. You would want the solutions to be isotonic, the same concentration.

18. False. The negative sign means energy is released. The system would warm up.

19. False. The more negative the exponent the less soluble the material would be.

20. False. This is known as the common ion effect and would decrease the tendency for solute to dissolve. This is expressed in Le Chatelier's principle.

21. True. There are several classes of colloids however.

22. False. Detergents are an entirely different type of material.

23. False. High BOD values indicate that large amounts of oxygen will be consumed which lowers water quality.

24. False. Many common household materials are hazardous.

Answers to Understanding of Concepts Questions:

1. If a solution is saturated at one temperature it would be supersaturated at a lower temperature. If the solution were cooled it would become supersaturated.

2. The extra gram would not dissolve. It would be added to the solid portion present at the bottom of the container.

3. When working with solids or gases in liquid solvents we use the term soluble. Liquid in liquid would involve the term miscible.

4. The oxygen atoms in the solute would be able to form some interactions with the hydrogen part of the water molecule so it would be hydrophilic.

5. Hydration is a special case of solvation. Hydration refers to water systems only and solvation would be applicable to any solvent, including water.

6. When a gas interacts chemically the quantitative aspects of Henry's law become unreliable. Some of the gas would be converted to another chemical.

7. A liter of solution is made by placing enough solvent with the solute to bring the total volume to one liter. A kilogram of solvent would vary in volume depending on the solvents density. A low density solvent, like alcohol, would have a total volume greater than one liter.

8. Pollution quantities tend to be very small. Molar amounts are not convenient at such small quantities.

9. The effect would be different because methanol is a molecule and a mole would provide one mole of particles. A mole of calcium chloride would provide three moles of ions so its effect would be greater by nearly three times.

10. The osmotic pressure is defined as the pressure needed to match that of the system. In this case unless the applied pressure was 560 torr you would not stop osmosis from occurring.

11. The value for AgCl is greater so it would be more soluble.

12. The emulsifier brings the two liquid components of a colloid together and keeps them from separating.

13. If the molecule has a polar end it would be hydrophilic and if the other end was a hydrocarbon it would be hydrophobic. This is how soap works. The acid end interacts with water and the hydrocarbon chain interacts with the grease or dirt.

14. Decaying materials will use oxygen so they would increase the demand for oxygen.

Answers to Testing the Analytical Concepts

1. This is a Henry's law problem. $S_g = K_H P_g$. The equation shows us that we must have an increase in the gas pressure to dissolve more. With the given data we can solve for K_H and then apply with value to the new set of conditions.

$$K_H = \frac{S_g}{P_g} = \frac{23.55 \text{ mL/L}}{1.00 \text{ atm}} = 23.55 \text{ mL/L atm}$$

$$P_{nitrogen} = \frac{100.0 \text{ mL gas per L}}{23.55 \text{ mL per L per atm}} = 4.24 \text{ atm}$$

2. Weight percent is a ratio of the weight of the item in question per weight of solution x100%. We need the weight of NaCl in 0.25 mol.

$$\text{g NaCl} = 0.25 \text{ mol x } 58.5 \text{ g NaCl/mol} = 14.6 \text{ g}$$

$$\text{Wt.\%} = \frac{\text{Wt. NaCl}}{\text{Total Wt}} \text{ x}100 = \frac{14.6 \text{ g}}{500.\text{g}} \text{ x } 100\% = 2.93\%$$

3. Molarity is defined as moles / liter of solution. First we must convert 100.0 g of $BaCl_2$ to moles.

$$\text{moles} = \frac{100.0 \text{ g}}{208.3 \text{ g/mol}} = 0.480 \text{ moles}$$

$$\text{M} = \text{Moles / Volume in liters} = \frac{0.480 \text{ mole}}{1.5 \text{ L}} = 0.320 \text{ M}$$

The second part is a molality problem since it involves mass of solvent. From above we know the number of moles. The solvent involves 1500 mL whose density if 0.993 g/ml so we can determine the Kg of solvent.

Kg solvent = 1500 mL x 0.993 g/mL x 1 Kg / 1000 g = 1.489 Kg

$$m = \frac{\#\ \text{moles}}{\text{Kg}} = \frac{0.480\ \text{moles}}{1.489\ \text{Kg}} = 0.322\ \text{m}$$

4. Relationships involved in the colligative property problem are:

Mol Wt = Wt involved / moles used

\# moles = molality x Kg solvent

We know the weight of solute and the Kg of solvent so we need a value for molality which will allow us to solve for number of moles. Molality can be determined from the equation $K_f\, m = \Delta T_f$

$$m = \frac{\Delta T_f}{K_f} = \frac{(10.0 - 6.5)°C}{11.8\ °Cm^{1-}} = \frac{3.5°C}{11.8\ °Cm^{1-}} = 0.297\ m$$

$$\#\ \text{moles} = m \times \text{Kg solvent} = \frac{0.297\ \text{mol} \times 10.84\ \text{g}}{\text{Kg}} \frac{\times 1\ \text{Kg}}{1000\text{g}}$$

$$= 3.22 \times 10^{-3}\ m$$

$$\text{Mol Wt.} = \frac{\text{Wt involved}}{\text{Moles}} = \frac{0.700\ \text{g}}{3.22 \times 10^{-3}\ m} = 218\ \text{g mol}^{1-}$$

5. The osmotic pressure relationship is $\Pi = c\,R\,T$. The gas constant is 0.082 $L{\bullet}atmK^{1-}mol^{1-}$ and the temperature in K is 273 + 25 or 298.

The osmotic pressure in the system is:

$$\Pi = 7.50\ \text{torr} \times \frac{1\ \text{atm}}{760\ \text{torr}} = 9.87 \times 10^{-3}\ \text{atm}$$

$$c = \frac{\Pi}{RT} = \frac{9.87 \times 10^{-3} \text{ atm}}{0.082\,1 \text{ atm K}^{1-} \text{ mol}^{1-} \times 298 \text{ K}} = 4.04 \times 10^{-5} \text{ mol/L}$$

$$\text{Mol Wt.} = \frac{\text{Wt.}}{4.04 \times 10^{-5} \text{ mol/L} \times 0.030\text{L}} = 6.511 \times 10^{4} \text{ g/mol}$$

6. We first need a balanced equation and from that we can set up the equilibrium expression.

	$La(OH)_3$ <=====>	La^{+3}	+ 3 OH^{-1}
Initial	excess	0	0
At equil	excess	x	3x

From the balanced equation we see that the concentration of La^{+3} is also the solubility of the compound.

$K_{sp} = [La^{+3}][OH^{-1}]^3 = 1.0 \times 10^{-19}$

$(x)(3x)^3 = 1.0 \times 10^{-19}$

$27x^4 = 1.0 \times 10^{-19}$

x = 7.8 x 10-6 mol / liter = solubility

7. This is a solubility problem that involves a common ion. The common ion effect will decrease the solubility of the salt in the solvent. Because of the excess K^{+1} ion Le Chatelier's principle tells us the equilibrium will prevent some solubility.

KBO_3 <=====> $K^{+1} + BO_3^{-1}$

$K_{sp} = [K^{+1}][BO_3^{-1}] = 1.3 \times 10^{-2}$

The value of the potassium ion is 0.5 plus what would dissolve. An for each K^{+1} ion that does dissolve we get one BO_3^{-1}.

$1.3 \times 10^{-2} = [0.50 + x][x]$

Since the salt is reasonably soluble as indicated by the large K value the value of x will be significant and the solution needs solving with the quadratic equation.

$1.3 \times 10^{-2} = 0.50x + x^2$

and solving gives x = 0.025 mole / liter

Answer to Chapter 15 Crossword

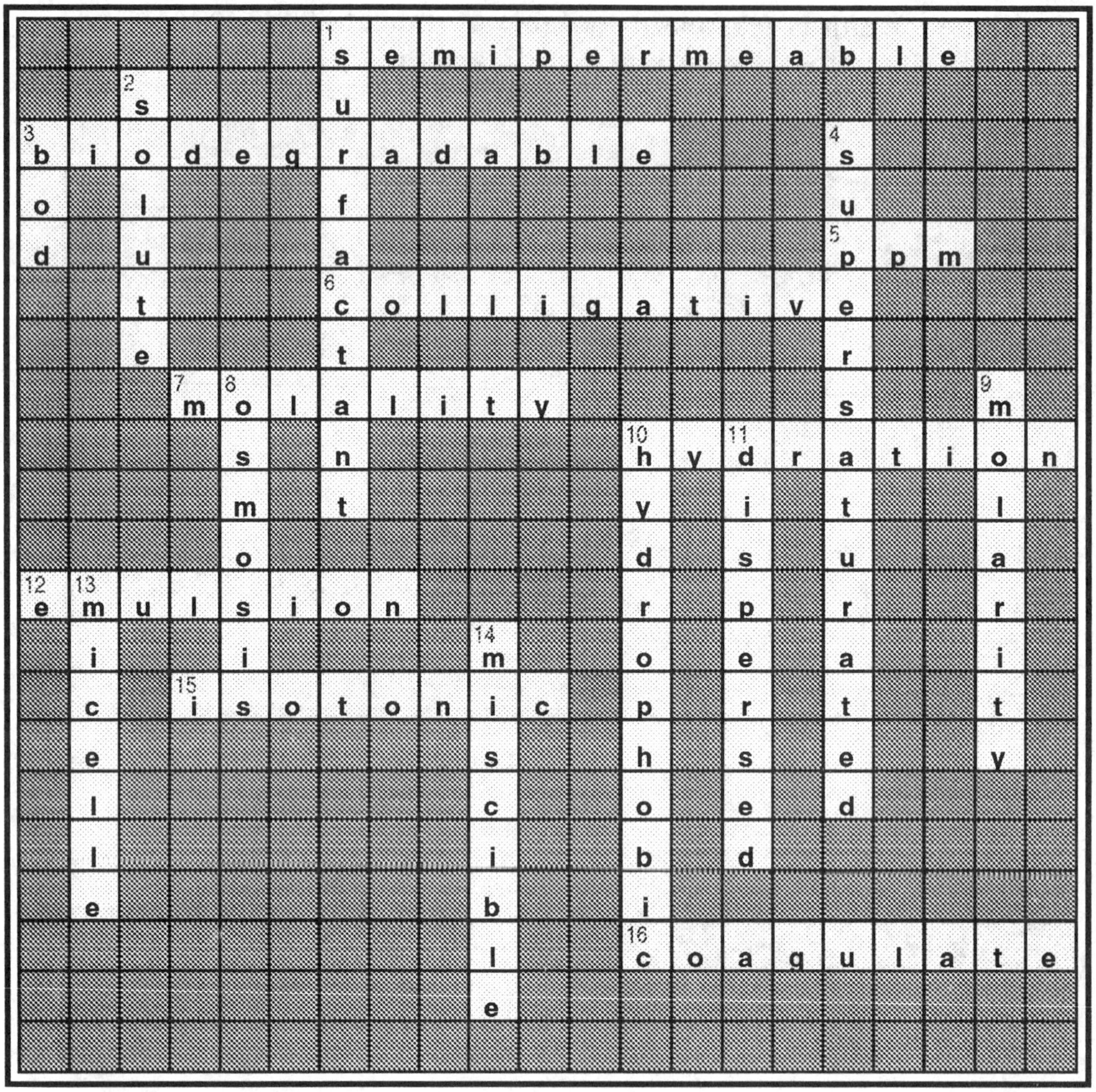

Some Supplementary Readings Relating to This Chapter:

1. Drinking water limits set for 23 more compounds, Chemical & Engineering News , June 1, 1992, page 19.

2. A New Blueprint for Water's Architecture, Science , Vol. 256, 26 June 1992, page 1764.

3. Tainted Water, Chem Matters , February 1988, page 13.

4. Waterborne radon linked to higher cancer risk, Chemical & Engineering News , September 17, 1990, page 23.

5. Saint's Blood, Chem Matters , February 1993, page 12.

6. Natural Dyes, Chem Matters , December 1986, page 4.

7. Denatured Alcohol, Chem Matters , December 1990, page 14.

8. Ferrofluids Are Making the Ball Bearing Look Passe´ , The New York Times , April 19, 1992.

Chapter 16 Acid and Bases

Acids and bases are some of the most common materials we come in contact with. They, or the products of their reactions, provide us with many important materials or the means to obtain materials. Acids, however, can be the cause of very expensive chemical changes, such as corrosion. Salts are the product of acid/base reactions. There are many different salts and they have many varying uses.

Focus Your Study Efforts

We can define acids and bases in several ways:

Objectives:

1. You should be able to identify acids and bases in the three common ways they are defined: Arrhenius, Bronsted, and Lewis.
2. You should be able to use the Bronsted concept to identify what materials may be amphoteric.
3. You should be able to identify conjugate bases of weak acids and conjugate acids of weak bases.

There are several ways to define acids and bases. The simplest is the Arrhenius definition, however it is valid only in a water solution. Acids are H^{+1} producers and bases are OH^{-1} producers.

The hydrogen ion, H^+, is not very stable in water because of its high charge density. For that reason you usually see it represented as H_3O^+ which means it is associated with a water molecule. This can actually consist of a group of water molecules interacting with a hydrogen ion.

The Bronsted concept of acids and bases includes all materials the Arrhenius does but it defines acids and bases in terms of proton action.

The acid donates a proton and the base accepts the proton. This concept is not limited to hydroxide ion for bases.

When ammonia, NH_3, is placed in water it dissolves and can be represented as $NH_{3(aq)}$. However, some of the ammonia reacts with water and a few ions are formed, so we have NH_4^+ and OH^-. You will also see it written as NH_4OH.

In the Bronsted definition some materials can either accept or donate a proton and they are said to be amphiprotic. An example would be HSO_4^{-1}.

HSO_4^{-1} =====> $H^{+1} + SO_4^{-2}$ as an acid
$HSO_4^{-1} + H^{+1}$ ===> H_2SO_4 as a base

The Bronsted definition also allows for the idea of autoionization which means two molecules react and one donates a proton to the other resulting in an acid and base being formed.

The Bronsted definition involves conjugate pairs. The following equation illustrates the meaning.

$H_2PO_4^{-1} + H_2O$ <===> $HPO_4^{-2} + H_3O^{+1}$

The $H_2PO_4^{-1}$ (an acid) and HPO_4^{-2} (its conjugate base) are a pair. The same is true for H_2O (a base) and H_3O^{+1} (its conjugate acid).

Identify the conjugate pairs in a reaction of nitrous acid and sodium carbonate.

Making use of table 16.2. Notice that Na_2CO_3 is not listed but CO_3^{-2} is. It is the carbonate ion that is the base.

From the table HNO_2 has NO_2^- as a conjugate base and CO_3^{-2} has a conjugate acid of HCO_3^{-1}.

Also note that HCO_3^{-1} is a base and strong enough to react with HNO_2 so a second reaction could occur with the conjugate acid of H_2CO_3 for HCO_3^{-1}.

A third, and much more inclusive definition of acids and bases, is the Lewis Concept. This concept identifies an acid as any material the can accept an electron pair to form a bond and the base is any material that can donate the electron pair to the bond. These bonds are coordinate covalent since both electrons come from the same material.

Using the Lewis definition, cations and neutral molecules can be acids.

The Lewis definition is not as common in the introductory course as the Bronsted concept. However, the Lewis ideas are used extensively in the area of organic chemistry and also in coordination chemistry.

Acids/bases may have several varying parameters:

Objectives:

1. You should be able to distinguish between binary acids and oxoacids.
2. You should be able to write ionization reactions for poly-protic acids or bases.

Acids may be binary, such as HCl, or oxoacids, such as H_3PO_4. The classification is based on the presence or absence of oxygen in the anion. Since some acids have more than one available hydrogen they are also classified on the basis of available hydrogens. HCl is monoprotic where phosphoric is polyprotic.

Bases are also capable of reacting with more than one hydrogen ion. Consequently the term polyprotic base is used.

The pH scale is a major tool for defining solutions:

Objectives:

1. You should be able to use $[H^+]$or $[OH^-]$ to calculate a pH value for a solution.
2. You should be able to explain the significance of the log scale as it applies to change in concentration of hydrogen or hydroxide ion.

Systems that have more H^{+1} than OH^{-1} are defined as acidic systems. Those with an excess of hydroxide ion are basic. Pure water has some autoionization in which the concentration of the hydrogen ion equals that of the hydroxide ion. This results in a neutral solution. At 25°C these values are 1.0×10^{-7} molar for each.

When written as an equilibrium for the reaction we get the water constant, $K_w = 1.0 \times 10^{-14}$. For convenience the values of these ions are expressed as the negative log and we have a scale known as the pH scale. (The p before a term means the negative log of the term.) We also can develop a scale based on the hydroxide ion, and this gives us a pOH value.

In pure water the $[H^{+1}] = 1.0 \times 10^{-7}$ so the pH value is 7. Likewise the $[OH^{-1}]$ is 1.0×10^{-7} and the pOH is 7. In pure water pH + pOH = 14.

$HClO_4$ is a strong acid. If 0.00030 moles is present in 250.0 mL of solution what is the pH?

$pH = -\log[H_3O^+]$. We need to know $[H_3O^+]$.

$[H_3O^+] = \frac{\text{moles}}{\text{Vol liters}} = \frac{3.0 \times 10^{-4}}{0.2500\text{L}} = 1.2 \times 10^{-3}$

$pH = -\log[H_3O^+] = -\log[1.2 \times 10^{-3}] = -(-2.92) = 2.92$

This general relationship can be represented in a diagram:

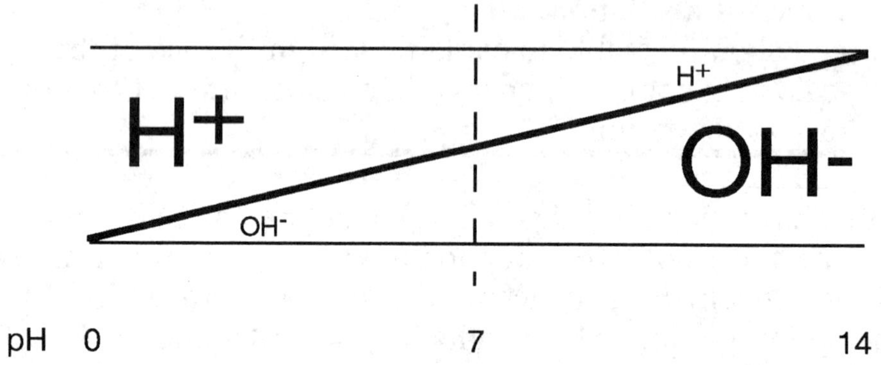

We define a strong acid as one with a low pH value. A strong base has a high pH value. HCl is a strong acid because it provides many hydrogen ions and has a pH of about 1 for a 0.1M solution. Acetic acid is much weaker since a 0.1M solution has many fewer hydrogen ions and a pH of about 3.

A solution has a hydrogen ion concentration of 1.0×10^{-4} M. What is the pH and pOH of this solution?

We know that $pH = -\log[H_3O^+]$.

$$pH = -\log[10^{-4}] = -(-4) = 4$$

$pH + pOH = 14.00$ so $pOH = 14.00 - pH = 14.00 - 4.00 = 10.00$

The pH scale is a log scale. This is very important to keep in mind. A change of 1 unit on the pH scale is a concentration factor change of 10.

A solution was made by dissolving 4.0 grams of NaOH in water to make 200.0 mL of solution. What is the pH?

This is a multiple step problem. We want pH but are working with a hydroxide so we calculate $[OH^-]$ first, then the pOH, and finally pH.

moles NaOH = Wt. Used / Mole Wt. = 4.0 g / 40.0 g mol^{-1} = 0.1
M = No. moles / Vol liters = 0.10 mol / 0.2000 L = 0.50 M

pOH = $-\log[OH^-]$ = -log [0.50] = -(-0.30) = 0.3
pH = 14.00 - pOH = 14.00 - 0.30 = 13.70

Always remember that the pH value does not tell how much total acid or base is present. It only identifies the amount of hydrogen ion present. A weak acid and a strong acid both can be 1.0 molar. However, the strong acid is nearly all in the ionized form, and the weak acid is more in the molecular (non-ionized) form.

Equilibrium constants are used to determine strengths:

Objectives:

1. You should be able to use K_a or K_b values to determine relative strengths of acids or bases.
2. You should be able to use the Bronsted chart to predict if reactions are product-favored or not.
3. You should be able to estimate the pH of a weak acid or base solution given concentration data and its K value.
4. You should be able to estimate K_2 or K_3 values for polyprotic acids if given a K_1 value.

We can determine the total amount of acid present by reacting the material with a base. If we react a known quantity of an acid with base of known concentration, and volume, the amount of acid can be calculated. This is a process known as titration. When this is done all of the acid, ionized and molecular, will get reacted.

Weak acids or bases are in equilibrium between the ionic and molecular states. When working with weak materials the actual quantity of ions varies considerably. We use ionization constants to represent the amount of the material present as ions.

Ionization constants are like any other equilibrium constant and are derived in the same way. Although you will see various forms of K, don't think of them as entirely different problem situations. Instead of K_{eq} we use K_a and K_b. You also have encountered K_{sp} in chapter 15.

Using the Bronsted system along with the relationship; $K_a \cdot K_b = K_w = 1.0 \times 10^{-14}$, we can determine the K value for the conjugate of any acid or base. Notice in table 16.2 of your text that if you multiply the K_a and K_b values for a conjugate pair you will get the value of 1.0×10^{-14}.

A particular 0.10M monoprotic acid solution had a pH of 3.31. What is the $[H_3O^+]$ and pH?

$pH = -\log[H_3O^+]$ so $[H_3O^+] = 10^{-pH}$

$[H_3O^+] = 10^{-3.31} = 4.90 \times 10^{-4}$ {to solve on your calculator enter 10 then y^x then 3.31 then ± and finally =}

Acid + H_2O ====> H_3O^+ + acid anion

Solving for a K_a value is aided by putting a data table together:

	[acid]	$[H_3O^+]$	[anion]
Init conc	0.10	≈0	0
Conc change	-x	+x	+x
Equil conc	0.10 -x	+x	+x

$$K_a = \frac{[H_3O^+][\text{acid anion}]}{[\text{acid}]} = \frac{x \cdot x}{0.1 - x} = 2.40 \times 10^{-6}$$

(assume x is negligible)

What is the K_b value for the conjugate base of the acid in problem above?

$$K_w = K_a \cdot K_b = 1.0 \times 10^{-14}$$

$$K_b = \frac{1.0 \times 10^{-14}}{K_a} = \frac{1.0 \times 10^{-14}}{2.40 \times 10^{-6}} = 4.17 \times 10^{-9}$$

Table 16.2 in the text can be very useful for predicting reactions of acids and bases. If you lay a ruler (or similar object) on the chart and connect the acid and base you wish to react you can predict the reaction. If the slope of the line is negative then the reaction is favored. The steeper the slope the more favored the reaction will be.

Will acetic acid and sodium dihydrogen phosphate react spontaneously?

This question is readily answered by referring to table 16.2 of the text. If we draw a line from CH_3COOH to $H_2PO_4^{-1}$ the dihydrogen phosphate ion, we have a positive slope. This would predict a non-spontaneous reaction.

Nitrous acid has a K_a value of 4.5×10^{-4}. What would the pH of a 0.070M solution be?

We first write the equation for the reaction and then determine the K_a , the $[H_3O^+]$ can then be calculated which allows us to determine pH from the equation, $pH = -\log[H_3O^+]$.

$HNO_2 + H_2O$ <===> $H_3O^+ = NO_2^-$

$K_a = \frac{[H_3O^+][NO_2^-]}{[HNO_2]}$

	$[HNO_2]$	$[H_3O^+]$	$[NO_2^-]$
Init conc	0.070	≈0	0
Conc change	-x	+x	+x
Equil conc	0.070 -x	+x	+x

$K_a = \frac{x \cdot x}{0.070 - x} = 4.5 \times 10^{-4}$

(a small x value will not affect 0.070)

$x = [H_3O^+] = 5.61 \times 10^{-3}$ with this particular problem the assumption that the value of x is negligible is questionable. It would probably be better to solve a quadratic equation.

$pH = -\log[H_3O^+] = -\log [5.61 \times 10^{-3}] = 2.25$

The way the text table 16.2 works is very similar to the way a redox table, which will be introduced in chapter 17 works. If you learn to use one properly you will use both properly.

When we have polyprotic acids or bases more than one ionization constant is required. Each proton (or hydroxide) has its own K value. As a general rule the K values change by a factor of 10^{-5} for each new ionization of the molecule. If K_{a1} is 10^{-6} then K_{a2} would be about 10^{-11}.

Reactions will proceed from strong acids and bases to weaker acids and bases. Unless the stronger items are reactants, the reaction will not be product favored.

Water has significant effects on acid/base systems:

Objectives:

1. You should be able to explain the leveling effect of water on strong acids.
2. You should be able to recognize which salts are expected to undergo hydrolysis reactions in water.
3. You should be able to write balanced equations for simple hydrolysis reactions.
4. You should be able to predict the effect of adding an acid or base on the solubility of insoluble salts.

Since water is our most significant solvent. It is important to realize that when an acid is placed in water it will always produce the hydronium ion, H_3O^+. This is as strong an acid as you can get in water. The result of this is given the name - Leveling Effect.

When salts are placed in water they may just go into solution or, undergo a reaction with the water. If the later occurs we say that a hydrolysis reaction has occurred.

Salts of strong acids and strong bases, such as NaCl, from HCl and NaOH, will not undergo hydrolysis. They will give a solution with a neutral pH value.

A salt of a strong base and weak acid, such as NaCN, from NaOH and HCN, the reaction of the CN^- ion with water will produce a basic solution. Since CN^- is the conjugate base of a weak acid, it is a strong base.

$CN^- + HOH \text{------>} HCN + OH^-$ results in many hydroxide ions and few acid ions. This is a basic solution.

What is the pH of a 0.060 M potassium nitrite solution?

When added to water KNO_2 will produce K^+ and NO_2^- ions. The potassium ion will not react with water but the NO_2^{-1} will undergo hydrolysis.

$NO_2^- + H_2O$ <====> $HNO_2 + OH^-$

The K_b value for nitrate ion can be found in table 16.2 and is 2.2×10^{-11}.

$$K_b = 2.2 \times 10^{-11} = \frac{[HNO_2]\,[OH^-]}{[NO_2^-]}$$

	$[NO_2^-]$	$[OH^-]$	$[HNO_2]$
Init conc	0.060	≈0	0
Conc change	-x	+x	+x
Equil conc	0.060 -x	+x	+x

$$\frac{(x)(x)}{0.060 - x} = 2.2 \times 10^{-11}$$

(assume x is negligible)

$x = [OH^{-1}] = 1.15 \times 10^{-6}$

$pOH = -\log [OH^{-1}] = -\log[1.15 \times 10^{-6}] = 5.94$

$pH = 14.00 - pOH = 14.00 - 5.94 = 8.06$

Strong acid and weak base salts, such as NH_4NO_3, from NH_4OH and HNO_3, will react with water to yield a solution with an acidic pH value. The NH_4^+ ion will react with water but the NO_3^- ion will not.

$NH_4^+ + HOH$ -----> $NH_4OH + H^+$.

Salts that are the result of reactions between weak acids and weak bases are more difficult to predict. Both of the ions will interact with water but unless the K_a and K_b values are identical one of the ions will undergo more hydrolysis than the other. The ion with the greater K value will determine the overall result.

If a salt is not water soluble it sometimes can be made soluble by adding an acid or base to the water. This may result in a reaction between the acid (or base) and the insoluble salt. These reactions often produce a gas which escapes from the solution. The net result, according to Le Chatelier's principle, is that when the gas escapes the reaction shifts to produce more gas and this results in the solid dissolving. When the gas escapes a reaction product is lost and equilibrium can not be maintained.

Buffer solutions resist pH changes:

Objectives:

1. You should be able to explain how to make a buffer solution and select an acid/base pair to produce a buffer of desired pH.
2. You should be able to calculate the pH of a buffer given information about the ratio of concentrations of the conjugate acid/base pair.

A buffer solution will resist a change in pH value as an acid and/or base is added to the solution. A buffer solution is made from a weak acid and the salt of that acid or a weak base and a salt of that base.

If we make a buffer from the acid HCN and a salt such as KCN the solution will have a pH greater than the acid system only. According to the common ion effect, if you add a material to a system which already has that ion an equilibrium shift will occur.

$HCN <===> H^+ + CN^-$ is an equilibrium system, when nearly 100% ionic KCN is added, the CN^- concentration goes way up, this causes the hydrogen ion to react with CN^- lowering the amount of that ion in solution. With less H^+, the higher, (closer to neutral) the pH value will be.

The Henderson-Hasselbalch equation can be used to predict the pH value for a buffer solution. This equation will also allow us to calculate the change in pH as the ratio of acid and salt in the system is changed.

What would be the pH of a buffer solution made from a solution of 0.10 M HCN and 0.25 M NaCN?

Solution of the problem is easily accomplished by use of the Henderson-Hasselbalch equation. This is a very useful and versatile equation that allows us to calculate several parameters.

$$pH = pK_a + \log \frac{[\text{anion}]}{[\text{weak acid}]}$$

$$HCN + H_2O <===> H_3O^+ + CN^-$$

$[CN^-] = 0.25M$ and $[HCN] = 0.10$ the large excess of cyanide ion will prevent nearly any HCN to ionize.

$$pH = 9.40 + \log \frac{(0.25)}{(0.10)} = 9.40 + \log 0.25 - \log 0.10$$

$$= 9.40 +(-0.60) -(-1.00) = 9.80$$

Titration is a process used to analyze acid/base solutions:

Objectives:

1. You should be able to define titration, equivalence point, and end point.
2. You should be able to select an appropriate indicator for an acid/base reaction.
3. You should be able to sketch the shape of titration curves for various combinations of strong and weak acids and bases.

An acid/base titration involves adding an acid (or base) to a base (or acid) solution in a controlled amount until the titrant (material being added) just uses up all the material being titrated. When this occurs the reaction is at the equivalence point. In this system the amount of H^+ added just matches the amount of OH^- that was present.

The equivalence points usually can not be seen without the aid of an indicator. Indicators are highly colored compounds which have one color at one pH value, and another color at some other pH value. Where the indicator changes color we have an end point. In a titration system we try to match the equivalence point and end point values.

When titrating strong acids and bases the equivalence point will be very near pH 7 and there will be a rapid change in pH value very close to the equivalence point.

What would be the pH of a solution after 35.0 mL of 0.150 M KOH was added to 50.0 mL of 0.145 M HCl?

Since KOH is a strong base and HCl is a strong acid both will ionize completely and the salt formed will not undergo hydrolysis.

$$HCl + KOH \Longrightarrow KCl + H_2O$$

We need to determine the moles of each material. From this we can determine the excess reagent and then calculate concentration and pH.

moles HCl $= 0.145$ m/L x 0.0500 L $= 7.25 \times 10^{-3}$
moles KOH $= 0.150$ m/L x 0.0350 L $= 5.25 \times 10^{-3}$

HCl is in excess by 2.00×10^{-3} moles.

$$[H_3O^+] = \frac{2.00 \times 10^{-3} \text{ moles}}{0.0850\text{L} \quad \text{(total volume)}} = 2.35 \times 10^{-2}$$

$pH = -\log [H_3O^+] = -\log 2.35 \times 10^{-2} = 1.63$

Titration of a weak acid (or base) will give a gradual rise in pH value and a much less abrupt pH change at the end point. The equivalence point will also not occur at pH 7.

Symbols found in Chapter 16

Page	Symbol	
698	H_3O^+	Hydronium ion, often seen as H^+
704	K_w	Ionization constant for water, 1.0×10^{-14}
705	pH	Used to represent acid/base character
711	K_a	Ionization constant for acids
711	K_h	Hydrolysis constant
711	K_b	Ionization constant for bases
727	pK_a	Log (negative) of K_a value

Terms in Chapter 16

The following terms are listed for your convenience. They should not become a list for you to memorize. You should have a working definition for each. If you have doubt about any particular term check its definition in the glossary and/or review it in the chapter. Some of these terms are included in the crossword puzzle later in the chapter. This chapter also makes use of several terms introduced in chapters 4, 5, 6 and 10.

Alkalis	Salt	Acid
Base	Hydronium ion	Bronsted Concept
Clathrate cage	Amphiprotic	Oxoacids
Conjugate acid-base pair	Monoprotic acid	Monoprotic base
Polyprotic acid	Polyprotic base	Ionization constant
Frasch process	Contact process	Titration
Acid ionization constant	Base ionization constant	
Leveling effect	Net ionic equation	Hydrolysis
Acidosis	Alkalosis	Buffer solution
Henderson-Hasselbalch equation		Equivalence point
End point	Lewis Acid	Lewis Base
Amphoteric		

Concept Maps for Acids and Bases

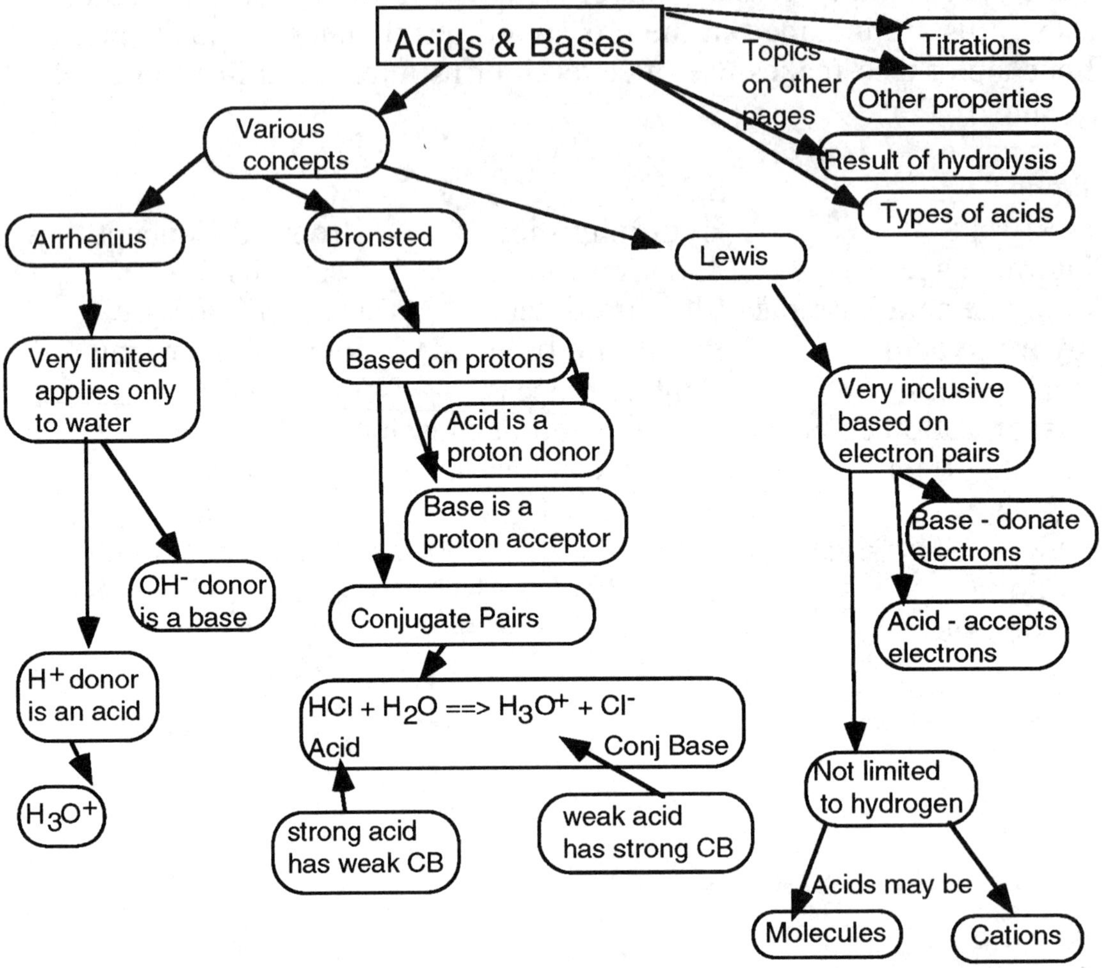

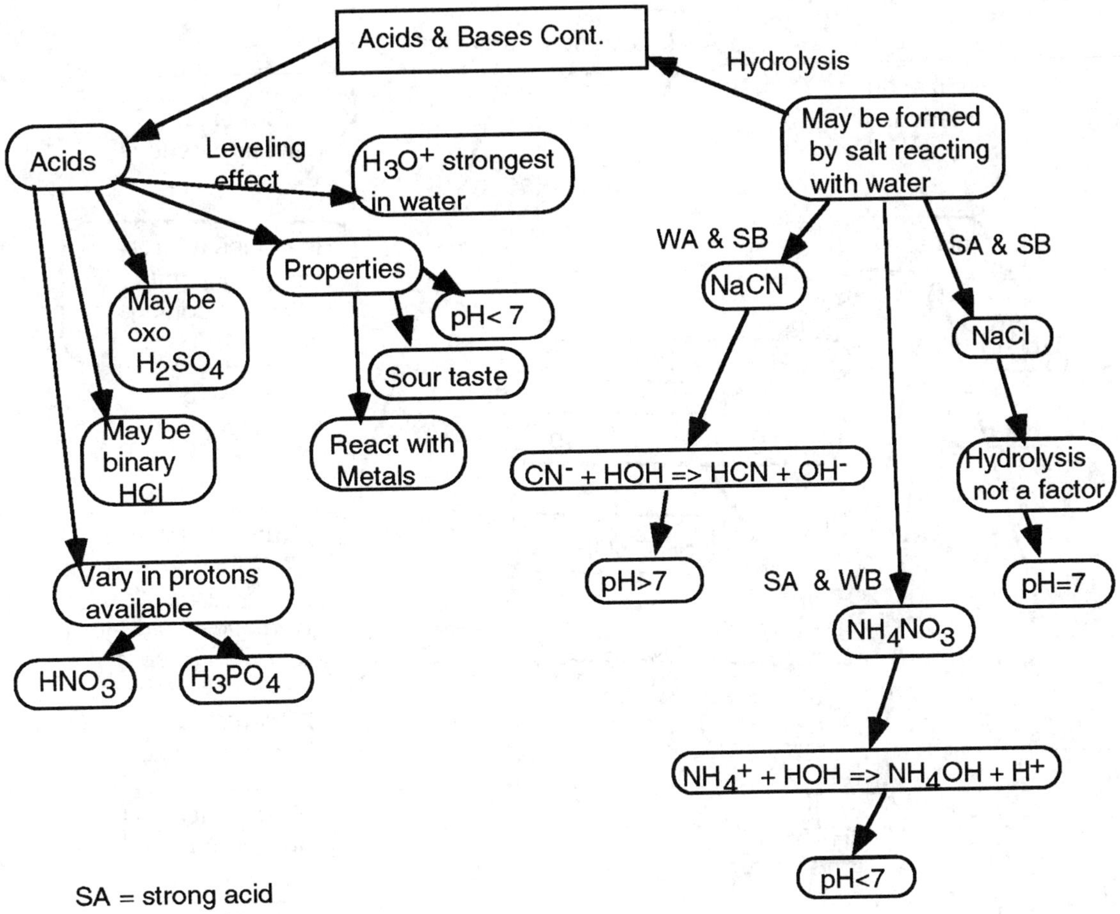

SA = strong acid
SB = strong base
WA = weak acid
WB = weak base

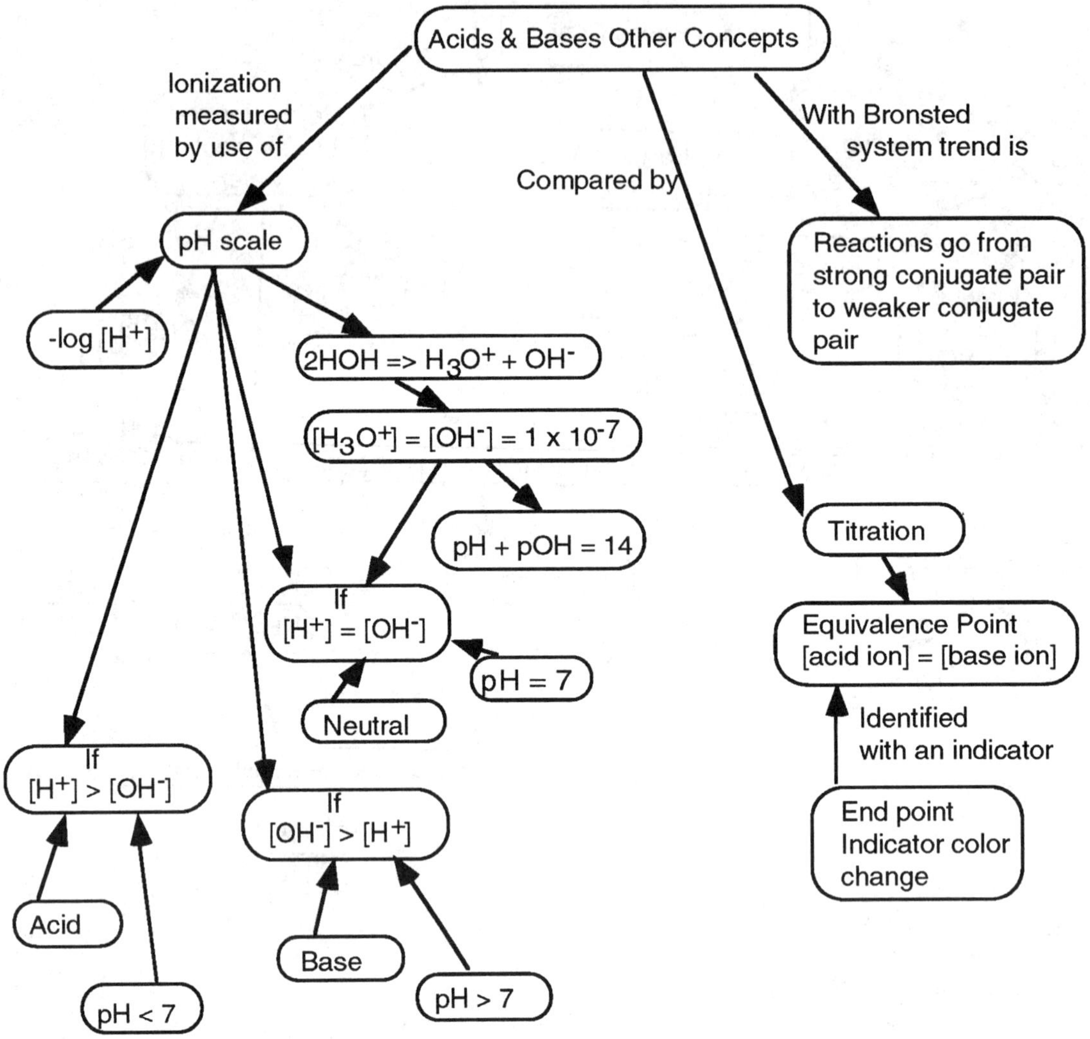
Acids & Bases Other Concepts
Ionization measured by use of
pH scale
-log [H+]
2HOH => H3O+ + OH-
[H3O+] = [OH-] = 1 x 10-7
pH + pOH = 14
If [H+] = [OH-]
pH = 7
Neutral
If [H+] > [OH-]
Acid
pH < 7
If [OH-] > [H+]
Base
pH > 7
Compared by
Titration
Equivalence Point [acid ion] = [base ion]
Identified with an indicator
End point Indicator color change
With Bronsted system trend is
Reactions go from strong conjugate pair to weaker conjugate pair

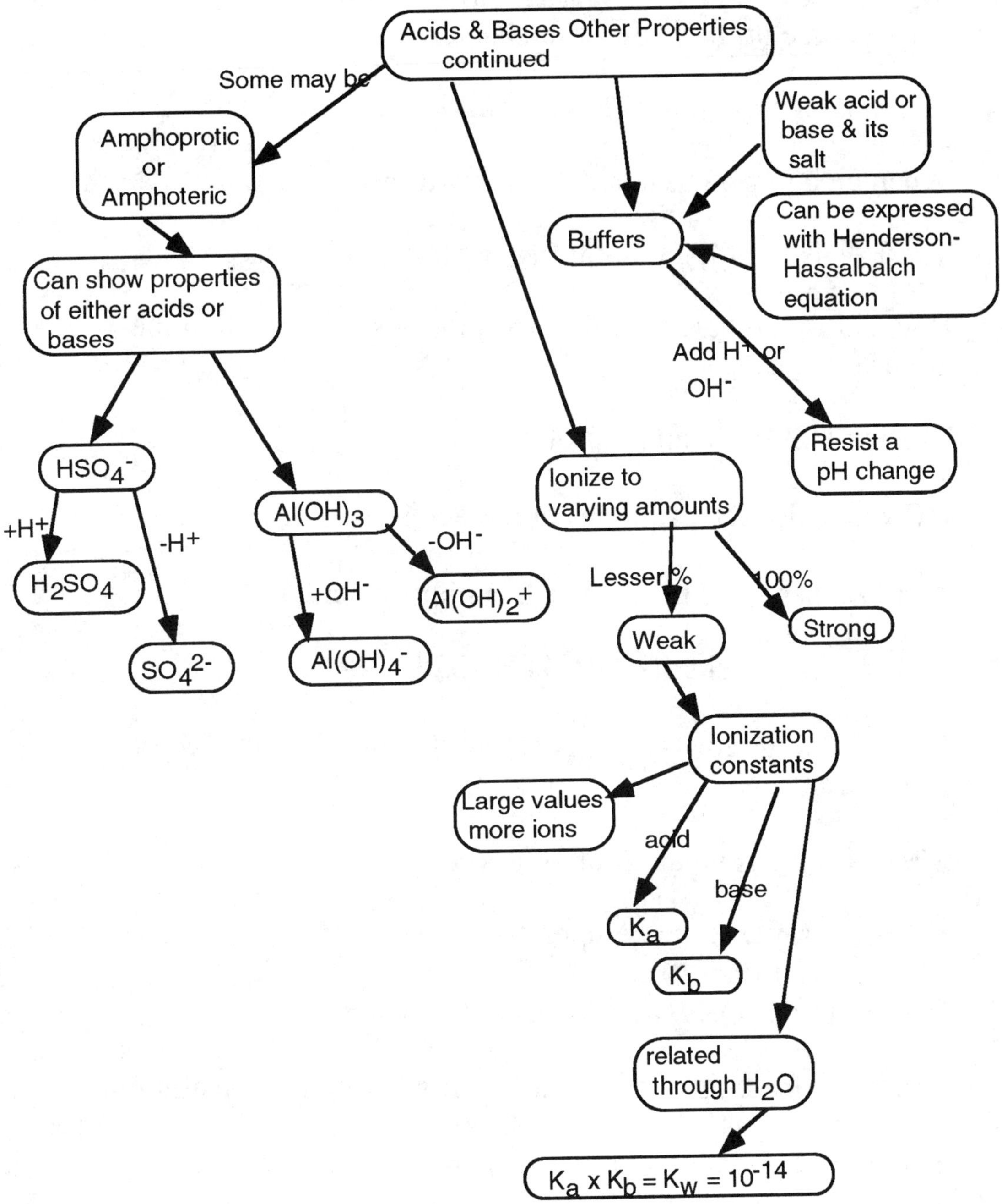
Acids & Bases Other Properties continued
Some may be
Amphoprotic or Amphoteric
Can show properties of either acids or bases
HSO_4^-
+H+
-H+
H_2SO_4
SO_4^{2-}
$Al(OH)_3$
-OH-
+OH-
$Al(OH)_2^+$
$Al(OH)_4^-$
Buffers
Weak acid or base & its salt
Can be expressed with Henderson-Hassalbalch equation
Add H+ or OH-
Resist a pH change
Ionize to varying amounts
Lesser %
100%
Weak
Strong
Ionization constants
Large values more ions
acid
base
K_a
K_b
related through H_2O
$K_a \times K_b = K_w = 10^{-14}$

Acids and Bases True/False Statements

Answers are at the end of the chapter.

1. The most inclusive acid/base definition is that of Arrhenius.

2. When an acid and base react a salt will be produced.

3. The H^{+1} is not the favored form of this ion in water.

4. Both acids and bases are defined in terms of protons in the Bronsted concept.

5. The ion HSO_4^{-1} is amphiprotic.

6. HCN and H_2S are examples of oxoacids.

7. A strong base will have a weak conjugate acid.

8. The conjugate acid of SO_4^{-2} is HSO_4^{-1}.

9. If the K_1 value for H_3PO_4 is about 10^{-4} the K_2 value would be about 10^{-9}.

10. The pH scale is value limited to a range of 0 to 14.

11. K_w is the same at any temperature.

12. If a solutions pOH was 8 its pH value would be 6.

13. The pH value tells us how much acid is present in a solution.

14. Strong acids or bases do not have a K_a or K_b value.

15. A base with a pH of 10 is stronger than a base of pH 8.

16. If an acid has a K_a of 1.4×10^{-6}, its conjugate base has a K_b of 7.0×10^{-9}.

17. Fe^{+3} could be a Lewis acid.

18. HCl is a stronger acid in water than HNO_3.

19. If a strong acid and strong base react the salt formed will undergo extensive hydrolysis in water.

20. When boric acid, $K_a = 7.3 \times 10^{-10}$, and sodium bicarbonate, $K_b = 2.4 \times 10^{-8}$, react the solution will be acidic.

21. An insoluble salt that contains a weak conjugate base will likely not dissolve in an acid solution.

22. HClO and NaClO would make a buffer solution.

23. An acid with a pH of 3 is 100 times stronger than an acid of pH 5.

24. According to the Henderson-Hasselbalch equation if the concentration of the salt in a buffer solution is increased the solution pH will decrease.

25. A strong acid has some ability to buffer when a base is added.

26. An end point value must match the equivalent point value.

27. The molecule, $:PH_3$ would make a good Lewis base.

Testing Your Understanding of Concepts in Chapter 16.
Answers are at the end of the chapter.

1. What are the two major contributors to acid rain? Since these are not acids themselves how do they make rain acidic?

2. Why doesn't H^+ exist in water?

3. Make a table to compare the three acid/base concepts discussed in this chapter.

4. Why is HSO_4^{-1} a base in the Bronsted system but not in the Arrhenius system of acids and bases?

5. What identifies an acid proton in organic molecules?

6. HCl is a strong acid. What is true about the strength of the conjugate base?

7. If K_1 for H_3AsO_4 is about 10^{-5} what would be reasonable estimates for K_2 and K_3?

8. Why is pure water neutral?

9. The K_a value for formic acid is 1.8×10^{-4} and for hypochlorous acid it is 3.5 x 10-8. Which acid is stronger? How much stronger? Which conjugate base is stronger?

10. When a chemist says an acid is strong what does that imply?

11. Why doesn't the pH value often tell how much acid is present in a solution?

12. When 50.0 mL of a strong base is titrated with a strong acid what happens to the solution pH during the first 5-10 mL of adding titrant?

13. Why is H_3O^+ the strongest acid in water?

14. Which of the following salts would not have a neutral pH in water? KCl, $Mg(ClO)_2$, $NaHCO_3$, Li_2SO_3

15. What material, when added to Na_2HPO_4, would make a buffer solution?

16. The pKa values for two acids are 4.74 and 7.46. Which is the stronger acid?

17. What is the difference between end point and equivalence point?

Testing You on Some Analytical Concepts.
Answers are at the end of the chapter.

1. Calculate the $[OH^-]$ and pH of a 0.250 M solution of $Ba(OH)_2$.

2. Milk has a pH near 6.50. What is the $[H_3O^+]$?

3. Benzoic acid, C_6H_5COOH, has a K_a value of 6.46×10^{-5}. What is the pH of a 0.250 M solution?

4. If $HF_{(aq)}$ has a K_a of 6.76×10^{-4}, what is the K_b value?

5. What is the pH of a 0.050 M CH_3COONa solution. At this temperature K_b for the acetate ion is 5.75×10^{-10}.

6. What pH would a buffer of 0.25 M formic acid and 0.15M KHCOO have? $K_a = 1.78 \times 10^{-4}$.

Crossword - Acids and Bases

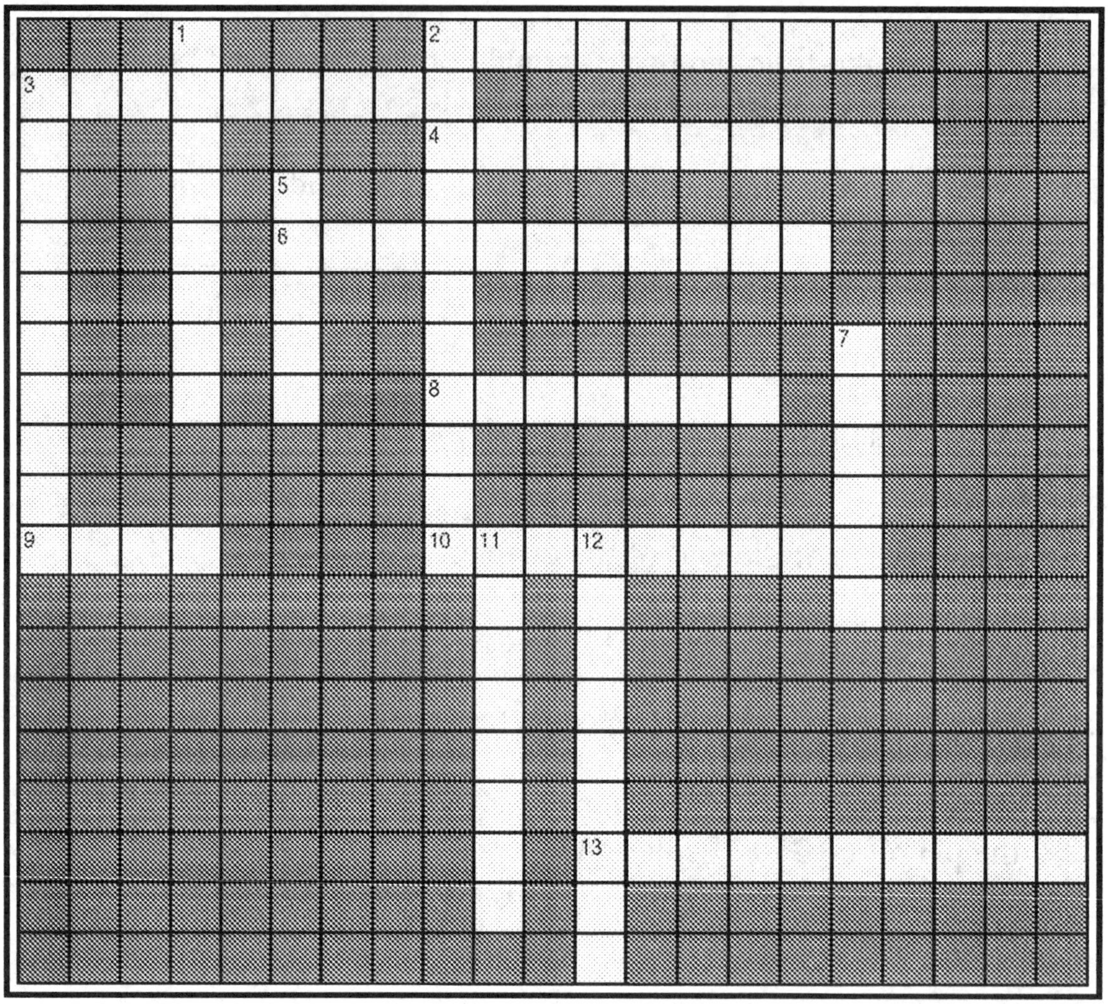

Crossword - Acids and Bases Clues

ACROSS

2 A blood condition that results when the blood pH is above 7.45.
3 This is the ion we used to represent a proton in water.
4 Some acids are able to provide more than one proton. They are said to be ____________.
6 At this point the amount of acid and base in a system have just been balanced.
8 An acid that contains oxygen is called an __________.
9 One of the products or any acid/base reaction.
10 The name given to the cage water molecules build around the hydrogen ion is __________.
13 Pure water does very little of this.

DOWN

1 Acid/base concept that defines each in terms of protons.
2 This type of material can either gain or lose protons.
3 We have this when a ion reacts with water.
5 This acid/base definition is the most comprehensive and involves electrons.
7 Solutions that resist a change in pH are called this.
11 The hydronium ion is the strongest acid possible in water. This is the _________ effect.
12 Acid or base solutions are often analyzed by this method.

Ties to Previous Chapters

We first encountered acids, bases, and salts in chapter 4. Some reactions of these materials were used in chapter 5. The principles of equilibrium introduced in chapter 8 are used extensively in this chapter. Organic acids were identified in chapter 13. Several sections of chapter 15 are background for topics in this chapter. Acids and bases are very important materials to a chemist and are related in many ways to other parts of chemistry.

Answers to True/False Statements:

1. False. This is the least inclusive. It is good only in water. The most inclusive used in introductory chemistry is the Lewis concept.

2. True. Salt and water are the result of acid/base reactions.

3. True. The favored form would be this ion attached to one or more waters.

4. True. Acids are proton donors and bases are proton acceptors.

5. True. It can lose a proton to form SO_4^{-2} or gain a proton to form H_2SO_4.

6. False. Oxoacids contain oxygen.

7. True. If the conjugate base is weak it will readily give up the proton. Strong acids are those that readily produce protons.

8. True. See number 5 above.

9. True. The rule of thumb is about a factor of 10^5 between each value of K in polyprotic acids or bases.

10. False. The pH scale is open ended. The practical limits are about 0 and 14 because of solubility factors.

11. False. K values are all temperature sensitive.

12. True. The pH + pOH = 14.00

13. False. The pH value tells us only how much hydrogen ion is present. A weak acid would have mostly non-ionized hydrogen which could produce acid conditions.

14. False. The K values are so large they have little meaning and are not used.

15. True. It has about 100 times more OH ion. The further from pH 7 the stronger.

16. True. $K_a \cdot K_b = 1.0 \times 10^{-14}$.

17. True. A Lewis acid can be any material that is capable of accepting an electron pair. This positive ion will readily do that.

18. False. In water all strong acids will produce the H_3O^+ ion. This is the Leveling Effect.

19. False. Neither of these ions is strong enough to react with water.

20. False. The K_b value is greater so it will predominate. Also if the two ions are compared in a Bronsted acid/base table the slope is positive rather than negative.

21. True. The weak base not tend to interact with the hydrogen ion of the acid.

22. True. Any weak acid and the salt of that acid will produce a buffer system. Not all are effective buffers however.

23. True. The pH scale is a log scale. Two pH value changes is 10^2 or 100.

24. False. The reverse would be true.

25. True. Strong acids will produce so many hydrogen ions that when some are used up there still are many present to keep the pH value constant. This is easily seen in a titration curve. The pH stays nearly constant for a good portion of the titration.

26. False. The closer together they are the better the result will be. They sometimes can not be made to match however.

27. True. The lone pair of electrons can be donated so this is a base.

Answers to Understanding of Concepts Questions:

1. SO_2 and NO_x, oxides of sulfur and nitrogen are the two major acid rain precursors. These materials are known as non-metal oxides or acid anhydrides. They will react with the rain to produce the acids. Refer back to chapter 5 for a review of metal and non-metal oxides.

2. The charge density of the bare H^{+1} is so high and the water molecules have a polar end which is negative so several water molecules will hydrate the ion.

3.

	Arrhenius	Bronsted	Lewis
acid	proton donor	proton donor	electron pair acceptor
base	hydroxide ion donor	proton acceptor	electron pair donor

4. HSO_4^{-1} cannot be an Arrhenius base since it does not contain any OH^{-1} ions to donate. It can however accept a proton and thus is a Bronsted base.

5. The organic carboxylic acid functional group is COOH. Only the hydrogen on the end of this group will be an acid proton.

6. According to the Bronsted concept an acid is strong because its conjugate base is weak. The weaker the conjugate base the stronger the acid.

7. A general rule used is that the change in K values is about 10^{-5} for each successive proton removed. K_2 then would be 10^{-10} and K_3 about 10^{-15}.

8. Pure water has no excess of either hydrogen ion or hydroxide ion. Water does autoionize a little but the concentrations of both ions are the same.

9. When ionization constants are less than one the closer to 1 the K value is the stronger the acid will be. Formic acid is stronger by about 10,000 times. The conjugate base of formic acid must be the weaker base.

10. A strong acid is one that provides many hydrogen ions to the system.

11. The pH value is only a measure of the free hydrogen ion present in the system. It does not measure the total amount of hydrogen that can eventually be supplied.

12. When a strong base is titrated with a strong acid there will be very little change in pH during the addition of the first few milliliters of base. Strong acids or strong bases have a self

buffering effect. In fact, very little change in pH would be noticed until well over 90% of the base was used up.

13. When any material that can donate protons is placed in water the conjugate acid of H_2O is H_3O^+. This is the strongest acid possible in water.

14. If the salt is made from the conjugate acid and base of strong acids and bases they will not hydrolyze. Of this group only the KCl is neutral since it is the salt of KOH and HCl.

15. A buffer is a combination of a weak acid and a salt of that acid. If Na_2HPO_4 is the salt then NaH_2PO_4 would be an acid.

16. The smaller the pK value the larger the K value. The pK value of 4.74 would mean a stronger acid.

17. The end point is where the indicator changes color. The equivalence point is where the amount of titrant added is just equal to the amount of material being titrated. We try to make these two values as close as possible when doing a titration.

Answers to Testing the Analytical Concepts

1. $Ba(OH)_2$ -----> Ba^{+2} + 2 OH^{-1} This is a strong base.
0.250 M 0.500 M

$pOH = -\log[OH^{1-}] = -\log (0.500) = 0.30$
$pH = 14.00 - pOH = 14.00 - 0.30 = 13.70$

2. $pH = -\log [H_3O^+]$

$[H_3O^+] = 10^{-pH} = 10^{-6.50}$
$[H_3O^+] = 3.0 \times 10^{-7}$

3. $$C_6H_5COOH + H_2O \Longleftrightarrow C_6H_5COO^{-1} + H_3O^+$$

	C_6H_5COOH	$C_6H_5COO^{-1}$	H_3O^+
init conc	0.250	0	≈ 0
conc change	-x	x	x
equil conc	0.250-x	x	x

$$K_a = \frac{[\ C_6H_5COOH\ ]\ [\ H_2O\ ]}{[\ C_6H_5COO^{-1}\ ]} = \frac{x \cdot x}{0.250 - x} = 6.46 \times 10^{-5}$$

assume x is negligible

$$x^2 = 6.46 \times 10^{-5} \times 0.250 = 1.62 \times 10^{-5}$$
$$x = 4.02 \times 10^{-3} = [H_3O^+]$$

$$pH = -\log[H_3O^+] = -\log(\ 4.02 \times 10^{-3}\ ) = 2.40$$

4. The equation below has been derived in the text.

$$K_a{\cdot}K_b = K_w = 1.00 \times 10^{-14}$$

$$K_b = \frac{K_w}{K_a} = \frac{1.00 \times 10^{-14}}{6.76 \times 10^{-4}} = 1.48 \times 10^{-11}$$

5. This salt will ionize completely:

$$CH_3COONa \longrightarrow CH_3COO^{-1} + Na^{+1}$$

The sodium ion will not react with water but the acetate ion will.

$$CH_3COO^{-1} + H_2O \Longleftrightarrow CH_3COOH + OH^{-1}$$

	CH_3COO^{-1}	CH_3COOH	OH^{-1}
Init conc	0.050	0	≈ 0
Conc change	-x	x	x
Equil conc	0.050 -x	x	x

$$K_b = \frac{[CH_3COOH]\ [OH^{-1}]}{[CH_3COO^{-1}]} = \frac{(x)(x)}{0.050 - x} = 5.75 \times 10^{-10}$$

$x^2 = 5.75 \times 10^{-10}\ (0.050) = 2.88 \times 10^{-11}$
$x = [OH^{1-}] = 5.36 \times 10^{-6}$

$pOH = -\log[OH^{-1}] = -\log\ (5.36 \times 10^{-6}) = 5.27$

$pH + pOH = 14.00$
$pH = 14.00 - 5.27 = 8.73$

6. This problem is conveniently solved with the Henderson-Hasselbalch equation.

$$pH = pK_a + \log \frac{[\text{anion}]}{[\text{acid}]}$$

$pK_a = -\log K_a = -\log 1.78 \times 10^{-4} = 3.75$

$$pH = 3.75 + \log \frac{[0.15]}{[0.25]} \quad 3.75 - 0.22 = 3.53$$

Answer to Chapter 16 Crossword

			1 b					2 a	l	k	a	l	o	s	i	s				
3 h	y	d	r	o	n	i	u	m												
y			o					4 p	o	l	y	p	r	o	t	i	c			
d			n		5 L			h												
r			s		6 e	q	u	i	v	a	l	e	n	c	e					
o			t		w			p												
l			e		i			r								7 b				
y			d		s			8 o	x	o	a	c	i	d		u				
s								t								f				
i								i								f				
9 s	a	l	t					10 c	11 l	a	12 t	h	r	a	t	e				
									e		i					r				
									v		t									
									e		r									
									l		a									
									i		t									
									n		13 i	o	n	i	z	a	t	i	o	n
									g		o									
											n									

Some Supplementary Readings Relating to This Chapter:

1. Salt, Chem Matters , December 1992, page 4.

2. Water Cluster Cradles H_3O^+ Ion In Stable Cagelike Structure, Chemical & Engineering News , April 6, 1991, page 47.

3. Precipitated Calcium Carbonate Finds Growing Niche in Paper Market, Chemical & Engineering News , April 29, 1991, page 9.

Chapter 17 Electrochemistry

Electrochemistry is the application of oxidation-reduction reactions. This may involve a battery to produce electric current from a chemical reaction or the use of electrical current to bring about a chemical change. Since energy is involved in the reaction either way, we can apply the rules of thermodynamics to the system.

Focus Your Study Efforts

Some ideas in this chapter are a review since they were introduced in chapter 5. This chapter takes these ideas into much more depth.

Items and processes in redox:

Objectives:

1. You should be able to distinguish between the processes of oxidation and reduction and oxidizing and reducing agents.
2. You should be able to identify when the process of electrolysis is occurring.
3. You should be able to assign oxidation numbers to various elements in a compound given the compound formula.

Every compound is electrically neutral. Various elements in the compound may have gained or lost electrons but there is no net gain or loss of electrons by the system. Knowing a few rules, oxidation states for each element in a given compound can be assigned.

Determine the oxidation numbers for the elements in Na_2HAsO_3.

Some of the elements are easy to determine. The major problem will be with As. Sodium is a 1A metal so it will be +1. Hydrogen is +1 except if it is bonded only to a metal. The oxygen is -2 except when it is a peroxide or bonded to fluorine.

+2 +1 -6
Na_2HAsO_3

This leaves a difference of -3, so the As has a charge of +3 which will result in a net charge on the molecule of 0.

The oxidation reduction process involves electron transfer. This also means ions may be formed. It is, however, also considered redox when electron control is transferred so that one atom gains more control than it would normally have.

Oxidation and reduction are processes, in contrast to oxidizing and reducing agents, which are things.

When a material gains an electron(s) we have the process of reduction. The item that gained the electron(s) is an oxidizing agent. Likewise, when a material loses an electron(s) we have an oxidation process and the item that lost the electrons is a reducing agent.

When we force the electron transfer to occur we have a process known as electrolysis.

Redox reactions involve two separate processes:

Objectives:

1. You should be able to write equations for oxidation and reduction half-reactions.
2. You should be able to use half-reactions to balance redox reactions.

The redox reaction is divided into half reactions. The oxidation half-reaction that involves a reducing agent (often a metal), for example,

$Mg \text{ ---> } Mg^{2+} + 2e^{1-}$. The reduction half-reaction involves the oxidizing agent (often a non-metal). An example is $S + 2e^{1-} \text{ ---> } S^{2-}$.

When half-reactions are combined we end up with a net reaction. No electrons will be shown in the net reaction since they are only transferred and thus conserved.

It is much less intuitive to balance redox reactions (unless they are a simple combination or replacement) than most other kinds. As a result of this, a series of steps are usually followed as we work our way through the reaction. These steps are outlined on pages 754 and 755 of the text. <u>You will find it best not to try and shortcut the procedure</u>. Also, remember you must conserve charge in the reaction. Always leave the balancing of hydrogen, oxygen, and water to the last.

Balance the redox reaction between H_2O_2 and $Cr_2O_7^{-2}$ knowing that $H_2O_{2(aq)} + Cr_2O_7^{-2}{}_{(aq)} \text{ ----> } Cr^{3+}{}_{(aq)} + O_{2(g)}$.

First assign oxidation numbers and determine what is oxidized and what is reduced.

$$\overset{-1}{H_2O_{2(aq)}} + \overset{+6}{Cr_2O_7^{-2}{}_{(aq)}} \text{ ----> } \overset{+3}{Cr^{+3}{}_{(aq)}} + \overset{0}{O_{2(g)}}$$

We see that Cr goes from +6 to +3 so it gains 3 electrons. The oxygen goes from -1 to 0 so it gives up 1 electron each atom.

Identify the half-reactions:

$Cr_2O_7^{-2}{}_{(aq)} \text{ ----> } Cr^{+3}{}_{(aq)}$ reduction half-rxn

$H_2O_{2(aq)} \text{ ----> } O_2$ oxidation half-rxn

Balance for atoms:

$Cr_2O_7^{-2}{}_{(aq)} \text{ ----> } Cr^{+3}{}_{(aq)} + 7\text{oxygen}$

$Cr_2O_7^{-2}{}_{(aq)} \text{ ----> } Cr^{+3}{}_{(aq)} + 7\ H_2O_{(l)}$

We balance the oxygen by converting it to water. We will take care of the hydrogen balance latter.

$H_2O_{2(aq)}$ ----> $O_2 + 2H^{+1}$

Balance for charges:

$Cr_2O_7^{-2}{}_{(aq)} + 14H^{+1} + 6e^{-1}$ ----> $Cr^{+3}{}_{(aq)} + 7\ H_2O_{(l)}$

$H_2O_{2(aq)}$ ----> $O_2 + 2e^{-1}$

Balance for electrons between half-reactions:

1x $Cr_2O_7^{-2}{}_{(aq)} + 14H^{+1} + 6e^{-1}$ ----> $Cr^{+3}{}_{(aq)} + 7\ H_2O_{(l)}$

3x $3H_2O_{2(aq)}$ ----> $3O_2 + 6e^{-1}$

Add the half-reactions:

$Cr_2O_7^{-2}{}_{(aq)} + 14H^{+1} + 6e^{-1} + 3H_2O_2$ ----> $2Cr^{+3}{}_{(aq)}$ + $7\ H_2O_{(l)} + 3O_{2(g)} + 6H^{+1}{}_{(aq)} + 6e^{-1}$

Cancel like items on each side:

$Cr_2O_7^{-2}{}_{(aq)} + 8H^{+1} + 3H_2O_2$ ----> $2Cr^{+3}{}_{(aq)} + 7\ H_2O_{(l)} + 3O_{2(g)}$

Check for charge and water balance.

Several factors affect the outcome of redox reactions:

Objectives:

1. You should be able to describe the parts of an electrochemical cell and explain the function of each part.
2. You should be able to define electrodes in terms of the redox reaction that occurs at them.
3. You should be able to explain the relationship between the electrode reactions and the potential developed by the cell.

The pH of a solution may have a significant effect on the products. The same oxidant or reductant may give very different results in an alkaline medium than it does in an acidic solution.

A cell consists of two electrodes immersed in electrolytes that are separated but connected via a salt bridge. These electrodes are then connected via an external conductor, such as a wire, which allows us to use the electrons that are transferred. We are able to use the electrons by directing them through a light bulb, motor, or some other devise.Because we can use the electrons they are able to do work for us.

It is best to forget the idea of which electrode is + or - . <u>It is better to define the electrode in terms of what process occurs at it</u>. Oxidation will always occur at the anode and reduction will occur at the cathode.

Electrodes are usually placed in an electrolyte containing their ion. For example, a zinc metal electrode would be placed in a solution of zinc nitrate. However, some materials do not have a solid form so a substitute electrode is used. If you are trying to reduce bromine, an electrode made of graphite may be used because there is no solid bromine available.

The voltage developed by a cell is determined by the electrodes or the reaction that occurs at them. The better they are as oxidizing and reducing agents the greater the potential. The size of the electrode plays no part in the cell voltage.

Work is done by electron flow:

Objectives:

1. You should be able to explain how time and electrode size affect current flow and/or the amount of electrical work done by the cell.

The amount of electrical work done by the cell is related to the flow of electrons. Those factors that make for a greater number of electrons transferred in a unit of time, such as electrode size, will increase the current flow.

The terms coulomb and ampere are related to measuring the amount of work. The terms EMF and volt are related to the cells potential energy.

> If we wanted to purify some copper we could electroplate the metal from an impure electrode to a pure one. What amount of copper would be deposited if we ran the reaction at 1.62 A for 90.0 minutes?
>
> We need to make the pure electrode the cathode. The reaction $Cu^{+2} + 2e^{-1} \text{----}> Cu^{0}$ shows a 2 electron change.
>
> $$\text{g Cu} = 90.0 \text{ min} \times \frac{60 \text{ sec}}{\text{min}} \times \frac{1.62 \text{ C}}{1 \text{ sec}} \times \frac{1 \text{ mole e}}{96500 \text{ C}} \times \frac{1 \text{ mol Cu}}{2 \text{ mol e}}$$
>
> $$\times \frac{63.55 \text{ g Cu}}{1 \text{ mole Cu}} = 2.88\text{g Cu}$$

Standard cell conditions are important for comparison:

Objectives:

1. You should be able to define standard conditions for electrochemical cells and the importance of the hydrogen electrode.
2. You should be able to use a table of reduction potentials to predict if a reaction is product-favored or reactant-favored.
3. You should be able to explain the relationship between E° and ΔG°.

A series of standards have been agreed upon so that cells can be compared. All potentials are measured against a standard hydrogen electrode whose potential is set as 0.00 volts when at a standard set of conditions. Standard conditions include: 1M solutions, 1 atmosphere of pressure for gases, and a temperature of 298K.

A table of potentials known as reduction potentials is available. All reactions in this table are written as reduction reactions and all are referenced to the standard hydrogen electrode (SHE). See table 17.1 in your textbook.

The convenient aspect of this table is that it can be used to predict cell potentials just like a Bronsted acid/base table can be used to predict the probability of the acid and base reacting. If a set of electrodes is such that a line drawn from left to right has a downward slope the reaction is product favored. The material on the right acts as the oxidizing agent and the one on the left is the reducing agent.

Note that one of the reactions has to be reversed when setting up the overall equation. When you do this the listed potential for the half-reaction also must have the sign changed. As an example look at a reaction between bromine and nickel:

$Br_{2(l)} + 2\ e^{-1} \text{ ---> } 2\ Br^{-1}_{(aq)}$ $\qquad E° = +1.36$ v

$Ni^{+2}_{(aq)} + 2e^{-1} \text{ ----> } Ni_{(s)}$ $\qquad E° = -0.25$ v

reverse the nickel reaction

$Ni_{(s)} \text{ -----> } Ni^{+2}_{(aq)} + 2e^{-1}$ $\qquad E° = +0.25$ v

Potentials are added algebraically

Net rxn $\quad Ni_{(s)} + Br_{2(l)} \text{ ------> } Ni^{+2}_{(aq)} + 2\ Br^{-1}_{(aq)}$ $\quad E° = +1.61$ v

Determine the feasibility of a reaction between $Cd_{(s)}$ and $Cu^{+2}_{(aq)}$. Calculate the reaction potential.

Checking table 17.1 we find that if a line is drawn between Cu^{2+} and $Cd_{(s)}$ is has a downward slope. From this we would predict a product favored reaction.

$Cu^{+2}_{(aq)} + 2e^{-1}$ -------> $Cu_{(s)}$ +0.337 v
$Cd^{+2}_{(aq)} + 2e^{-1}$ -------> $Cd_{(s)}$ -0.40 v

Since $Cd_{(s)}$ is a reactant we need to reverse the second reaction.

$Cd_{(s)}$ --------> $Cd^{+2}_{(aq)} + 2e^{-1}$ +0.40 v

Net $Cu^{+2}_{(aq)} + Cd_{(s)}$ -------> $Cd^{+2}_{(aq)} + Cu_{(s)}$ +0.74 v

When net potential values are (+) the reaction is product favored. Reaction potentials can be related to Gibbs free energy (see chapter 7). Both potentials and free energy values will indicate if a reaction is product favored or not. However, do not let the signs confuse you. A (+) E° value and a (-) ΔG° value indicate product favored reactions.

Mathematically potential and free energy are related through the Faraday constant, F. $\Delta G° = -nFE°$.

What is the potential and free energy change for the reaction of NaOCl and NaI ?

$NaOCl + H_2O + 2\ NaI$ -----> $NaCl + 2\ NaOH + I_2$

From the half reaction table we get the following:

$OCl^{-1} + H_2O + 2e^{-1}$ -----> $Cl^{-1} + 2\ OH^{-1}$ +0.890 v
$2\ I^{-1}$ ----> $I_2 + 2e^{-1}$ -0.535 v
This is a two electron change.

$E° = 0.89\ v - 0.535\ v = +0.35\ v$

$\Delta G = -nFE°$

$$\Delta G = -(2e^{-1}) \times \frac{9.65 \times 10^4 \text{ J}}{\text{V mol e}^{-1}} \times 0.35 \text{ v} = -67550\text{J} = -68 \text{ kJ}$$

Batteries and cells are not the same:

Objectives:

1. You should be able to explain the difference between cells such as the primary, secondary, and fuel cell.
2. You should be able to explain the chemical reactions that occur when a secondary cell is being used and recharged.

Technically a battery is a collection of cells. You will, however, often find a cell being called a battery.

A primary cell (battery) is one that is not rechargeable. They are made of materials that eventually are consumed. There are several kinds of these batteries made from different materials and providing different voltages.

The secondary battery is one that is capable of being recharged. The charging and discharging cycles are the reverse of each other. With these batteries the anode and cathode are reversed during the use and recharge cycles. Again, there are several different kinds of these batteries.

Fuel cells are becoming more important in overall use. These cells are different, in that, a constant supply of reactant materials is provided to them from some external source.

Redox reactions have many uses:

Objectives:

1. You should be able to explain the process of corrosion and some ways in which the corrosion of iron can be controlled.
2. You should be able to explain the processes involved in electrolysis.

Oxidation-reduction is not always a useful reaction. Corrosion is an extremely common redox reaction that costs enormous amounts of money and material each year.

Control of corrosion is addressed by two common methods. The anodic inhibition is accomplished by placing a protective surface on the material needing protection. Cathodic protection is really a substitute corrosion in which a sacrificial metal is applied to the metal you wish to preserve.

Electrical energy can be used to bring about a chemical decomposition. This process is known as electrolysis. If an electric current is passed through molten sodium chloride sodium metal and chlorine can be produced.

Metals can be plated on any number of objects. This process is electroplating. In the process of electroplating the object to be plated is made the cathode.

You need to remember that these reactions will occur only if sufficient voltage is applied to bring about the chemical reaction, and, if enough current flows to make it occur in a reasonable time.

Symbols found in Chapter 17

Page	Symbol	
760	EMF	Electromotive force
761	E°	Standard conditions potential value
766	V	Volts
768	$E°_{ox}$	Standard potential for anode reaction
768	$E°_{red}$	Standard potential for cathode reaction
768	E_{net}	Standard potential for the total reaction
768	ΔG°	Gibbs Free Energy value
769	F	Faraday constant 96500 coulomb/mol
769	C	Coulomb

Terms in Chapter 17

The following terms are listed for your convenience. They should not become a list for you to memorize. You should have a working definition for each. If you have doubt about any particular term check its definition in the glossary and/or review it in the chapter. Some of these terms are included in the crossword puzzle later in the chapter. This chapter also makes use of several terms introduced in chapters 6 and 10.

Reduction	Oxidation	Electrolysis
Oxidizing agent	Reducing agent	Half-cell
Half-reaction	Voltaic cell	Electrochemical cell
Battery	Electrode	Anode
Cathode	Salt bridge	Porous membrane
Electromotive force	Coulomb	Volt
Standard conditions	Standard EMF	Faraday constant
Standard hydrogen electrode		Primary battery
Standard reduction potential		Secondary battery
Rechargeable battery	Alkaline battery	Mercury battery

Lithium battery	Lead storage battery	Ni-cad battery
Fuel cell	Corrosion	Anodic inhibition
Cathodic protection	Galvanized	Sacrificial anode
Electroplating		

Concept Maps for Electrochemistry

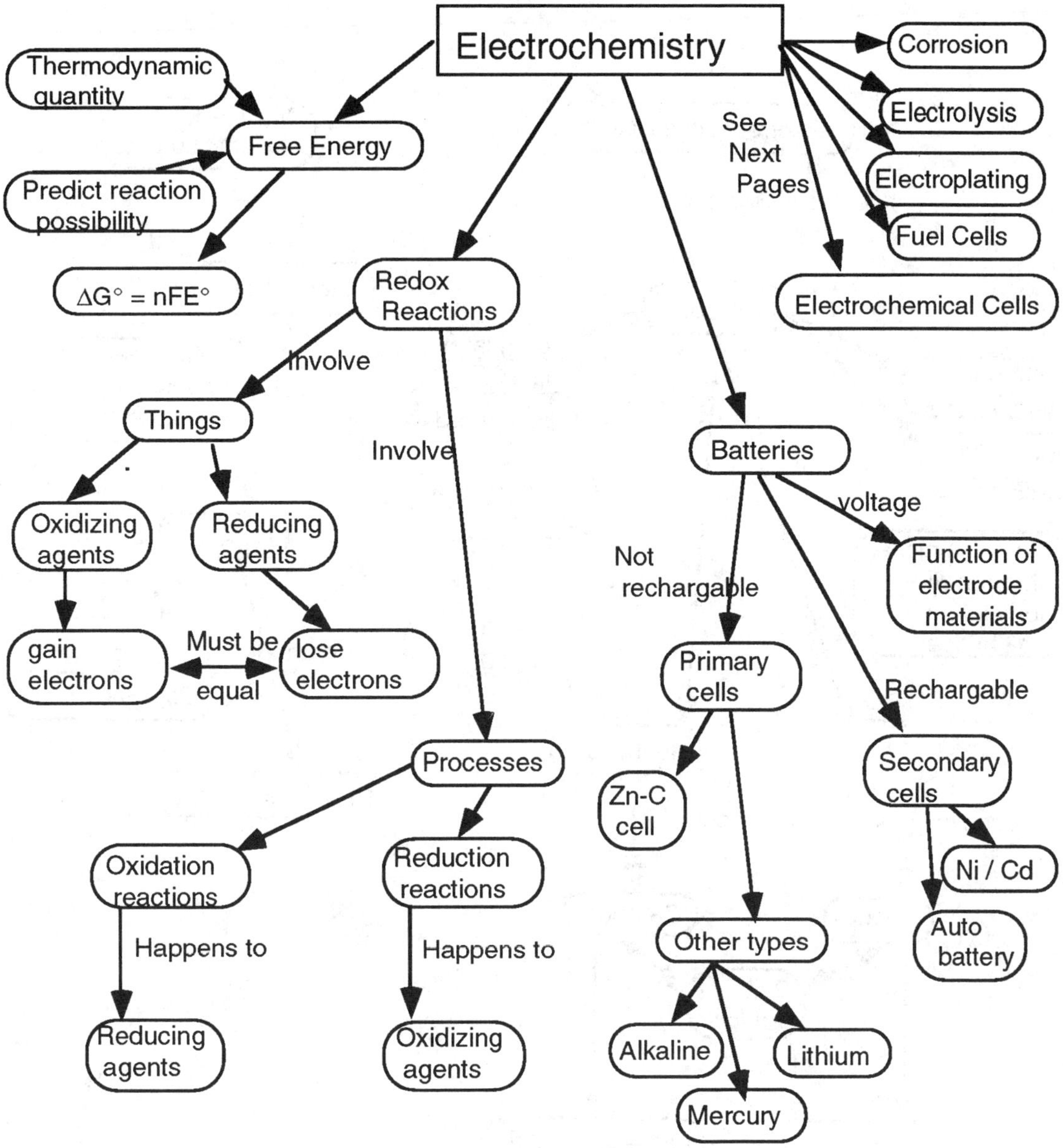

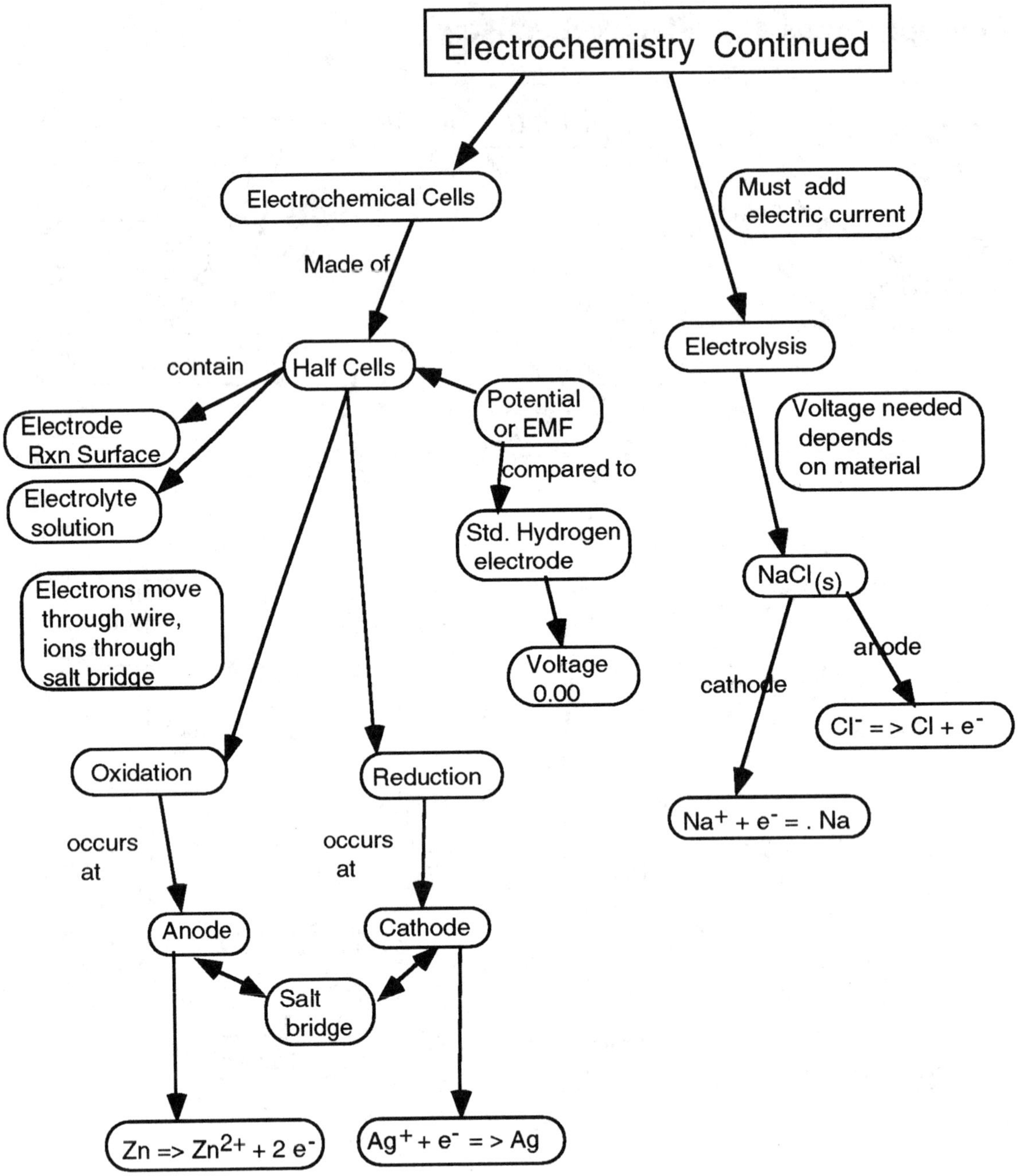
Electrochemistry Continued
Electrochemical Cells
Made of
Half Cells
contain
Electrode Rxn Surface
Electrolyte solution
Potential or EMF
compared to
Std. Hydrogen electrode
Voltage 0.00
Electrons move through wire, ions through salt bridge
Oxidation
Reduction
occurs at
occurs at
Anode
Cathode
Salt bridge
Zn => Zn2+ + 2 e-
Ag+ + e- = > Ag
Must add electric current
Electrolysis
Voltage needed depends on material
NaCl(s)
anode
cathode
Cl- = > Cl + e-
Na+ + e- = . Na

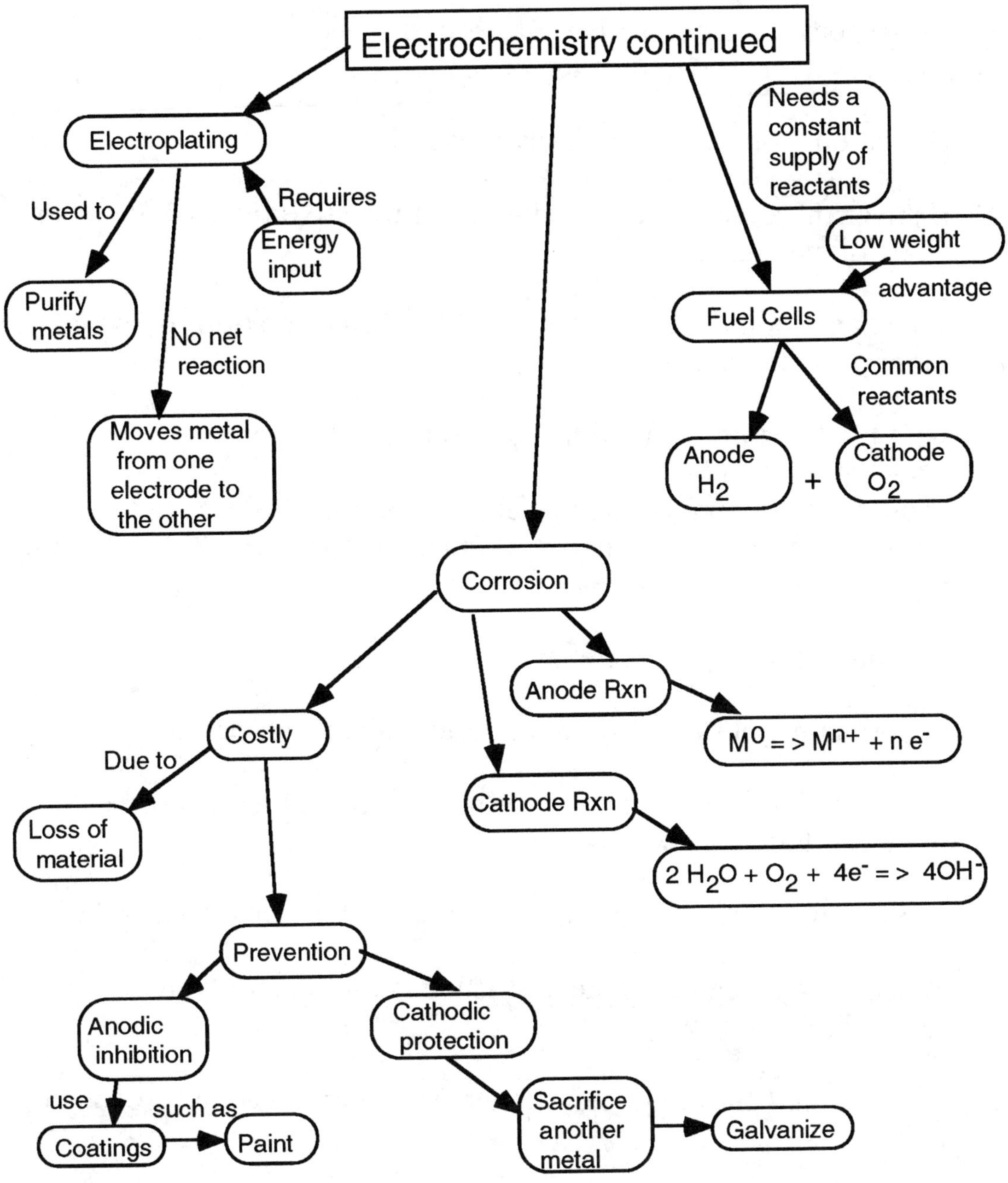
Electrochemistry continued
Electroplating
Used to
Purify metals
No net reaction
Moves metal from one electrode to the other
Requires
Energy input
Needs a constant supply of reactants
Low weight
advantage
Fuel Cells
Common reactants
Anode H_2
+
Cathode O_2
Corrosion
Anode Rxn
$M^0 => M^{n+} + n e^-$
Cathode Rxn
$2 H_2O + O_2 + 4e^- => 4OH^-$
Costly
Due to
Loss of material
Prevention
Anodic inhibition
use
Coatings
such as
Paint
Cathodic protection
Sacrifice another metal
Galvanize

Electrochemistry True/False Statements

Answers are at the end of the chapter.

1. Oxidation is more common than reduction.

2. If silver metal acts as a reducing agent the silver ion would act as an oxidizing agent.

3. When an oxidation reaction occurs the material oxidized becomes more positive in charge.

4. Fe^{2+} could be both an oxidizing and reducing agent.

5. If one half-cell in a complete cell has a atom that loses two electrons then the other half-cell must have an atom that will gain two electrons.

6. Cathodes have a positive charge.

7. A porous membrane may function just like a salt bridge.

8. Larger electrodes in a battery will allow the battery to produce a higher voltage.

9. Cell voltages are relative values compared to some standard.

10. A cell with a negative potential value is a product favored reaction.

11. The standard flashlight cell (battery) is an example of a secondary cell.

12. Both electrodes in the lead storage battery are lead metal.

13. Iron will corrode faster in a dry climate than a damp one.

14. Electric water heaters often have a magnesium block connected to the iron heating element. This magnesium block provides cathodic protection to the heating element.

15. To electroplate an object it must be made the cathode.

Testing Your Understanding of Concepts in Chapter 17.
Answers are at the end of the chapter.

1. What is the difference between oxidation and oxidizing agent?

2. What are several ways the electrolysis process can be used?

3. Using the reaction: $2\ H_2 + O_2$ ----> $2\ H_2O$ identify the oxidizing agent, the reducing agent, the oxidation reaction and the reduction reaction.

4. Why don't all redox reactions act like a battery and produce and electric current?

5. Discuss electrodes in comprehensive format.

6. Why doesn't sugar water make a good salt bridge material?

7. Why are there so many sizes of dry cell batteries since they all develop the same voltage?

8. What kind of material makes a strong oxidizing agent?

9. What are some basic differences between primary batteries, secondary batteries and fuel cells?

10. What materials are essential for the corrosion of iron to occur?

11. Would copper be a good metal to use for cathodic protection of iron? Explain.

12. How are electrolysis and electroplating similar?

Testing You on Some Analytical Concepts.
Answers are at the end of the chapter.

1. Determine the oxidation state of:
   Cr in Na_2CrO_4
   S in MgS_2O_8

2. Balance the reaction for a cell of PbO_2 in sulfuric acid with aluminum metal.

3. What is the potential of the cell in question number 2?

4. What is the Gibbs free energy change in an alkaline battery as outlined on page 771 of the text?

5. If you wanted to electrolyze $MgBr_{2(s)}$ what potential would have to be applied?

6. What mass of Ni would be formed if 0.200 Amps was passed into a solution of $NiSO_4$ for 30 minutes?

Crossword - Electrochemistry

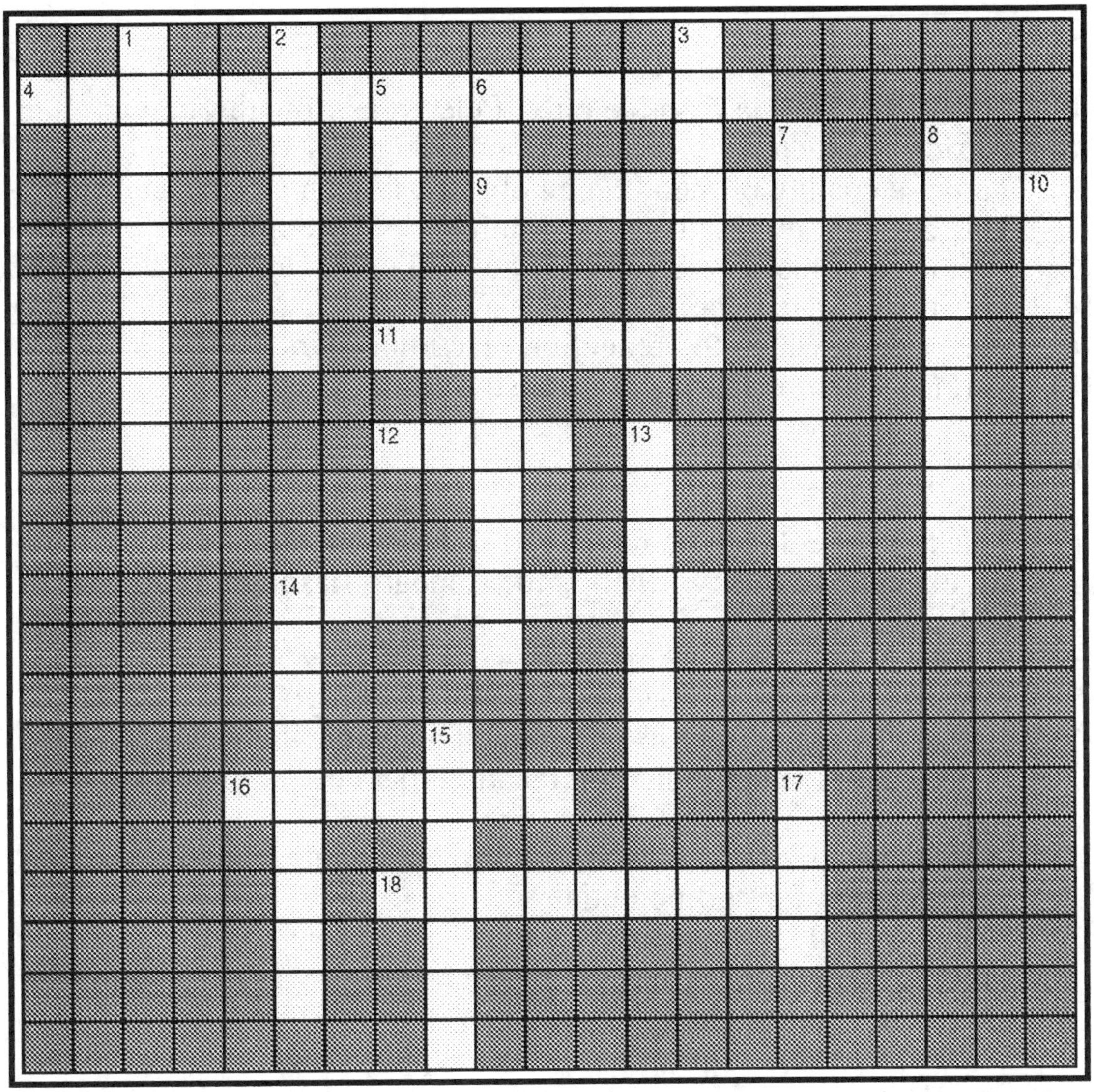

Crossword - Electrochemistry Clues

ACROSS

4 This type of cell is possible only if electrons flow through an outside conductor.
9 To put a coating of metal on an object using an electric current.
11 This constant represents a fixed number of coulombs.
12 Unit used to measure electrical potential energy differences.
14 This type of agent gains electrons.
16 Electrode at which the reduction reaction occurs.
18 Surface where electrons are transferred.

DOWN

1 This type of cell can be recharged. Like in an automobile.
2 An example of this type cell is the standard flashlight unit. Not rechargeable.
3 A collection of cells forms this.
5 Consists of electrodes, salt bridge and electrolyte.
6 Use of electric current to bring about a chemical change.
7 A natural oxidation reduction reaction that costs a great deal of money.
8 Process using zinc to reduce reaction with another metal.
10 Abbreviation for cell potential.
13 This potential is symbolized by E°.
14 Process that occurs at the cell anode.
15 This is the unit of charge.
17 This type of cell requires a constant supply of reactants. Used in space craft.

Ties to Previous Chapters

The material in this chapter is closely related to chapters 5, 7 and 10.

Oxidation reduction is extremely common in nature and often is present even when we are looking at a reaction from another aspect. For example we classify some reactions as combination or decomposition yet they more often than not are also redox reactions.

Many organic reactions are oxidation-reduction reactions even though oxygen may not be involved in the reaction. We often think of oxygen as the only oxidizing agent but we need to remember that any number of materials can be oxidizing agents. Any material that will gain an electron is a potential oxidizing agent.

Answers to True/False Statements:

1. False. Oxidation and reduction occur together. The number of electrons gained must equal the number lost.

2. True. The silver ion would undergo the reverse reaction so it would gain an electron. We define oxidizing agents as those materials that gain electrons.

3. True. For example Ag has a charge of 0 the product as a result of its reaction has a charge of +1.

4. True. Fe^{+2} could lose another electron (reducing agent) and become the Fe^{+3} ion or it could gain two electrons (oxidizing agent) and become the iron atom.

5. False. The atoms do not have to undergo equal reactions just the net result must be equal numbers of electrons. Cu can lose 2 electrons but it could be in a cell with silver whose ion gains only 1 electron.

6. False. This is true sometimes but not always. The same electrode may be either + or - depending on if it is providing energy or being recharged.

7. True. Either would allow some ions to pass through to equalize charge concentration in the electrolyte solutions.

8. False. The voltage is determined by the materials that make up the electrodes. It has nothing to do with the size.

9. True. The hydrogen electrode at standard conditions is the reference for all other half-cells.

10. False. The potential value must be positive to be product favored.

11. False. The standard battery can not be recharged so it is a primary cell. The automobile battery is of the secondary type.

12. False. One of the electrodes is lead (IV) oxide.

13. False. Moisture is necessary for the reaction to occur. Dry areas such as Arizona are used to store much military hardware.

14. True. This is just like galvanizing nails or a garbage can.

15. True. The cathode is where electrons are available to convert the ion to the metal.

Answers to Understanding of Concepts Questions:

1. Oxidation is a process. This process involves a reducing agent. An oxidizing agent is a material that is capable of gaining electrons.

2. Electrolysis is a many faceted process. It is important to plating of metals on objects, it is important in batteries, it can be used to purify metals and is important in manufacturing several metals.

3. We recognize oxygen as an oxidizing agent that makes hydrogen the reducing agent. We also know that oxidizing agents undergo reduction reactions just as reducing agents undergo oxidation reactions.

Red Agt   Oxid Agt

$2\ H_2 \quad + \quad O_2 \quad \text{---->} \quad 2\ H_2O$

reduction rxn (O_2 → $2\ H_2O$)

oxidation rxn ($2\ H_2$ → $2\ H_2O$)

4. In order for a redox reaction to produce energy like a battery does, the reaction must be controlled, and the electron flow directed. This is usually not done in regular chemical reactions.

5. Electrodes are a surface when a chemical reaction occurs via the transfer of electrons. Electrodes are classified as anodes, where oxidation reactions occur, and cathodes, where the reduction reaction occurs. Usually electrodes are placed in an electrolyte solution that contains an ion of the metal.

6. Salt bridges need to conduct ions. Sugar water does not have ions present so it does not allow for completion of the circuit in the cell.

7. The size of the cell determines the amount of materials that are present and can react. Larger cells have more material and larger surface areas so they will chemically last longer and the larger surfaces make it possible to have higher electron flow meaning greater current.

8. Strong oxidizing agents are materials that are easy to reduce.

9. Primary cells are not rechargeable, the materials get entirely consumed in the reaction. Secondary cells are rechargeable meaning the reaction is easily reversed. Fuel cells differ in that you must constantly supply reactants to them.

10. Since iron is a reducing agent you need an oxidizing agent, usually oxygen in air. You also need a solvent to keep the oxygen in contact with the metal for a period of time, water is the prime material. Finally you need a material present to provide some electrolyte, often a salt.

11. To provide cathodic protection the material must be more reactive than the material being protected. Copper is less reactive than iron so it would not work.

12. Electroplating is just one form of electrolysis.

Answers to Testing the Analytical Concepts

1. The oxidation states of Na and O are fixed. The sodium being a 1A metal is +1 and oxygen will be -2 since it is not a peroxide or reacted with fluorine. This gives a total of +2 for Na and -8 for oxygen.The compound must be neutral so Cr is +6.

   Mg is a 2A metal so it has an oxidation state of +2. As above the oxygen is -2 for a total of -16. The two sulfur have a total charge of +14 to make the compound neutral. Each sulfur is +7.

2. Start with the half-reactions from table 17.2.

$PbO_2 + SO_4^{-2} + 4H_3O^{+1} + 2e^{-2} \text{--->} PbSO_4 + 6H_2O$ E° = +1.685 v
$Al^{+3} + 3e^{-1} \text{---->} Al$ E° = - 1.66 v

We need to reverse the second reaction
$Al \text{---->} Al^{+3} + 3\ e^{-1}$ E° = + 1.66 v

Balance for electrons

(3x) $PbO_2 + SO_4^{-2} + 4H_3O^{+1} + 2e^{-2} \text{--->} PbSO_4 + 6H_2O$
(2x) $Al \text{---->} Al^{+3} + 3\ e^{-1}$

$3PbO_2 + 3SO_4^{-2} + 12\ H_3O^{+1} + 6\ e^{-1} \text{--->} 3PbSO_4 + 18H_2O$
$2Al \text{---->} 2Al^{+3} + 6\ e^{-1}$

Combine the reactions by adding

$3PbO_2 + 3SO_4^{-2} + 12H_3O^{+1} + 2Al \text{--->} 3PbSO_4 + 18H_2O + 2Al^{+3}$

3. The net cell potential will be the sum of the two voltages:
$E°_{net} = +1.685 + 1.66 = + 3.345$ v

4. $\Delta G° = -nFE°$

The potential for the reaction is given as 1.54 volts. We also know from the reactions that this is a 2 electron change.

$$\Delta G° = -(2e) \times \frac{9.65 \times 10^4\ J}{V\ mol\ e} \times 1.54\ V \times \frac{1kJ}{1000J} = 297\ kJ\ mol^{-1}$$

5. The process will require a potential that exceeds the cell potential. The Mg half-cell needs to be reversed so we will do that in one step. The half-cell potentials are:

$Br_2 + 2e^{-1}$ -----> $2Br^{-1}$ $E° = + 1.08$

Mg ---> $Mg^{+2} + 2e^{-1}$ $E° = + 2.37$

Net Mg + Br_2 ----> $MgBr_2$

A voltage of at least 3.45 volts must be applied in order to cause the reaction to occur. It must be applied since we are attempting to reverse the reaction as shown.

6. The mass of the metal produced is determined by the time and the current flow. Also 1 amp is a coulomb per second. A mole of electrons is 96500 coulombs.

$$\text{Mass Ni} = 30\text{ min} \times \frac{60\text{ sec}}{\text{min}} \times \frac{0.200\text{ C}}{\text{sec}} \times \frac{1\text{ mol e}}{96500\text{ C}} \times$$

$$\frac{1\text{ mol Ni}}{2\text{ mol e}} \times \frac{58.7\text{ g}}{1\text{ mol Ni}} = 0.11\text{ g Ni}$$

Answer to Chapter 17 Crossword

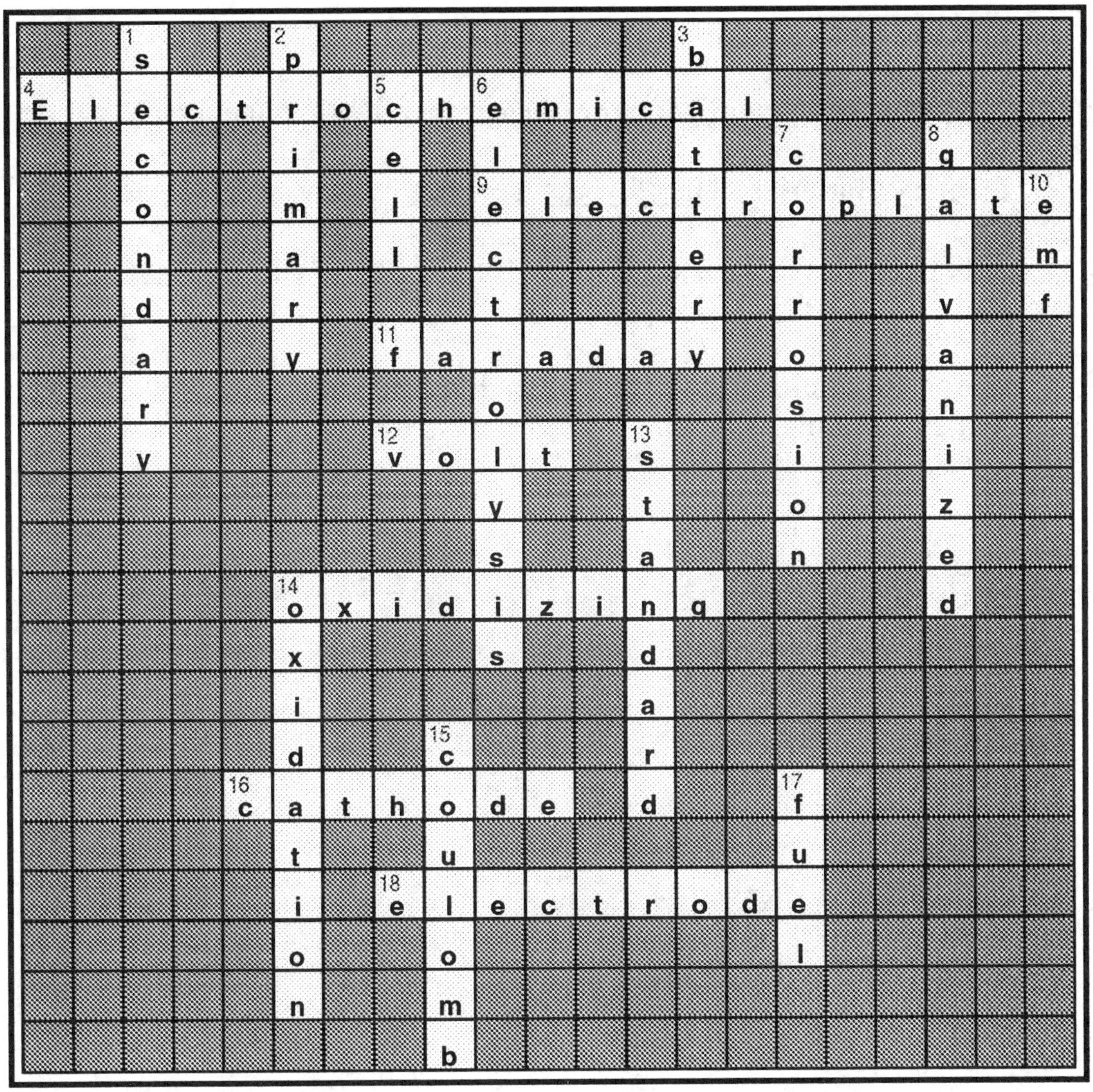

Some Supplementary Readings Relating to This Chapter:

1. Electric Cars, Chem Matters , April 1993, page 4.

2. Leaf Jewelry, Chem Matters , December 1987, page 14.

3. Primary Batteries, The Physics Teacher , January 1981, page 12.

4. A Nickel Metal Hydride Battery for Electric Vehicles, Science , Vol.260, 9 April 1993, page 176.

5. Going Beyond Batteries to Power Electric Cars, The New York Times , March 3, 1993.

Chapter 18 Elements from Land, Sea & Air

This chapter discusses how several materials are obtained from their native state. Several important industrial processes are outlined. There is considerable use of materials from previous chapters especially the oxidation-reduction chapter.

Focus Your Study Efforts

Electrolysis is used for purification and extraction.

Objectives:

1. You should be able to describe the electrochemical process as it is used to produce pure elements.
2. You should be able to explain reasons for the need of pre-electrolysis work-up of some materials.
3. You should be able to write mathematical conversions for various relationships.

The use of electric current to produce both elements and compounds is extensive. It is important that you recognize how the process works and why additional materials, such as $CaCl_2$ with NaCl, are used along with the material being subjected to change.

Either a melt or water solution can be used. The process is simpler if a melt of the pure material is used rather than a water solution. When a solution is electrolyzed you have the potential for electrolysis of the solvent as well as the solute. This is where potential tables help determine what reaction will actually occur.

The higher the charge on a metal ion the more electrons it takes to convert it to the metal atom. A mole of various ions, having different charges, will yield varying numbers of moles of atoms in relation to the ion charge for the same amount of electric current.

> If the same amount of electric current, $6.75 \times 10^{+3}$ kWh, was used to produce Al and Cu, which would be produced in the greater number of moles?
>
> This question can be answered on the basis of electrons involved. $Al^{+3} + 3e^{-1} \longrightarrow Al$ and $Cu^{+2} + 2e^{-1} \longrightarrow Cu$. It takes 3 moles of electrons to produce the same number of moles of Al as 2 moles of electrons will produce moles of Cu. Thus 1.5 times as much Cu will be produced.

Electrolysis methods are very energy intensive. The amount of energy consumed can readily be calculated for any given reaction. This is a fact industry is very interested in and may be the determining factor in what process will be used.

> Write mathematical relationships for the following:
>
> Current
> Faraday
> Joules
> Watt
> Kilowatt hours
> Coulombs
>
> $$\text{Current} = \text{amps} = \frac{\text{coulombs}}{\text{seconds}}$$
>
> $$\text{Faraday} = \frac{\text{Coulombs}}{\text{electron}} \times N$$
>
> $$\text{Joule} = \text{coulombs} \times \text{volts}$$
>
> $$\text{Watt} = \frac{1\ \text{joule}}{\text{second}}$$
>
> $$\text{kWh} = 1000\ \text{watts} \times 1\ \text{hour} \times \frac{3600\ \text{sec}}{\text{hour}} \times \frac{1\ \text{joule/ sec}}{\text{watt}}$$

$$\text{Coulomb} = \frac{1 \text{ joule}}{\text{volt}}$$

Electrolysis must involve extensive thought and care in the design of the process so that electrolyzed materials can be collected and kept pure.

Often an extensive pre-electrolysis work up is required to provide a material that can be efficiently electrolyzed. Obtaining magnesium from sea water is an example.

Special electrodes are often used in electrolysis. Sometimes because pure solids of the desired material, such as chlorine, are not available. In other cases the material being produced may be very corrosive. Consequently "inert" electrodes, such as graphite or platinum, are used.

A person wanted to plate an object with silver and determined it would take 1.35 g of Ag to do it. If the plating solution had 1.50 amps flow for 15 minutes, would enough silver get deposited?

The reaction $Ag^{+1} + 1e^{-1} \text{ ----> } Ag$, shows that only one electron is required per silver ion.

Since we need grams of silver we can determine what relationships are required to solve the problem.

Current x time ---> Charge (coulombs) ----> moles of e^{-1} --->
moles of ions ---> moles of metal

$$\text{g Ag} = 1.50 \text{ amps} \times 15 \text{ min} \times \frac{60 \text{ sec}}{\text{min}} \times \frac{1 \text{ mole } e^{-1}}{96500 \text{ C}} \times \frac{1 \text{ mole Ag}}{1 \text{ mole } e^{-1}}$$

$$\times \frac{107.9 \text{ g Ag}}{1 \text{ mol Ag}} = 1.51 \text{ g Ag}$$

This is in excess of the amount of silver needed.

Water is an unlimited resource for us. It also is a potential source of hydrogen gas which can be obtained by electrolysis. If an acid-water solution is electrolyzed, how many kWhrs would be required to produce 500.0 g of H_2? What minimum voltage would be required? If the cell were only 80% efficient how many kWhrs would be required?

Cell potential is required to solve the problem so we start there. What is the minimum cell potential we need? This information is available in tables of reduction potentials.

$O_2 + 4\ H^{+1} + 4\ e^{-1}$ ----> $2\ H_2O$ +1.23 V
$2\ H^{+1} + 2\ e^{-1}$ ----> H_2 0.00 V

balance for electrons by doubling the second reaction
reverse the first equation to match our system

$4\ H^{+1} + 4\ e^{-1}$ ----> $2\ H_2$ 0.00 V

net rxn $2\ H_2O + 4\ H^{+1} + 4\ e^{-1}$ ---> $2\ H_2 + O_2 + 4\ H^{+1} + 4\ e^{-1}$
the voltage is negative 1.23

$2\ H_2O$ ----> $2\ H_2 + O_2$ -1.23V
we must supply a minimum of 1.23 volts

The acidity allows the solution to conduct and facilitates the reaction. $2\ H_3O^{+1} + 2\ e^{-1}$ ---> $H_2 + H_2O$ shows that a mole of hydrogen requires 2 moles of electrons.

We can determine the charge required for 500 grams of H_2.

$$\text{Charge} = 500\text{g } H_2 \times \frac{1 \text{ mole } H_2}{2.016 \text{ g } H_2} \times \frac{2 \text{ mol } e^{-1}}{1 \text{ mol } H_2} \times \frac{96500 \text{ C}}{1 \text{ mol } e^{-1}} =$$

$$= 4.79 \times 10^7 \text{ coulombs}$$

Charge can be converted to kWhrs

$$\text{kWhr} = 4.79 \times 10^7 \text{ C} \times 1.23 \text{ V} \times \frac{1 \text{ kWhr}}{3.60 \times 10^6 \text{ J}} = 16.4 \text{ kWhr}$$

If the system is less than 100% efficient we need to ratio up the power value. We would expect that less efficiency would require more power to accomplish the same task.

$$\text{kWhr} = 16.4 \text{ kWhr} \times \frac{100}{80} = 20.5 \text{ kWhr at } 80\%$$

Metallurgical Processes.

Objectives:

1. You should be able to define the terms minerals, ores, and metallurgy.
2. You should be able to explain why metals require a purification process after being extracted from their ore.
3. You should be able to identify the products of fractionation of air.

Minerals, ores, and metallurgy are all related. The mineral is a naturally occurring substance that contains a material, usually a metal, that we are interested in. If the mineral is concentrated enough to be economically workable it is called an ore. The process of separating the specific metal from the ore is known as metallurgy.

Most metals are found as compounds in the oxidized state. Most metallurgical processes involve a reduction of the ore to yield the pure metal. Sulfide ores are volumetrically a large class, but involve a relatively few elements. Insoluble sulfides are common in nature.

Most metals when prepared from their ore have a considerable impurity content. As a result, purification processes must be developed. These processes are often specific for a given metal.

Iron is an excellent example of how the addition of other materials can significantly change the properties of a metal. Steel, as shown in table 18.2 of the text, is not a single material. There are many types of steel.

Usually air is not thought of as a source of materials. It is however, our major source of oxygen, nitrogen, and argon. The process used for separation is distillation; however, at extremely low temperatures.

Nitrogen is an extremely important resource. The element, however, is very non-reactive. Plant life is very dependent upon nitrogen compounds since it is a primary nutrient. Before a plant can use nitrogen it must be converted to soluble compounds. This conversion process is known as nitrogen fixation.

Symbols found in Chapter 18

Page	Symbol	
796	C	Coulomb which is an amps x seconds
796	J	Joule of energy
796	A	Amps

Terms in Chapter 18

The following terms are listed for your convenience. They should not become a list for you to memorize. You should have a working definition for each. If you have doubt about any particular term check its definition in the glossary and/or review it in the chapter. Some of these terms are included in the crossword puzzle later in the chapter. This chapter also makes use of several terms introduced in chapters 6 and 17.

Downs process	Chlor-alkali process	Amalgam
Faraday	Watt	Nafion
Hall-Heroult process	Sacrificial anode	Ore
Metallurgy	Blast furnace	Slag
Pig iron	Cementite	Cast iron
Steel	Carbon steel	Refractory tube
Alnico V	Roasting	Floatation process
Superplastic	Blister copper	Anode slime
Mineral	Joule-Thompson effect	Linde process
Cryogens	Primary nutrients	Nitrogen fixation
Legumes	Black smokers	Tectonic plate

Concept Maps for Elements from Land, Sea & Air

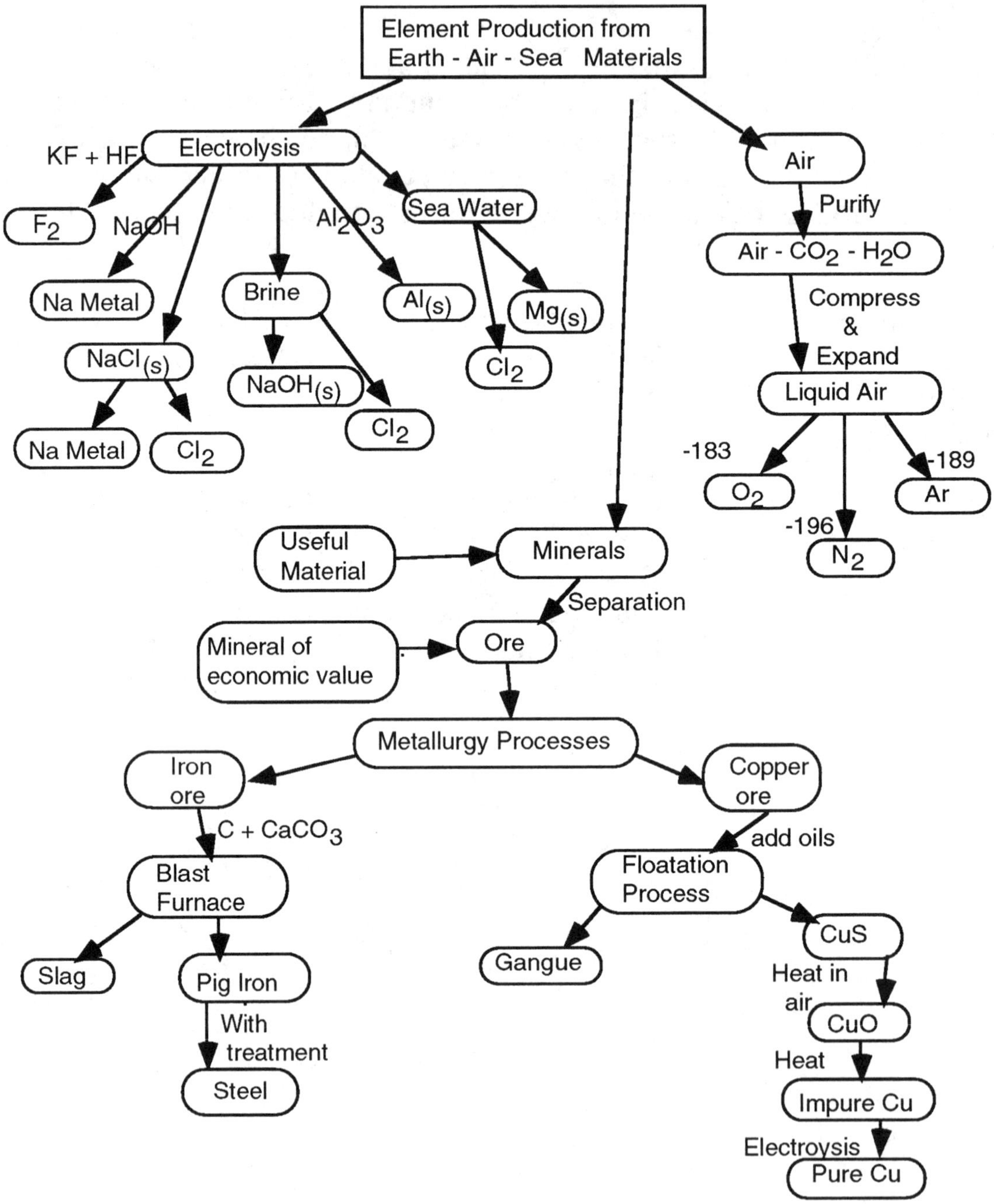

Elements from Land, Sea & Air True/False Statements

Answers are at the end of the chapter.

1. Electrolysis, as a method of element production, is often used because it is a low cost procedure.

2. Electrolytic production of elements often does not produce the exact products that are preferred.

3. It is not uncommon to add materials to an electrolytic cell in an effort to lower the melting point.

4. The greater the charge on the metal ion we would like to reduce to the metal, the more energy it will consume.

5. Most electrolytic cells work at high voltage and low currents.

6. A major use of some metals is in the production of other metals.

7. Ores consisting of nearly pure metals are uncommon.

8. A blast furnace provides a reducing environment to reclaim iron from its ore.

9. Pig iron and cast iron are the same material.

10. To make steel all the carbon must be removed from the pig iron.

11. Blast furnaces are lined with a refractory material.

12. The roasting process converts metal sulfides directly to the pure metal.

13. Some processes, such as the floatation process, depend upon the hydrophobic and hydrophilic character of materials.

14. Anode slime is a worthless by-product that has major environmental problems.

15. The production of cryogens uses very little energy.

16. Nitrogen gas is an excellent fertilizer.

17. Geologic processes currently are producing deposits of some minerals.

Testing Your Understanding of Concepts in Chapter 18.
Answers are at the end of the chapter.

1. Why is it essential to consult a reduction potential table before predicting anode and cathode products during the electrolysis of solutions?

2. Why are different products produced when you electrolyze molten NaCl or a brine of NaCl?

3. Why does the Hall-Héroult process for producing aluminum metal run at such high currents?

4. $MgCl_2$ is electrolyzed without any added material but Al_2O_3 has cryolite, Na_3AlF_6, added before electrolysis. Why the difference in procedures?

5. What functions do coke and limestone have in the chemistry occurring in a blast furnace?

6. Why is roasting used in the recovery of Cu from its ores?

7. What is the basis for separation of the components of air?

Testing You on Some Analytical Concepts.

Answers are at the end of the chapter.

1. If a molten NaCl bath was subjected to 20.0 amps of current for 6.00 hours what mass of sodium metal would be produced?

2. How many kWhrs is required to plate 10.0 grams of Cr from CrO_3 in an acid medium? The cell is run at 4.5 volts for 308 minutes at 6.00 amps. The cathode reaction is:

   $CrO_3 + 6H^{+1} + 6e^{-1}$ ----> $Cr + 3\ H_2O$

Crossword - Elements from Land, Sea & Air

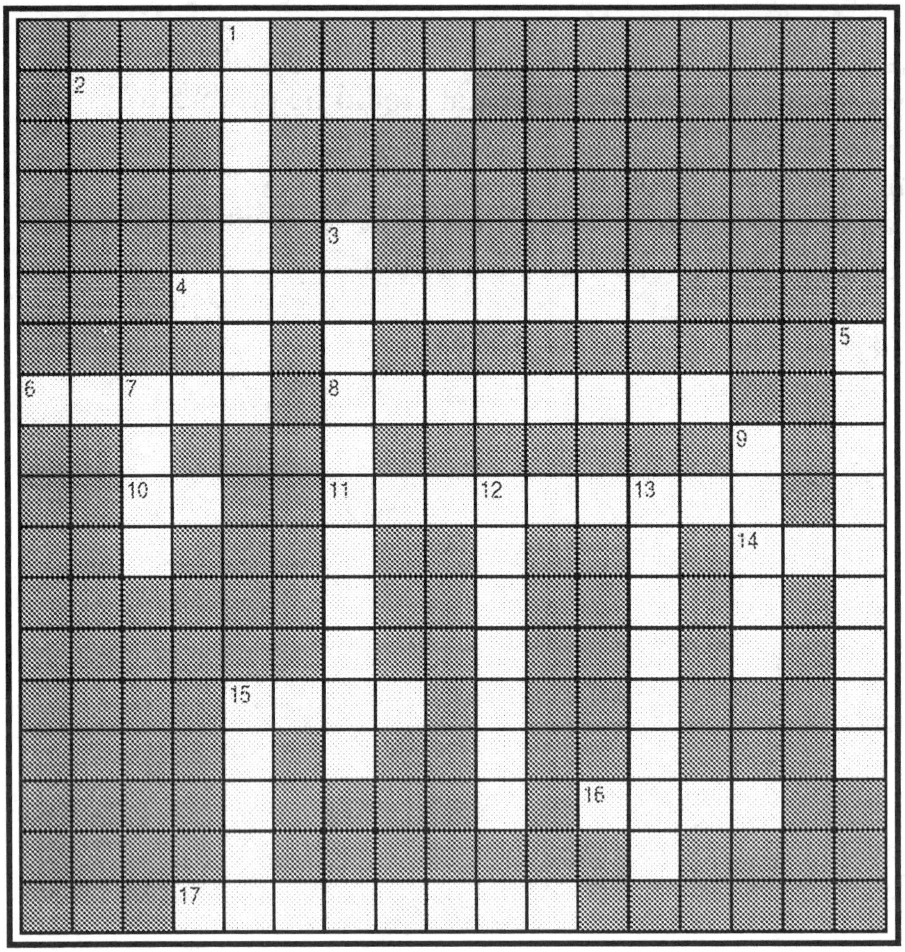

Crossword - Elements from Land, Sea & Air Clues

ACROSS

2 Element essential for plant growth but needs to undergo fixation.
4 Chemical reduction process to separate a metal from its ore.
6 This type of cell is used to obtain elements from molten NaCl.
8 Process used to convert ores to the oxide form. Often used with copper.
10 Symbol for the element titanium
11 This process of separation involves the hydrophobic and hydrophilic properties of ores.
14 A mineral that contains an element in sufficient quantity to make it profitable to extract.
15 Waste material from the blast furnace.
16 Compound used to convert magnesium ions to the insoluble hydroxide.
17 Electrolytic process to extract this metal developed in both France and the U.S.A.

DOWN

1 Term applied to very cold materials such as liquid nitrogen.
3 This type of anode is used to prevent corrosion of other metals.
5 The chief source of this element is sea water.
7 One joule per second.
9 The slime (residue) from this electrode may be of value.
12 Combination of a metal and mercury.
13 Geologic plates that may indicate where chemical reactions are occurring.
15 A group of alloys made with the element iron.

Ties to Previous Chapters

Processes used in obtaining pure elements from materials in nature make use of many chemical principles. This chapter requires that you consider information first presented in chapter 4, Chemical Compounds. You must understand the aspects of stoichiometry presented in chapter 6. Energy aspects from chapter 7 are used as well as ideas from the gases in chapter 12 and properties of solutions from chapter 13. Aspects of electrochemistry developed in the last chapter, chapter 17, are used here.

Answers to True/False Statements:

1. False. Electrolytic processes consume large amounts of electric energy. Electrolytic plants usually are located only where electric energy is the least expensive.

2. True. Sometimes the electrolysis is done on solutions and as a result the water may be electrolyzed instead.

3. True. This done with such materials as aluminum and sodium chloride.

4. True. The objective is to add electrons to convert the ion to the atom. The electrons are provided by the current and since more electrons are needed, the cost is increased for more highly charged ions.

5. False. The reverse is true. The reactions are not difficult to cause, but large amounts of current flow are needed to make the product at a rapid rate.

6. True. Pure metals can make good reducing agents.

7. True. Most metals are reactive in the atmosphere and have been converted to oxides or sulfides or other compounds.

8. True. To make iron the oxide must be reduced to the metal.

9. True. Cast iron is pig iron put into specific molds.

10. False. Nearly all steel, as shown in table 18.2 of the text, contains some carbon. However, it is usually much less than what is present in pig iron.

11. True. The refractory material is essential to prevent the furnace walls from melting or reacting.

12. False. The roasting process often converts the sulfide to the oxide. The oxide must then be reduced.

13. True. The soil and rock components get wetted easily by water so they can be separated.

14. False. It may be an environmental problem, but it often contains small amounts of valuable metals such as gold and silver.

15. False. This process involves a lot of energy in the compression of the gases. Cold doesn't necessarily mean that energy is not consumed.

16. False. The nitrogen is very unreactive. It is nitrogen compounds that act as the fertilizer.

17. True. This is especially true where tectonic plates are active.

Answers to Understanding of Concepts Questions:

1. When solutions are involved in electrolysis the possibility of solvent electrolysis also occurs. The reduction potential table can provide information regarding the relative chance of solute or solvent being reacted.

2. The answer to this question is related to question # 1 above. Since Na^{+1} is so difficult to reduce the cathode reaction involves water instead. The potential for water reaction is much less than Na^{+1} reduction.

3. There are probably several reasons for this, however two would be very reasonable. High current flow will produce more product in the same time as a small amount of current and secondly, it takes 3 electrons to produce the metal atom from the ion.

4. The melting point of $MgCl_2$ is relatively low and easily obtained. Al_2O_3 has a very high melting point. The Na_3AlF_6 is much lower and when aluminum oxide is dissolved in it, the melting temperature drops even more. This would be expected as a consequence of the colligative property aspect studied in chapter 15, solution chemistry.

5. Coke is carbon which is converted to CO in the furnace. CO is a very good reducing agent and removes the oxygen from Fe_2O_3. The limestone has an entirely different function. Limestone reacts with the silica present and converts it to a material known as, slag, that is easily removed from the process.

6. Roasting is the conversion of Cu_2S to Cu_2O by heating. This combination will, upon further heating, yield copper and SO_2.

7. When liquified, the various gases found in air have different boiling points so they can be fractionally boiled off.

Answers to Testing the Analytical Concepts

1. The reaction at the cathode is: $Na^{+1} + 1\ e^{-1} \rightarrow Na$. This gives us the electron relationship. From this, the charge can be determined.

$$C = 20.0 \text{ amp} \times 6.00\text{hr} \times \frac{3600 \text{ sec}}{\text{hr}} = 4.32 \times 10^5 \text{ C}$$

This is then converted to moles of electrons consumed:

$$\text{moles of } e^{-1} = \frac{C}{F} = \frac{4.32 \times 10^5 \text{ C}}{96500 \text{ C/ mol } e^{-1}} = 4.48 \text{ mol } e^{-1}$$

$$\text{Moles of Na} = \text{Moles of } e^{-1}$$

$$\text{g Na} = 4.48 \text{ mol } e^{-1} \times \frac{1 \text{ mol Na}}{1 \text{ mol } e^{-1}} \times \frac{23.0 \text{ g Na}}{1 \text{ mol Na}} = 103 \text{ g Na}$$

2. From the given reaction it is seen that each mole of Cr produced will require 6 moles of electrons.

$$\text{mol of Cr} = 10.0 \text{ g Cr} \times \frac{1\text{mol Cr}}{52.0 \text{ g Cr}} = 0.192 \text{ mol Cr}$$

$$\text{mol of } e^{-1} \text{ needed} = 0.192 \text{ mol Cr} \times \frac{6 \text{ mol } e^{-1}}{1 \text{ mol Cr}} = 1.15 \text{ mol } e^{-1}$$

$$C = 308 \text{ min} \times \frac{60 \text{ sec}}{\text{min}} \times 6.00 \text{ amp} \times 1.15 \text{ mol } e^{-1} = 1.28 \times 10^5 \text{ C}$$

$$\text{joules} = 1.28 \times 10^5 \text{ C} \times 4.5 \text{ V} = 5.74 \times 10^5 \text{ joules}$$

$$\text{kWhr} = 5.74 \times 10^5 \text{ joules} \times \frac{1 \text{ kWhr}}{3.6 \times 10^6 \text{ J}} = 0.159 \text{ kWhr}$$

Answer to Chapter 18 Crossword

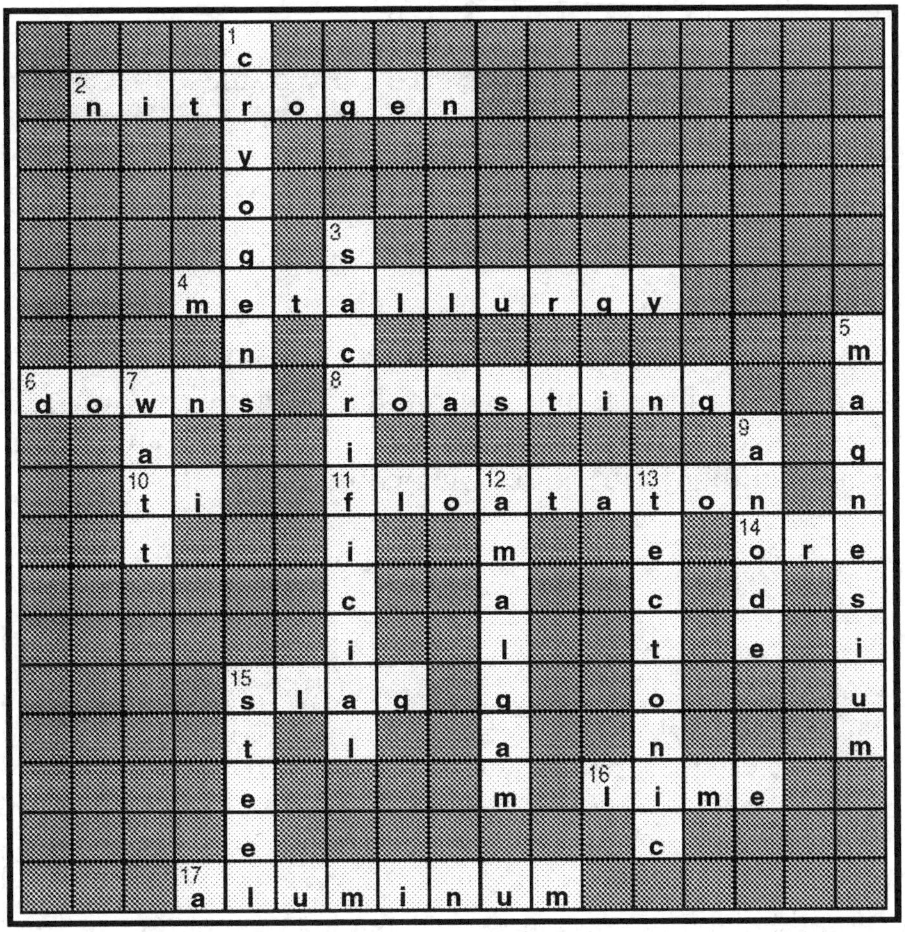

Some Supplementary Readings Relating to This Chapter:

1. The New Gold Rush, Chem Matters , October 1989, page 4.

2. Greens Pick an Enemy: Chlorine, the Everywhere Element, The New York Times , December 20, 1992.

Chapter 19 Nuclear Chemistry

Nuclear changes in atoms and how chemists make use of these is the emphasis in this chapter. Natural and synthetic changes in the elements, production of new isotopes, uses and hazards of various isotopes, and reasons for variations in the stability of atomic nuclei are discussed.

Focus Your Study Efforts

Development of the field of nuclear science involved many people. The progress seen is the result of work building on that of many previous workers. Within a few years major progress was made in this area. You should be somewhat familiar with the people and time chronology.

Radioactive processes are natural and random:

Objectives:

1. You should be able to identify sources of radioactive decay, both natural and synthetic.
2. You should be able to characterize the three major types of radiation observed in radioactive decay processes.
3. You should be able to write a balanced equation for a nuclear reaction.
4. You should be able to explain why a nucleus undergoes a decay process.

Radioactivity is a natural phenomenon that results from incoming radiation from space and the natural changes of materials all over the earth.

There are several forms of radioactive decay but three are very common as far as chemists are concerned. They are alpha, beta, and gamma. Each type of decay causes specific changes in the originating material.

We express nuclear changes in the form of nuclear reactions. These reactions must show charge and mass conservation.

Complete the following neuclear reactions:

(A) $^{218}_{84}Po = ^{214}_{82}Pb + \_\_\_\_$

(B) $^{210}_{83}Bi = ^{0}_{-1}e + \_\_\_\_$

(C) $^{99}_{43}Tc = ^{99}_{43}Tc + \_\_\_\_$

(D) $^{14}_{7}N + \_\_\_\_ = ^{17}_{8}O + ^{1}_{1}H$

(E) $^{235}_{92}U + ^{1}_{0}n = 2\,^{1}_{0}n + ^{97}_{40}Zr + \_\_\_\_$

With any reaction of this type we must conserve charge and mass. Also the symbol needed will be determined by the atomic number. We look at the atomic numbers to determine the missing value.

(A) We need 2 charge units and 4 mass units. This corresponds to He.

$^{218}_{84}Po = ^{214}_{82}Pb + ^{4}_{2}He$

(B) We need 84 charge units to make the product side 83. No mass change occurs.

$^{210}_{83}Bi = ^{0}_{-1}e + ^{210}_{84}Po$

(C) This shows no mass or charge change. The only item that fits this radiation criteria is the release of a gamma ray.

(D) The missing item is a particle that is used to collide with the nitrogen. The right side has 9 charge units and 18 mass units. The missing item needs 2 charge and 4 mass units.

$^{14}_{7}N + ^{4}_{2}He = ^{17}_{8}O + ^{1}_{1}H$

(E) With this problem be sure to account for 2 neutrons on the product side. Total mass and charge on reacted side:
Charge = 92 + 0 = 92
Mass = 235 + 1 = 236

The product side shows:
Charge = 0 + 0 + 40 = 40
Mass = 1 + 1 + 97 = 99

Needed:
Charge 92 - 40 = 52
Mass 236 - 99 = 137

$^{235}_{92}U + ^{1}_{0}n = 2\ ^{1}_{0}n + ^{97}_{40}Zr + ^{137}_{52}Te$

The radioactive decay process is totally unpredictable for a specific atom. The entire process is predicated on a statistical basis involving large numbers of atoms. We can not predict when a specific nucleus will change, but we can predict that within a given time, a certain percentage of a large sample will undergo change.

Radioactivity is the result of a nucleus attempting to adjust its makeup to reach a more stable, lower energy state.

Nuclear stability has some favored combinations:

Objectives:

1. You should be able to interpret a decay series and write a decay series if given the decay sequence.
2. You should be able to explain binding energy and how it relates to the process of fission and fusion.

There appears to be a favored range in the ratio of protons to neutrons in the nucleus. Within a nucleus this ratio is usually adjusted via either alpha or beta decay. Other processes such a K-capture or positron emmision are also possible.Gamma radiation has no effect on the proton/neutron ratio.

Often a single decay step will not reach the optimum stability. When this occurs a set of sequential decay steps will take place. This results in what is known as a radioactive decay series.

Energy is involved in holding any nucleus together. The greatest nuclei stability occurs near that of an iron isotope. Nuclei lighter than this tend to combine (fusion) and those heavier tend to split (fission).

An isotope of manganese with a mass of 54.938 amu would have what binding energy for manganese - 55?

We need the mass for the constituent particles:

25 protons at 1.0078 amu = 25.195 amu
30 neutrons at 1.0087 amu = 30.261 amu

Total mass of particles = 55.456 amu
Actual mass = 54.938 amu
Mass defect = 0.518 amu

Energy change is calculated from Einstein's equation:

$\Delta E = \Delta m \cdot c^2 = 0.518$ amu x 1.66×10^{-27} kg/amu x $(3.00 \times 10^8)^2$
$= 7.738 \times 10^{-11}$ Joules per particle

Activity and half-life are not the same:

Objectives:

1. You should be able to explain the differences between the half-life of a nucleus and the activity of a radioactive sample.
2. You should be able to calculate the half-life of a radioactive isotope if the sample activity is known.
3. You should be able to use half-life data to calculate the time for an isotope to decay to a particular activity.

Be careful not to confuse half-life and activity. Half-life is independent of sample size but activity is directly proportional to the size of the sample.

Activity relationship formulas have symbols with a subscript 0. Those with a 0 subscript refer to initial values at the start time.

Cobalt-60 has a $t_{1/2}$ of 5.27 years. If we started with a 100.0 gram sample now what fraction would be left after 20 years?

The amount present now is N_0 and after 20 years it will be N.

$$\ln \frac{N_0}{N} = \frac{0.693\ t}{t_{1/2}} = \frac{(0.693)\ (20 \text{ years})}{5.27 \text{ years}} = 2.63$$

$\ln N_0 - \ln N = 2.63$
$\ln 100 - \ln N = 2.63$

$\ln N = 1.975$
$N = 7.21$ grams

The activity relationship is useful when there is interest in how much material remains. It also allows us to calculate the $t_{1/2}$ for a given sample.

A sample of phosphorous-32 weighing 2.00 mg was analyzed after 33.3 days and found to contain 0.40 mg of phosphorous-32. What is the $t_{1/2}$ of the isotope?

$$t_{1/2} = \frac{0.693\ t}{\ln(N_0/N)}$$

the ratio of $N_0/N = 2.00 / 0.40 = 5.00$

$$t_{1/2} = \frac{0.693\ (33.3\ \text{days})}{\ln 5.00} = \frac{23.08}{1.61} = 14.3\ \text{days}$$

Radiochemical dating is dependent upon having a history of isotope concentrations. It also assumes that the isotopic ratio has remained constant over very long time periods.

There are no elements beyond uranium that have half-lives long enough to exist by natural means. These elements are all made in laboratories. This may seem unlikely when we look at a $t_{1/2}$ of about 24000 years for a plutonium isotope. We must, however, remember that on a geologic time scale that is a very short period of time. Even for that amount of time, many half-lives have gone by since the earth was formed.

Fission is currently useful - fusion is not controlled:

Objectives:

1. You should be able to describe nuclear fission, nuclear fusion, and a nuclear chain reaction.
2. You should be able to identify and explain the function of the components in a nuclear reactor.

Fission is dependent upon a critical mass and a chain reaction. A critical mass is dependent upon the geometry of the material being used. A sphere is the most favored since it compacts the greatest number of particles in the smallest volume.

The key to a chain reaction is the release of several neutrons (3 in one case) for each fission that occurs. The exponential effect of this is a very large number of nuclei splitting within a very short time period.

It is the ability to control the number of neutrons that makes it possible to have a neuclear reactor. Some elements are good at absorbing neutrons and when inserted in the reactor can regulate the total neutron flux. If neutrons are absorbed the reaction will slow down.

Nuclear reactors vary considerably in how they are constructed and controlled. However, all reactors have certain common components: fuel, shielding, control rods, primary and secondary heat exchange, and a coolant.

All commercially usable nuclear reactions are of the fission type. Fusion has yet to be contained in sufficient quantity to be useful. In fact, after billions of dollars in research, the first break-even point with respect to energy input to the energy output was just reached.

The interest in fusion lies in the tremendous energy potential that is available. The low cost and abundant fuel supply, as well as the fact that long term waste is minimal, make it hard to discount the process.

Radiation is part of our environment:

Objectives:

1. You should be able to describe some sources of background radiation.
2. You should be able to describe some units used to measure the amount of radiation.
3. You should be able to discuss some uses for radioisotopes.

Everyone must be aware that we can not escape from radiation. There is a natural background radiation everywhere. Our lifestyle and medical technology add about half again to normal background effects.

Several radiation units are in use. They vary in how they can effectively be applied. You should be aware of their differences and when they can be used.

The curie, Ci, unit is a measure of activity. This is different than units like the rem, which measure radiation effects on a system.

Applications of radiochemical technology are present in several places. Food preservation, medical treatments, medical diagnostics, industrial tracers, and use in chemical and biological research to follow reaction systems and biological pathways are a few.

Symbols found in Chapter 19

Page	Symbol	
826	α	Alpha radiation particle

826	β	Beta radiation particle
827	γ	Gamma radiation
831	$_{+}\beta$	Positron - postitive electron
834	E_b	Binding energy
835	Δm	Mass defect
836	$t_{1/2}$	Half-life
838	A	Activity of a material at some time
839	A_o	Activity at time zero or start time
839	t_o	Time zero
839	N	Number of radioactive atoms at some time
839	N_o	Number of radioactive atoms at time zero
839	k	Decay constant
851	R	röentgen - radiation dosage
851	Bq	bequerel - 1 disintegration per second
851	Ci	Activity unit for radiation
853	ALARA	As low as reasonably achievable radiation limit
858	PET	Positron emission tomography
858	υ	neutrino - decay product

Terms in Chapter 19

The following terms are listed for your convenience. They should not become a list for you to memorize. You should have a working definition for each. If you have doubt about any particular term check its definition in the glossary and/or review it in the chapter. Some of these terms are included in the crossword puzzle later in the chapter. This chapter also makes use of several terms introduced in chapters 3 and 9.

Phosphorescence	Radioactivity	Alpha radiation
Beta radiation	Gamma radiation	Nuclear reactions
Nuclear transmutations	Nucleons	Radioactive series
Positron	Electron capture	K-capture
Antielectron	Deuterium	Tritium

Binding energy	Mass defect	Fission
Binding energy per nucleon		Fusion
Half-life	Activity	Decay constant
Radiochemical dating	Transuranium element	Nuclear fission
Atomic reactor	Nuclear fusion	Plasma
Microballoon	röentgen	rad
rem	bequerel	curie
Background radiation	Tracers	Nuclear medicine
Imaging	Metastable	Positron emitters
Neutrino		

Concept Maps for Nuclear Chemistry

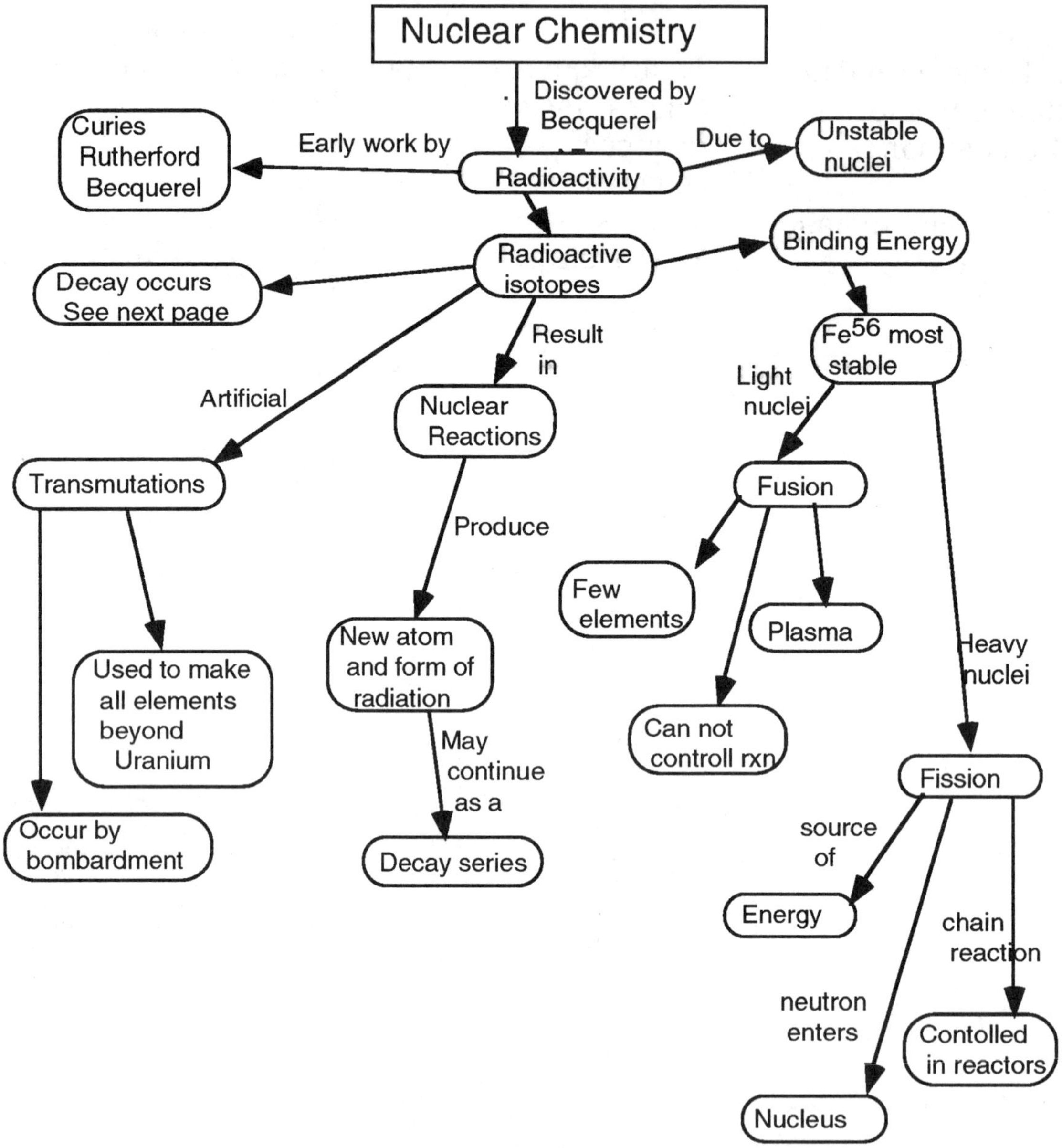

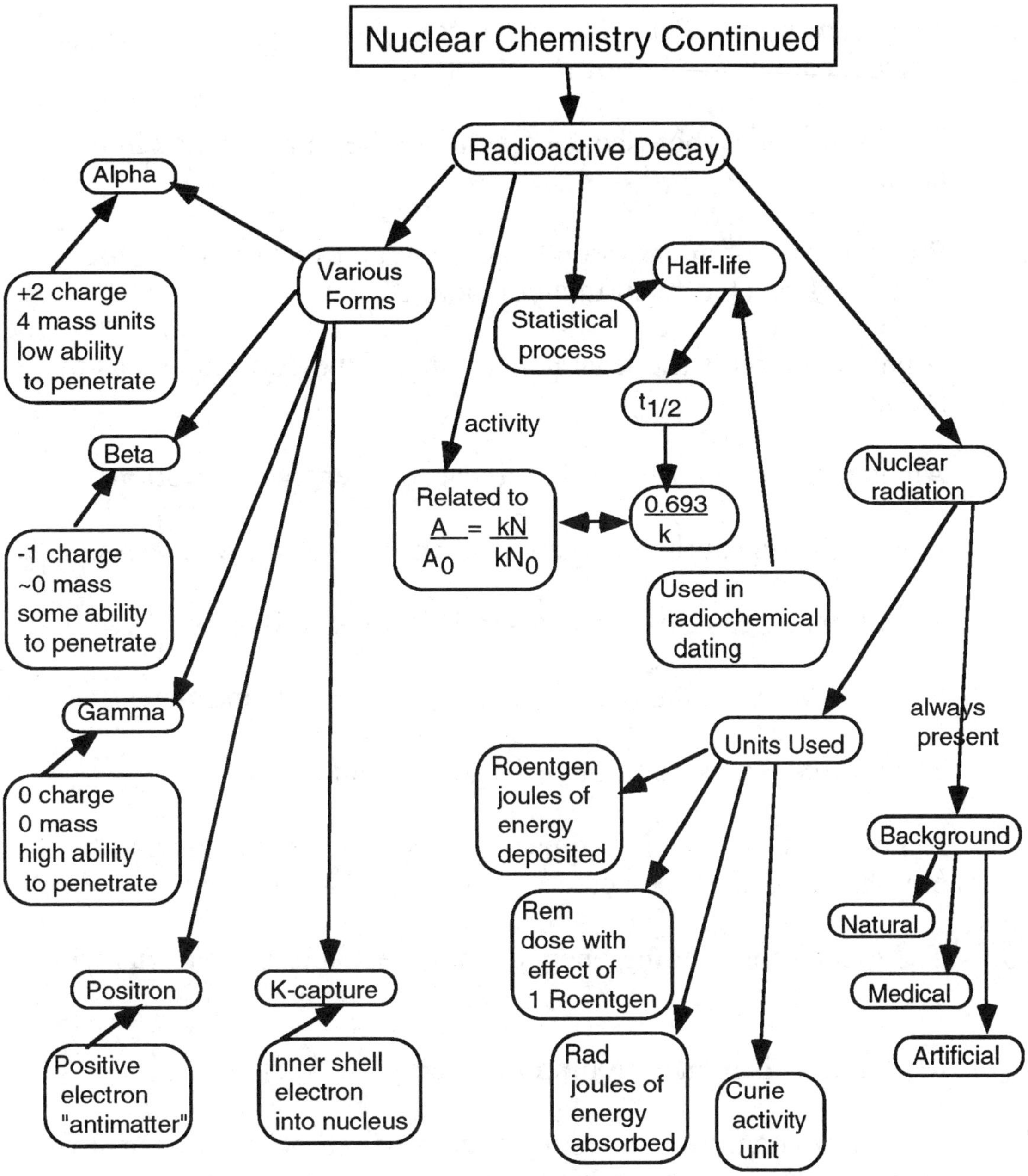
Nuclear Chemistry Continued
Radioactive Decay
Various Forms
Alpha
+2 charge
4 mass units
low ability to penetrate
Beta
-1 charge
~0 mass
some ability to penetrate
Gamma
0 charge
0 mass
high ability to penetrate
Positron
Positive electron "antimatter"
K-capture
Inner shell electron into nucleus
activity
Related to
A / A0 = kN / kN0
Statistical process
Half-life
t1/2
0.693 / k
Used in radiochemical dating
Nuclear radiation
Units Used
Roentgen joules of energy deposited
Rem dose with effect of 1 Roentgen
Rad joules of energy absorbed
Curie activity unit
always present
Background
Natural
Medical
Artificial

Nuclear Chemistry True/False Statements

Answers are at the end of the chapter.

1. X-rays and phosphorescence are due to the same change in the atom.

2. Radioactivity of an atom undergoes a major change when the element is reacted to form a compound.

3. Alpha radiation is the least penetrating of the common types of radiation.

4. Alpha radiation has the ability to cause extensive ionization.

5. Beta radiation results in a new isotope that has a lower mass.

6. The ultimate result of a decay series is a stable isotope.

7. Positrons are an example of what is often called antimatter.

8. The radioactive isotope of hydrogen is deuterium.

9. If a proton to neutron ratio has an excess of neutrons beta emission would be expected.

10. The greater the binding energy the more stable the nucleus will be.

11. Mass defect is related to binding energy.

12. Both fission and fusion result in the production of a material with more thermodynamic stability.

13. Stated half-life values refer only to a specific nuclei.

14. The greater the activity of a radioactive sample the shorter the half-life of that material.

15. Large values for a decay constant result in a long half-life.

16. When carbon-14 decays by beta emission the new isotope is of the element nitrogen.

17. Transuranium elements are common in nature.

18. The fission process is more energetic than the fusion process.

19. Nearly pure fissionable fuel is required for the nuclear reactor.

20. A plasma occurs at extremely high temperatures and can facilitate the fusion process.

21. The radition unit “rad” is a measure of radiation absorbed.

22. Background radiation is readily controlled and can be lowered with little effort.

23. Radon is a background radiation component in many places.

24. Radioactive tracers are useful only in medicine.

25. The neutrino is the result of a proton conversion to a neutron.

Testing Your Understanding of Concepts in Chapter 19.
Answers are at the end of the chapter.

1. Compare alpha and beta radiation in their ability to penetrate and ionize.

2. When writing nuclear reactions what must be conserved?

3. Why do radioactive series exist?

4. What is one factor that determines if an isotope will be radioactive or not? What is meant by band of stability?

5. Which elements would it be logical to predict as having the more stable nuclei?

6. Discuss and compare the terms activity and half-life.

7. What are some of the properties an isotope must posess to be used as a medical tracer?

8. It has been stated that since the time of open air nuclear testing the use of Carbon-14 as a dating material has had to be adjusted. Explain why this statement may be true.

9. What is a good material to use to cause nuclear transformations? Why is this so effective?

10. How is control of a nuclear reactor maintained?

11. What are some things you could do to lower your exposure to background radiation?

12. Why is nuclear fusion so difficult to accomplish in the laboratory?

13. How do “rads” and “curies” differ?

Testing You on Some Analytical Concepts.
Answers are at the end of the chapter.

1. What type of radioactive decay would one expect from Calcium-47? Explain.

2. If alpha particles are used to bombard Plutonium-239 a reaction results producing a neutron and another element. Write a reaction for the process.

3. An unknown beta emitting isotope had a radioactive decay constant of 1.0×10^{-13} sec^{-1}. What is its $t_{1/2}$ in years?

4. The luminous dial from an old watch was found to contain tritium. The half-life of tritium is 12.3 years. What is the value of k, the decay constant?

5. A piece of wood from an archeological site had a Carbon-14 activity of 7.0 per minute per gram of carbon. Assume there has been no change in the ratio of carbon isotopes because of nuclear testing. Current carbon samples give 15.3 disintegrations per minute per gram of carbon. The half-life of Carbon-14 is 5730 years. How old is the wood from this site?

Crossword - Nuclear Chemistry

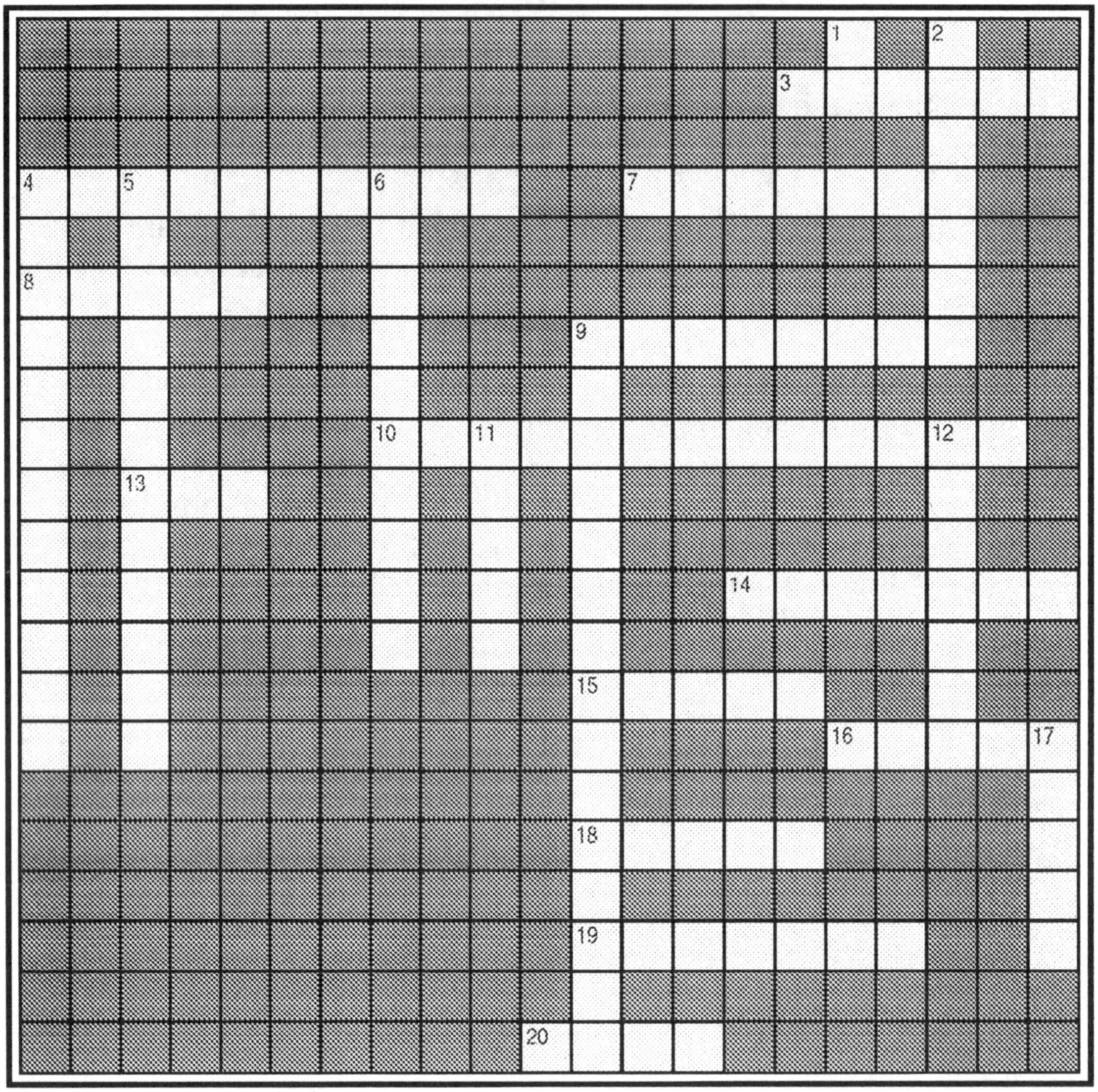

Crossword - Nuclear Chemistry Clues

ACROSS

3 Nuclear process we have yet to control that involves small nuclei.
4 Radioactive isotopes that are in transition to a more stable isotope. Often used in medicine.
7 These are radioactive isotopes used to follow a material in a chemical or biological system.
8 This type of reaction is necessary to have a bomb or nuclear reactor.
9 A low mass anti-matter particle.
10 The result of nuclei going to a more stable form and as a result, a new element may be produced.
13 A unit of radiation that is a measure of the amount of radiation absorbed.
14 This is a devise where a controlled nuclear reaction occurs.
15 A radioactive gas that we worry about in homes.
16 A very penetrating form of radiation. It is often a secondary effect in nuclear decay.
18 The unit of activity for radioactive materials.
19 Often thought of as being the same as isotopes. The protons and neutrons.
20 Radiation type that is equivalent to an electron.

DOWN

1 Symbol for transuranium fissionable element.
2 Nuclear process that involves atom splitting.
4 Glass spheres used to hold deuterium and tritium for fusion.
5 Those elements in the periodic table after uranium.
6 This type of radiation is always present but will vary from place to place.
9 Term applied to the glow that results after the radiation of a source is discontinued.

11 All radioactive elements will undergo this.
12 An isotope of hydrogen with two neutrons.
17 A high mass, low penetration type radiation.

Ties to Previous Chapters

Kinetics is an important aspect of nuclear chemistry. Kinetics is also important to chemical systems and was introduced chapter 8. The same ideas apply to both chemical and nuclear problems.

The periodic properties of the transuranium elements make it easy to place them in the periodic table. This relationship goes back to chapters three and nine.

Answers to True/False Statements:

1. False. X-rays are much more energetic than phosphorescence. Visible or Ultraviolet light can cause phosphorescence.

2. False. The radioactive process is not affected to any great extent by chemical reaction or change in physical conditions.

3. True. Alpha is stopped by a single sheet of paper. It also can not penetrate the skin.

4. True. Alpha has a very high charge so it can easily ionize materials it comes in contact with.

5. False. Beta has no mass so its loss does not change nuclear mass effects.

6. True. The series results when an element starts the process of changing to a more stable material. This often requires several changes until the proper proton to neutron ratio is obtained.

7. True. Positrons are positive electrons. Since they are not normal in our environment and are the reverse of normal, the term antimatter is applied.

8. False. The only radioactive hydrogen isotope is tritium. One proton and two neutrons.

9. True. Beta emission results in a neutron being converted to a proton.

10. True. The binding energy is the energy difference required to hold individual particles together compared to the energy actually in the nucleus. The greater this value the more stable.

11. True. The mass defect is what was converted to energy as a result of mass differences. The energy is the binding energy.

12. True. Either process results in material closer to the material with maximum binding energy - a nucleus of iron.

13. False. Changes are purely statistical. No specific nuclei is identified.

14. False. Activity is related to the amount of material present.

15. False. The smaller the K value the greater the half-life. The equation: $t_{1/2} = 0.693/k$ shows the relationship.

16. True. Beta results in a nucleus with one more proton.

17. False. They are very uncommon. They are the result of production in laboratories. They have short half-lives.

18. False. Fusion has the greater energy per unit of material. However fusion has not been controlled.

19. False. The fuel must be enriched with U-235 but need not be pure.

20. True. The plasma is a very high energy system. Fusion can occur only as the result of high energy.

21. True. Compare this to the other units so you know what each are.

22. False. There is little that can be done to stop the background radiation. Some factors such as shielding can help control it however.

23. True. This is one that is showing up in homes in many areas of the country.

24. False. Radioactive tracers are very important in medicine but they are also used in industry and research.

25. True.

Answers to Understanding of Concepts Questions:

1. Although alpha radiation has very low penetrating ability, it does have a high ability to cause ionization damage. Beta radiation is considerably more penetrating but not as effective at causing ionization.

2. Charge must be conserved. Mass is not totally conserved but as far as the chemist is concerned there is no noticeable mass change.

3. A series of changes will occur when nuclei undergo radiation emission but have still not reached the maximum stability possible. The process continues until eventually a stable isotope forms. It may take several steps.

4. Nuclear stability appears to be related to the ratio of protons to neutrons in the nucleus. The band of stability is a group of isotopes that are not radioactive and the radioactive isotopes change to approach this band.

5. The most stable isotope is one of iron. Those materials that are close to it would be expected to have stability.

6. Activity is the number of disintegrations that occur per unit of time. The more material present the greater the activity. Half-life is the time it takes for half of the remaining material to undergo a change. The amount of material does not change the half-life.

7. Medical isotopes must be specific for certain areas of the body, such as a material that will concentrate in the bones. The material must have a short half-life and it must not be so active or penetrating that it will cause damage to the body. These represent some, but not all, properties of importance.

8. The use of a material for dating depends on the ratio of the radioactive isotope and other isotopes being constant. If the open air testing increased the amount of radioactive carbon, the ratio will have changed and this would have to be accounted for.

9. To cause a nuclear change the material must enter the nucleus. To get past electrons and protons the material should be of neutral charge. The neutron would work well.

10. Control rods are used. These rods are made of material that is capable of absorbing neutrons. Since it is neutrons that cause the nuclear splitting, reducing their number would slow the reaction.

11. Some materials have higher radiation than others, concrete for example is greater than wood. Live in a wood house. Don't live in areas that have high quanties of radioactive minerals near by. Live at lower elevations so the atmosphere can screen out more cosmic radiation. Have fewer X-rays and don't fly in airplanes often.

12. Fission requires very high energy to initiate. This can not be obtained under laboratory conditions.

13. Rads measure the radiation dosage absorbed by tissue. Curies measure the activity of a radioactive sample.

Answers to Testing the Analytical Concepts

1. This question relates to the band of stability. Calcium-47 has 20 protons and 27 neutrons. This is on the left side of the stability band which would lead us to expect beta emission.

2. We need a mass and charge balance between reactants and products.

$$^{239}_{94}Pu + ^{4}_{2}He \Longrightarrow ^{1}_{0}n + ^{242}_{96}Cm$$

3. The relationship we can use is $t_{1/2} = 0.693 / k$. We also need to convert from seconds to years.

$$t_{1/2} = \frac{0.693}{1.0 \times 10^{-13}\ s^{-1}} \times \frac{1\ hr}{3600\ s} \times \frac{1\ day}{24\ hr} \times \frac{1\ yr}{365.25\ day} =$$

$$= 2.02 \times 10^{5}\ yr$$

4. The decay constant is expressed in sec^{-1} units.

$$t_{1/2} = \frac{0.693}{k} \quad \text{or} \quad k = \frac{0.693}{t_{1/2}}$$

First convert years to seconds.

$$\# \text{ sec} = 12.3 \text{ yr} \times \frac{365.25 \text{ day}}{\text{yr}} \times \frac{24 \text{ hr}}{\text{day}} \times \frac{3600 \text{ sec}}{\text{hr}}$$

$$= 3.88 \times 10^{8} \text{ sec}$$

$$k = \frac{0.693}{3.88 \times 10^{8} \text{ sec}} = 1.79 \times 10^{-9} \text{ sec}^{-1}$$

5. We have the relationship $\ln \frac{N}{N_0} = -kt$

We also know that $t_{1/2} = \frac{0.693}{k}$ or $k = \frac{0.693}{t_{1/2}}$

substitute for k in the ln relationship

$$\ln \frac{N}{N_0} = \frac{-0.693\ t}{t_{1/2}}$$

rearrange for t and convert t to a positive value

$$t = \ln \frac{N_0}{N} \times \frac{t_{1/2}}{0.693}$$

The N_0 / N ratio is 15.3 / 7.0 = 2.2

$$t = \ln 2.2 \times \frac{5730 \text{ yr}}{0.693} = 6.52 \times 10^{3} \text{ years}$$

Answer to Chapter 19 Crossword

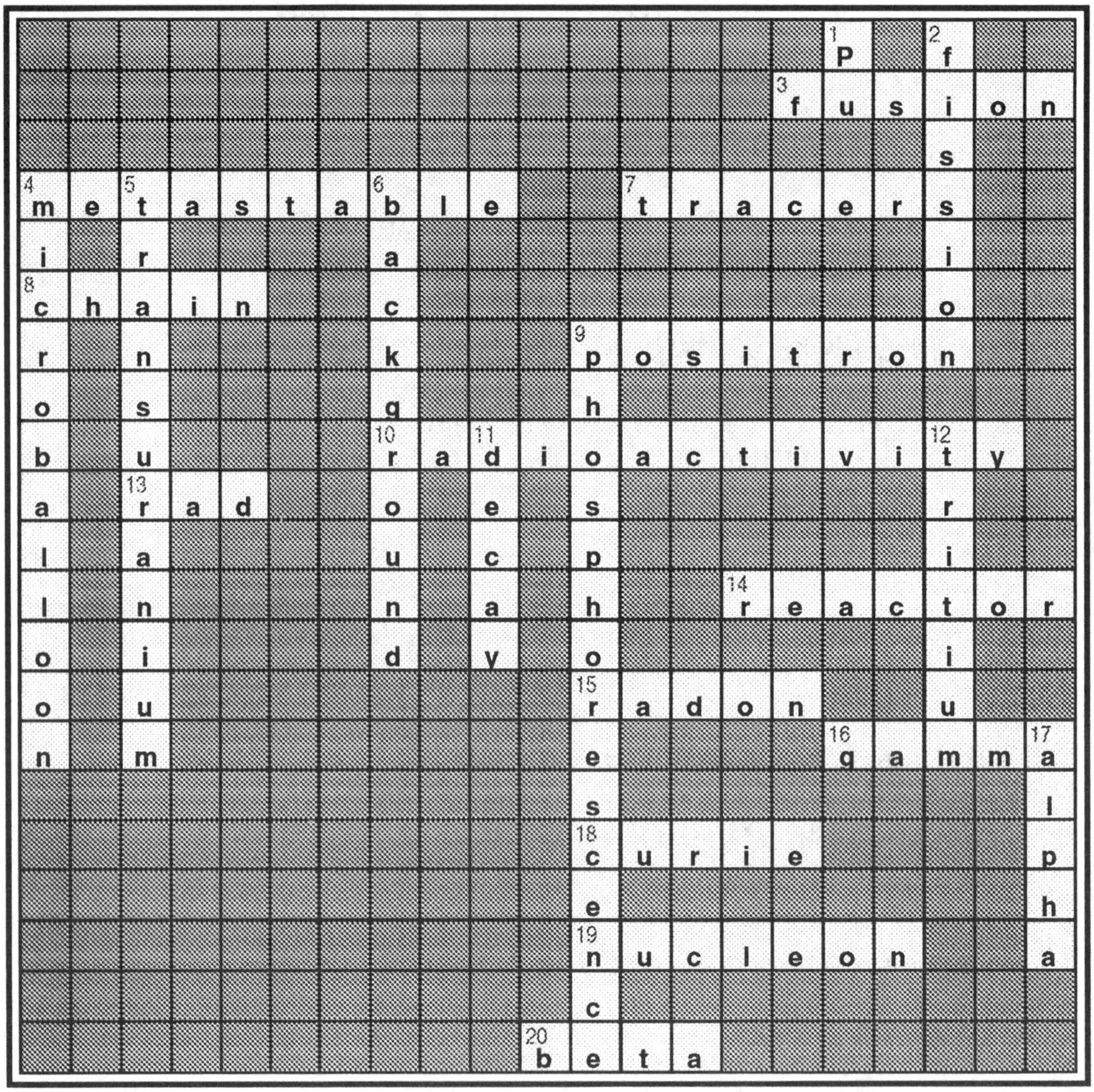

Some Supplementary Readings Relating to This Chapter:

1. Managing Risks from Indoor Radon, Chemical & Engineering News , October 11, 1993, page 35.

2. Researchers Take Up Environmental Challenge at Hanford, Chemical & Engineering News , June 21, 1993, page 9.

3. Europe: Betting Heavily on Fusion, Science , Vol. 250, 14 December 1990, page 1500.

4. U.S. Fusion Program: Struggling to Stay in the Game, Science , Vol. 250, 14 December 1990, page 1501.

5. Concern rises over nuclear waste explosion risk., Chemical & Engineering News , October 29, 1990, page 20.

6. Blinded by the light, Time , December 20, 1993, page 54.

7. Nuclear Plants Held Hostage to Old Fuel, The New York Times , October 3, 1992.

8. Getting Nuclear Plants Down to Size, The New York Times , July 12, 1992.

Chapter 20 Chemistry of the Atmosphere

The materials that occur in the atmosphere and some of the chemistry of these materials will be introduced in this chapter. Various air pollutants, types of pollutants, and some of the effects they have on us and our environment are identified. The relationship of our life style to various pollutants is discussed. One of the factors emphasized in this chapter is to make you aware of how interactive parts of our environment are.

Focus Your Study Efforts

The atmosphere is easily affected:

Objectives:

1. You should be able to give a description of the atmosphere and what its components are.
2. You should be able to discuss the measurement values used to express impurity concentrations.

The atmosphere, while a huge volume, is very much subject to localized disturbances. Often the tendency is to consider the atmosphere to be a homogeneous mixture, when in reality, it is far from that. Local influences often are not dissipated due to environmental, geographical, or concentration factors.

Atmospheric impurities are present in very small quantities, such as parts per million or billion. Our inability to comprehend values of this small size often make it difficult for us to consider these quantities to be hazardous. You need to recalibrate measurement skills when working at this level. The discussion of SO_2, text page 867, will help with this.

You should keep in mind that many of the problems in this chapter are useful for providing you with relationships about relative amounts. It would be well worth your time to think about the answers and their significance in a relative way rather than as absolute values.

> The quantity of SO_2 present in the atmosphere is about 2.00×10^{-8} percentage by volume. How many liters of SO_2 gas are present in one cubic mile of air?
>
> The first step in solving the problem is to convert to common units. Convert a cubic mile of volume to cubic centimeters (mL).
>
> $$\text{cm} = 1.00 \text{ mile} \times \frac{5280 \text{ feet}}{1.00 \text{ mile}} \times \frac{12 \text{ inches}}{1 \text{ foot}} \times \frac{2.54 \text{ cm}}{1.00 \text{ inch}}$$
>
> $$= 161{,}000 \text{ cm}$$
>
> A cubic mile is then $(161000)^3 = 4.17 \times 10^{15} \text{ cm}^3 = 4.17 \times 10^{12}$ L
>
> Of this total volume $2.00 \times 10^{-8}\%$ is SO_2
>
> Thus $4.17 \times 10^{12} \text{ L} \times 2.00 \times 10^{-10} = 830$ liters of SO_2 is present

Of all the atmospheric levels shown in Figure 20.3 of the text, the lower two are extremely significant to us. The lowest level, the troposphere, is where humans have the greatest impact. This lowest level is shallow, has a small volume, and does not readily mix with the level above it. What goes into the troposphere may circulate the globe several times and end up continents away, or the pollutants may be forced out, due to weather patterns, within a few miles. The number of variables is very large and they are not always predictable.

Pollutants can be classified:

> **Objectives:**
>
> 1. You should be able to describe what the term air pollutant means.

2. You should be able to classify pollutants into primary or secondary categories.
3. You should be able to name the major air pollutants.

Any material that degrades air quality can be classified as a pollutant. It is essential to recognize those factors that can be controlled. Some items can be addressed and they include: quantity of pollutant material, proportion of pollutants relative to the background level, how and in what form the material is released to the environment, and sometimes the chemical composition of the pollutant.

According to data in Table 20.2 of the text what percent of NO_x emissions is due to burning of fossil fuels compared to other sources?

Total emissions = 74×10^6 tons
Total from fossil fuels only = 66×10^6 tons

$$\% \text{ from fossil sources} = \frac{\text{Wt from fossil fuels}}{\text{Wt from total}} \times 100\% =$$

$$= \frac{66 \times 10^6}{74 \times 10^6} \times 100\% = 89\%$$

The primary pollutants, those that directly enter the atmosphere, sometimes may not be as great of concern as those resulting from atmospheric reactions, the secondary pollutants. Primary pollutants include oxides of sulfur and nitrogen, hydrocarbons, oxides of carbon, and chlorofluorocarbons.

Pollutant Classifications

Primary	Secondary
Particulates	N_2O_4
NO_x	R-O-O-R'
Hydrocarbons	O_3
SO_2	·OH
CO	PAN
CO_2	PAH
CFCs	
Volatile organic solvents	

Pollutants often change once they enter the atmosphere due to photochemical reactions resulting in free radical formation and/or photochemical dissociation.

> The following set of reactions is proposed as being possible in creating air pollution. Identify the underlined materials.
>
> $O_3 + \underline{RCH{=}CHR} \Longrightarrow \underline{R\text{-}C(H){=}O} + RO\cdot + HOC\cdot$
>
> $O_3 + h\nu \Longrightarrow O_2 + \underline{O}$
>
> $O + R\text{-}C(H){=}O + h\nu \Longrightarrow R\text{-}C(=O)\cdot + \underline{OH\cdot}$
>
> $O_2 + R\text{-}C(O)\cdot \Longrightarrow \underline{R\text{-}C(=O)\text{-}O\text{-}O\cdot}$

The first material is an unsaturated hydrocarbon.
The second material in the first reaction is an aldehyde.
The material in the second reaction is an oxygen atom.
The third reaction has a hydroxyl free radical.
The last item is an organic peroxide free radical.

Some pollutants will interact with water droplets to form aerosols and may eventually result in acid rain. Other materials may be concentrated on particulates. Aerosols and particulates are of major concern because they can carry large amounts of pollutants into the lungs to be deposited where they may be very hazardous. This most certainly occurs with carcinogen materials.

Smog is of two major types:

Objectives:

1. You should be able to distinguish between the two kinds of smog.
2. You should be able to discuss the conditions conducive to forming photochemical smog.

Smog often tends to be the foremost thought when the topic of air pollution is discussed. There are two major classifications of smog with significantly different characteristics. Industrial smog contains reducing components and several primary pollutants. Photochemical smog is rich in oxidizing agents and, as a result of reactions initiated by sunlight, contains several secondary pollutants.

As shown in Figure 20.7 of the text, photochemical smog is variable in components with the time of day. The concentrations also vary with time. Looking at the figure it is easy to see that a relationship exists between sunlight, pollutant levels, and population movement.

Photochemical smog is rife with free radicals and the reaction products they produce.

Free radical reactions are extensively outlined in the text (pages 874-880). You should look at these carefully paying special attention to similarities in many of these reactions.

Many components affect the atmosphere:

Objectives:

1. You should be able to describe the two roles ozone may play in the atmosphere.
2. You should be able to explain how acid rain is produced and the effects it may have on the environment.
3. You should be able to discuss what CFCs are and their role in ozone depletion.
4. You should be able to describe the "greenhouse effect".
5. You should be able to identify major chemicals that contribute to air pollution.

Ozone is involved in two separate roles in the atmosphere. When in the troposphere it affects plants, animals, and products in adverse ways. Its presence in the stratosphere is useful since it absorbs high energy radiation, which if it reaches the earth, can cause serious medical problems.

Hydrocarbons have many natural and human produced sources. As a pollutant they undergo several reactions via oxidation and/or with NO_2. This extensive chemistry makes hydrocarbons difficult to control.

Sulfur compounds are a major concern because of the cost to remove sulfur from fuels before it compounds reach the atmosphere. Also, the fate of sulfur in the atmosphere may be sulfuric acid, a very strong and corrosive acid. This acid is also a health concern because it aerosols into the lungs. Sulfuric acid corrodes metals, rocks, or objects made from rocks, and it also impacts on plant life.

The natural sources of carbon monoxide, CO, are many times greater than the human sources. The concern with human generation is that it very often is concentrated into very small environmental areas.

Acid rain is the consequence of oxides of sulfur and nitrogen interacting with moisture. Acid rain provides evidence with respect to how far reaching pollutions problems can go. Evidence shows that acid rain can result from pollution emissions hundreds of miles away.

The quantity of SO_2 present in the atmosphere is about 2.00×10^{-8} percentage by volume. Assume all of the SO_2 found in a cubic mile of air (about 830 liters) was converted to SO_3 by reacting with molecular oxygen and then converted to H_2SO_4. Assuming standard conditions are present for temperature and pressure what mass of acid would be produced?

This is a stoichiometry problem which means the place to start is with a balanced chemical equation or equations.

$$2\ SO_2 + O_2 ===> 2\ SO_3$$
$$2\ SO_3 + 2\ H_2O\ ====> 2\ H_2SO_4$$

Net $$2\ SO_2 + O_2 + 2\ H_2O ====> 2\ H_2SO_4$$

$$g\ H_2SO_4 = 830\ L\ SO_2 \times \frac{1\ mol\ SO_2}{22.4\ L\ SO_2} \times \frac{2\ mol\ H_2SO_4}{2\ mol\ SO_2} \times \frac{98\ g\ H_2SO_4}{1\ mol\ H_2SO_4} = 3.65 \times 10^3\ grams$$

The concerns with chlorofluorocarbons are an example of a material once thought to be very safe because they are so unreactive. However, because of this nonreactive character, CFCs have a long survival time and have been able to penetrate from the troposphere to the stratosphere. The high energy environment in the stratosphere enables CFCs to react rather well. The chief reactant, unfortunately, is ozone.

The "greenhouse effect" is a natural and essential phenomenon. It is the human perturbation that causes problems. While the major emphasis is on CO_2, other materials also contribute to the effect. The most important after carbon dioxide, are water vapor, methane, and ozone.

Carbon dioxide in the environment is a classic lesson about equilibrium and the ease at which a balance can be upset. As shown in Figure 20.23 of the text, the equilibrium cycle is very complex between CO_2 producers and CO_2 absorbers. Current evidence suggests that CO_2 is being produced at a rate greater than the absorbers can accept it. It may even be that large receptors, such as the oceans, are at equilibrium capacity.

Air pollution is due to many products in addition to those that receive major attention. Many industrial chemicals also contribute, to concerns, at least at the local level. Table 20.7 in the text identifies a few of the significant chemicals released.

When considering air pollution it is worthwhile to take into account what we do to ourselves within our own homes and work places. It brings to focus the question, when is pollution acceptable and what makes it so?

Symbols found in Chapter 20

Page	Symbol	
871	NO_2*	The * indicates an excited molecule
872	NO_x	Used to represent various oxides of nitrogen
873	R-O-O-R'	Representation for organic peroxides
878	FEV_1	Forced expiratory volume
879	•OH	Hydroxyl radical
880	PAN	Peroxyacetylnitrates - part of photochemical smog
880	PAH	Polynuclear aromatic hydrocarbons - smog material
888	CFC's	Chlorofluorocarbons
892	HCFC's	CFC substitutes
899	^{222}Rn	Radon gas air pollutant

Terms in Chapter 20

The following terms are listed for your convenience. They should not become a list for you to memorize. You should have a working definition for each. If you have doubt about any particular term check its definition in the glossary and/or review it in the chapter. Some of these terms are included in the crossword puzzle later in the chapter. This chapter also makes use of several terms introduced in chapters 10 and 11.

Troposphere	Stratosphere	Air pollutant
Adsorption	Absorption	Aerosols
Particulates	Mutagenic	Carcinogenic
Adsorb	Absorb	Free radical
Photochemical reaction	Electronically excited molecule	
Photodissociation	Smog	London type smog
Photochemical smog	Urban smog	Primary pollutant
Secondary pollutant	Hydroxyl radical	Acid rain
Polynuclear hydrocarbon		Slaked lime

Polynuclear aromatic hydrocarbon
Electrostatic precipitator
Greenhouse effect
Chlorofluorocarbons
Ozone holes

Concept Maps for Chemistry of the Atmosphere

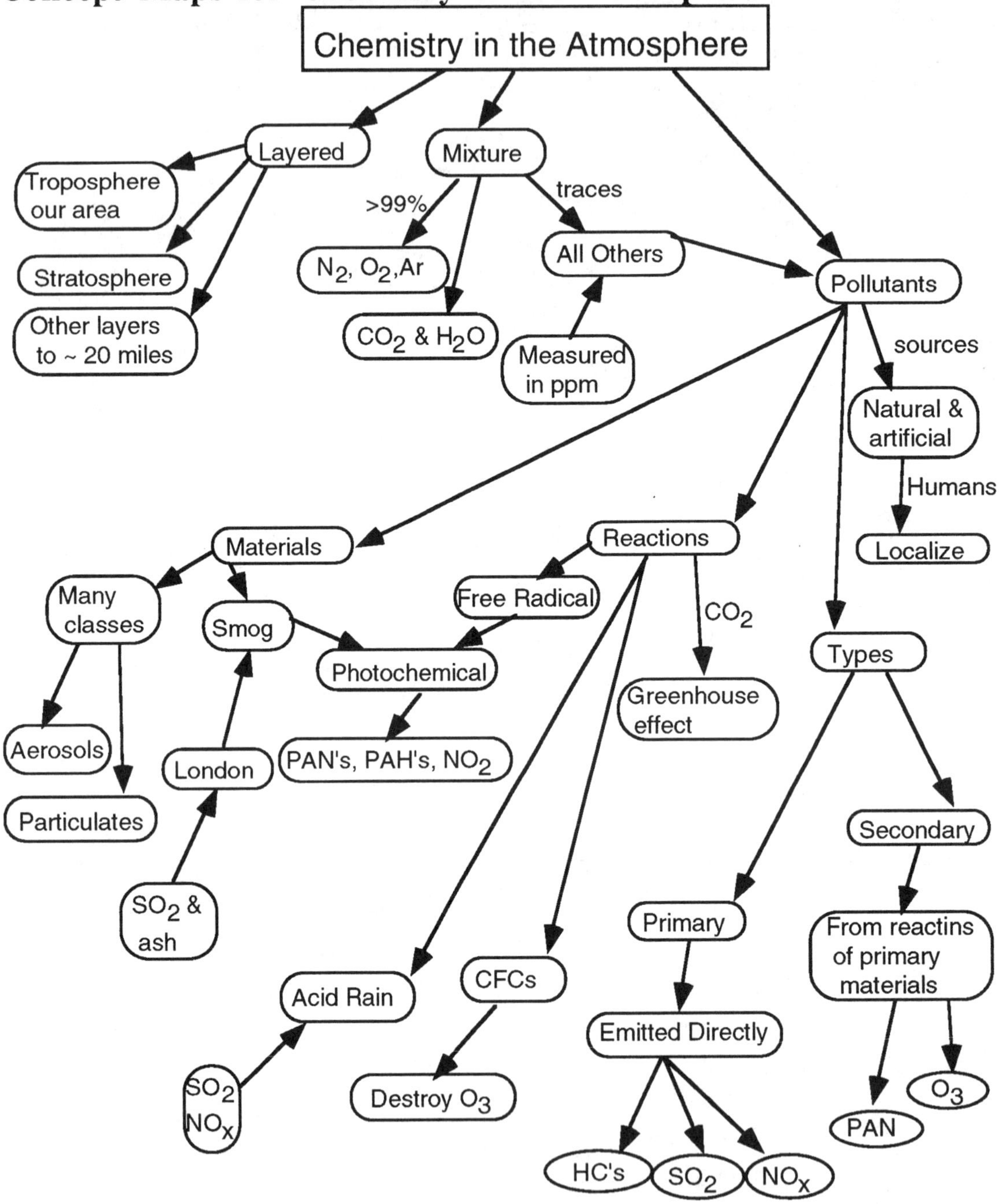

Chemistry of the Atmosphere True/False Statements

Answers are at the end of the chapter.

1. The atmosphere is homogeneous.

2. Atmospheric pollution is often region specific.

3. Expressing pollution concentrations by using percentages is not useful because of the small concentrations involved.

4. Air pollution can apply to the home as well as the troposphere.

5. Aerosols involve particles larger than those classified as particulates.

6. Aerosols may pick up pollutants by both adsorption and absorption processes.

7. Particulates can be controlled, at least to some extent, by physical methods.

8. Photochemistry is important in the process of generating London type smog.

9. Free radicals are very reactive species and are important in atmospheric reactions.

10. Sulfuric acid is a constituent of London type smog.

11. Ozone is a major component in generating organic peroxides.

12. Primary pollutants are the consequence of photochemical reactions.

13. Temperature is an important component in smog chemistry.

14. NO_2 has an effect on humans but is only a very minor problem with plants.

15. Ozone may be detrimental or beneficial depending upon where it is in the atmosphere.

16. The ·OH free radical is the only one of significance in atmospheric chemistry.

17. PAN and PAH are minor variations of the same pollutant.

18. Use of tall smoke stacks is an effective way to minimize SO_2 pollution.

19. Nearly all rain has a pH value less than 7.

20. CFCs are a major problem in the troposphere.

21. A major factor in the effectiveness of CFCs in the destruction of ozone is their ability to generate chlorine atoms.

22. There are some potential substitute compounds for CFCs.

23. It would be advantageous to totally eliminate the "greenhouse effect".

Testing Your Understanding of Concepts in Chapter 20.
Answers are at the end of the chapter.

1. Discuss factors that may affect local air quality.

2. What evidence exists that indicates that mixing can and does occur between atmospheric layers?

3. What are the differences between aerosols and particulates?

4. Identify several items that result from photochemical reactions.

5. How do London smog and Photochemical smog differ?

6. What are the basic differences between primary and secondary pollutants?

7. If the same nitrogen oxide pollutants were generated in the same amounts, on January 21 of any year, in Los Angeles, California and Minneapolis, Minnesota, how would the effects compare?

8. Why do more hydrocarbons appear in auto emissions if the vehicle is running fuel-rich? What are the implications of either fuel-rich or fuel-lean systems?

9. How does stratospheric ozone protect us from radiation?

10. What general class of particles is very common in various forms of photochemical reactions?

11. Methane is a very common hydrocarbon pollutant yet, relative to other hydrocarbons, it is not that abundant in the air of cities. Explain.

12. Why is there so much concern about PAH molecules as an air pollutant?

13. At one time tall exhaust stacks were thought to be the answer to SO_2 pollution. That is not the case today. Why the change in attitude?

14. What condition of pH is necessary before precipitation can really be classified as acid precipitation?

15. Identify several ways “acid rain” can be a hazard to the natural environment.

16. The materials known as CFCs were used because they were so chemically unreactive in the troposphere. Whey have they turned out to be such a problem since they are unreactive?

17. Are there any advantages to the “greenhouse effect”?

Testing You on Some Analytical Concepts.
Answers are at the end of the chapter.

1. The average adult requires about 8500 liters of air daily. If the concentration of SO_2 in the air inhaled is 10.0 ppm how much SO_2 will be inhaled in a day?

2. The amount of SO_2 emitted in a given area was determined to be 29.5×10^9 Kg. If this was removed by the limestone process what amount of calcium sulfate would need disposal?

$$CaCO_3 \Longrightarrow CaO + CO_2$$
$$2CaO + 2\,SO_2 + O_2 \Longrightarrow 2\,CaSO_4$$

3. Methane, CH_4, composes about 0.00015 percent of a certain atmosphere. What is that concentration expressed in ppm?

4. Hexane is an automobile fuel component. Hexane has a density of 0.664 g/mL and constitutes 5.0% of this gasoline sample. What amount of CO_2, at STP, would be produced when one liter of gasoline is burned to CO_2 and H_2O.

Crossword - Chemistry of the Atmosphere

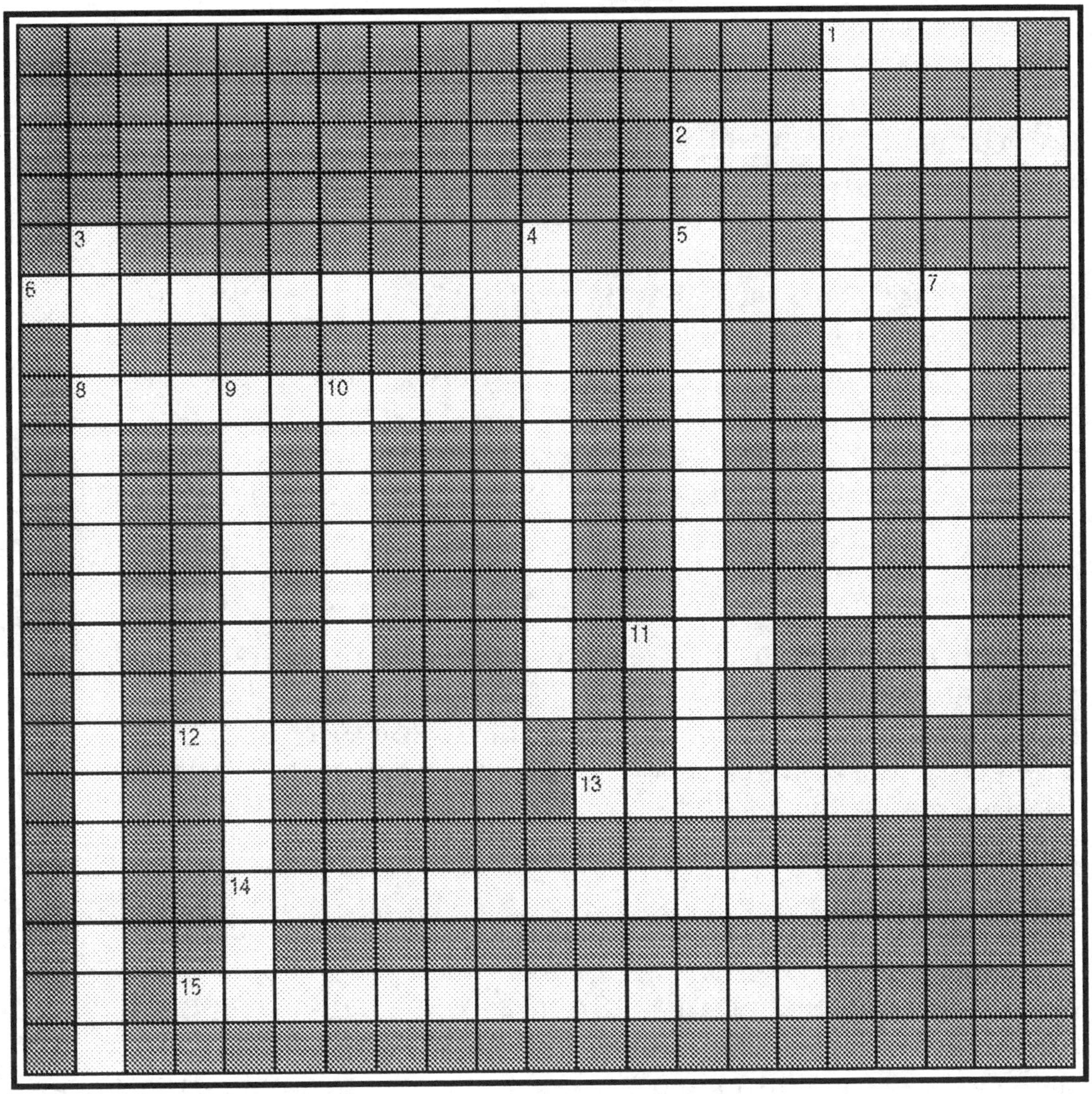

Crossword - Chemistry of the Atmosphere Clues

ACROSS

1 This type of air pollution is often divided into two major classes.
2 The name for the ·OH radical.
6 These are freon compounds and are identified as CFCs.
8 This is the region of the atmosphere nearest us. It is also the region where weather occurs.
11 Short for peroxyacetylnitrate pollutants.
12 The result of pollutants attaching to water droplets in the air.
13 Process of attaching to the surface rather than being within the material.
14 Materials capable of causing cancer are classified as this.
15 A method of removing particulate matter from industrial stacks.

DOWN

1 This is the region of the atmosphere where ozone is being destroyed.
3 Breakdown of a chemical due to the action of light.
4 This effect is largely due to excess amounts of the gas carbon dioxide.
5 The solid material found in the atmosphere is given this general name.
7 This class of pollutant is the result of a reaction in the atmosphere.
9 These type reactions result from light activation. These reactions lead to a type of smog found in the Los Angeles area
10 The gas from this element is one of the chief components of acid rain.

Ties to Previous Chapters

The atmosphere has many reactions occur in it and they involve oxidation-reduction (chapter17), acid formation, and reactions that acids cause (chapter 16). Since the atmosphere is mostly the gaseous state the study of gases from chapter 12 will be used. The atmosphere is a complex solution and the study of it will involving several of the items initially presented in chapter 15 on solution chemistry.

Answers to True/False Statements:

1. False. There are many variations in the atmosphere. Some are very local but others may be between hemispheres. Also the levels of the atmosphere are different and do not exhibit much mixing.

2. True. This may be due to natural or human causes. Los Angeles may be affected much more than the rest of California. A volcanic eruption in the northern hemisphere may have little effect in the southern hemisphere.

3. True. Pollution concentrations are very small. The use of values like percentages would require values like $2 \times 10^{-8}\%$. Values such as 2.0 ppm are more meaningful ways of expression.

4. True. When we paint or stain materials in the house or put out air fresheners we are adding something to the atmosphere that may degrade the quality.

5. False. Aerosols are very small water droplets that have materials either within them or on their surface.

6. True. Some materials may be dissolved into the droplet and others may just reside on the surface of the droplet.

7. True. Page 870 outlines several methods that can be used.

8. False. Photochemistry is important in photochemical smog. Some processes in London fog may be enhanced but it is primarily not a factor in London smog.

9. True. Free radicals are very unstable and quickly react with another species to stabilize themselves.

10. True. Sulfur dioxide is a major component of London smog and this is converted to sulfuric acid.

11. True. This is discussed on page 873 of the textbook.

12. False. Photochemical reactions use the primary pollutants and convert them to other materials.

13. True. Temperature is a critical component of many chemical reactions. Smog is no different. A temperature increase tends to increase the rate of a reaction.

14. False. It is true it affects humans but it also has an impact on plant life. This is discussed in the text on page 876.

15. True. Ozone in the troposphere has a negative impact on the environment. Ozone in the stratosphere is critical in stopping high energy radiation.

16. False. There are several free radicals involved in atmospheric chemistry. The hyroxyl radical is important but so are several others.

17. False. These are very different materials. See page 880 of the text for the structures of the materials.

18. False. The tall stacks minimize the effect in the local area, however this may just move the problem further downwind of the stack.

19. True. The fact that CO_2 dissolves in the rain droplet causes a pH in the range of 5.7. Carbon dioxide is always present.

20. False. CFCs are a stratosphere problem gas. That is where the ozone layer is and that is what CFCs attack.

21. True. The chlorine atom is extremely reactive and can cause a reaction and then get regenerated. A single chlorine atom can destroy thousands of ozone molecules.

22. True. Substitutes for CFCs are being produced. They do not always work the same and may not be compatible with current uses. They involve hydrogen in place of some of the halogen atoms. See page 892 in the text.

23. False. We are dependent upon the "greenhouse effect" to provide a liveable temperature on the earth. The concern is that we may over produce CO_2 and let the effect get out of balance.

Answers to Understanding of Concepts Questions:

1. Local air quality may be very different from the regional quality due to localized concentrators such as industry or automotive traffic, there may be local weather conditions, such as inversions, or the local topography may prevent good air mixing.

2. One good example is the fact that CFCs get into the stratosphere. This appears to take a long time but since they are emitted in the troposphere there must be mixing.

3. Aerosols are the result of pollutants being attracted to water droplets in the atmosphere. Particulates may be the direct result of stack emissions or other small particles that enter the atmosphere by various means.

4. Two major products of photochemical reactions include molecules in an excited state and free radicals. Other materials include organic peroxides, nitrogen oxides, and materials referred to as PAN's.

5. London smog is reducing in character where the photochemical type has many oxidizers present. London smog has particulates, SO_2, and materials such as soot. The photochemical type is susceptible to changes caused by sunlight and also has many hydrocarbons.

6. Primary pollutants are those emitted directly into the atmosphere where the secondary pollutants are produced by reactions of primary pollutants with oxidizers and activation by sunlight.

7. These reactions are temperature dependent. Higher temperatures make them go faster. Since Minnesota is much colder in January the effects would be less there than in California.

8. Fuel-lean would have an excess of air present so all the hydrocarbon should be used. Fuel-rich has an excess of the fuel so not all will be combusted. If the system is at the correct stoichiometry, optimum combustion occurs and pollutant levels should be minimized.

9. In the stratosphere, the ozone absorbs radiation in the ultraviolet region and is converted to oxygen atoms and molecules. This prevents the high energy radiation from reaching people. This

radiation can cause skin cancer and other medical problems if it reaches the surface of the earth.

10. Various free radicals are produced.

11. Other hydrocarbons are released in much greater abundance in the cities. Burning of fuels in industry and automobiles which do not produce methane add higher concentrations of the other components.

12. PAH molecules contain some forms that are known to cause cancer and many are toxic chemicals. These also make their way into the lungs where they can be very effective in causing problems.

13. Tall exhaust stacks merely place the pollutants higher in the atmosphere where they take longer to settle out. This gives the SO_2 time to interact and form sulfuric acid. In the delay of settling, the pollutants get carried farther downwind and create problems for other areas.

14. The natural pH value of water in contact with air is about 5.6. Unless the pH value is less than this it is considered normal.

15. Acid rain contains both sulfuric and nitric acids. These are strong acids and can readily react with many materials. Some plants are susceptible to acids or nitrogen oxides. Animals may get respiratory problems. The acidity can lower the pH of lakes and streams which has an adverse effect on fish or the food they eat. The high acidity may lower the pH of soils and release toxic metals into the water.

16. While unreactive in the troposphere they will last a long time. Eventually they are transported to the stratosphere. At this higher

elevation the ultraviolet light is capable of breaking C-halogen bonds. The free halogen atom, usually Cl, can then readily react with ozone.

17. Some "greenhouse effect" is essential. This helps keep the global temperature in a habitable range. The concern comes when the warming gets excessive.

Answers to Testing the Analytical Concepts

1. 10 ppm means that out of every million, 10^6, liters of air,10 will be SO_2.

$$L\ SO_2 = 8500\ L\ air \times \frac{10\ Liters\ SO_2}{1000000\ L\ air} = 0.085\ L\ or\ 85\ mL$$

2. The set of balanced equations shows that for each mole of SO_2 converted, a mole of calcium sulfate will be produced.

$$Kg\ CaSO_4 = 29.5 \times 10^9\ Kg\ SO_2 \times \frac{1\ mol\ SO_2}{64\ Kg\ SO_2} \times \frac{2\ mol\ CaSO_4}{2\ mol\ SO_2} \times$$

$$\frac{136.1\ Kg\ CaSO_4}{1\ mol\ CaSO_4} = 62.7 \times 10^9\ Kg$$

3. First convert the percent to a fraction, since percent is parts per 100 we have:

$$\frac{0.00015}{100} = 0.0000015$$

This can be converted to parts per million by multiplying the value by 1,000,000.

$0.0000015 \times 1{,}000{,}000 = 1.5$

4. The first step is to write a balanced equation for the combustion of hexane in oxygen:

$$2\ C_6H_{14} + 19\ O_2 \Longrightarrow 12\ CO_2 + 14\ H_2O$$

A liter of fuel has 5.0% hexane:

$$1000 \text{ mL fuel} \times \frac{5.0 \text{ mL hexane}}{100 \text{ mL fuel}} = 50.0 \text{ mL hexane}$$

$$\text{g of hexane} = 50.0 \text{ mL hexane} \times \frac{0.664 \text{ g hexane}}{1 \text{ mL hexane}} = 33.2 \text{ g}$$

$$\text{moles of hexane} = 33.2 \text{ g} \times \frac{1 \text{ mole hexane}}{86 \text{ g hexane}} = 0.39 \text{ mole}$$

$$\text{mol } CO_2 = 0.39 \text{ mol } C_6H_{14} \times \frac{12 \text{ mol } CO_2}{2 \text{ mol } C_6H_{14}} \times \frac{22.4 \text{ L } CO_2}{1 \text{ mol } CO_2}$$

$$= 51.9 \text{ L } CO_2$$

Answer to Chapter 20 Crossword

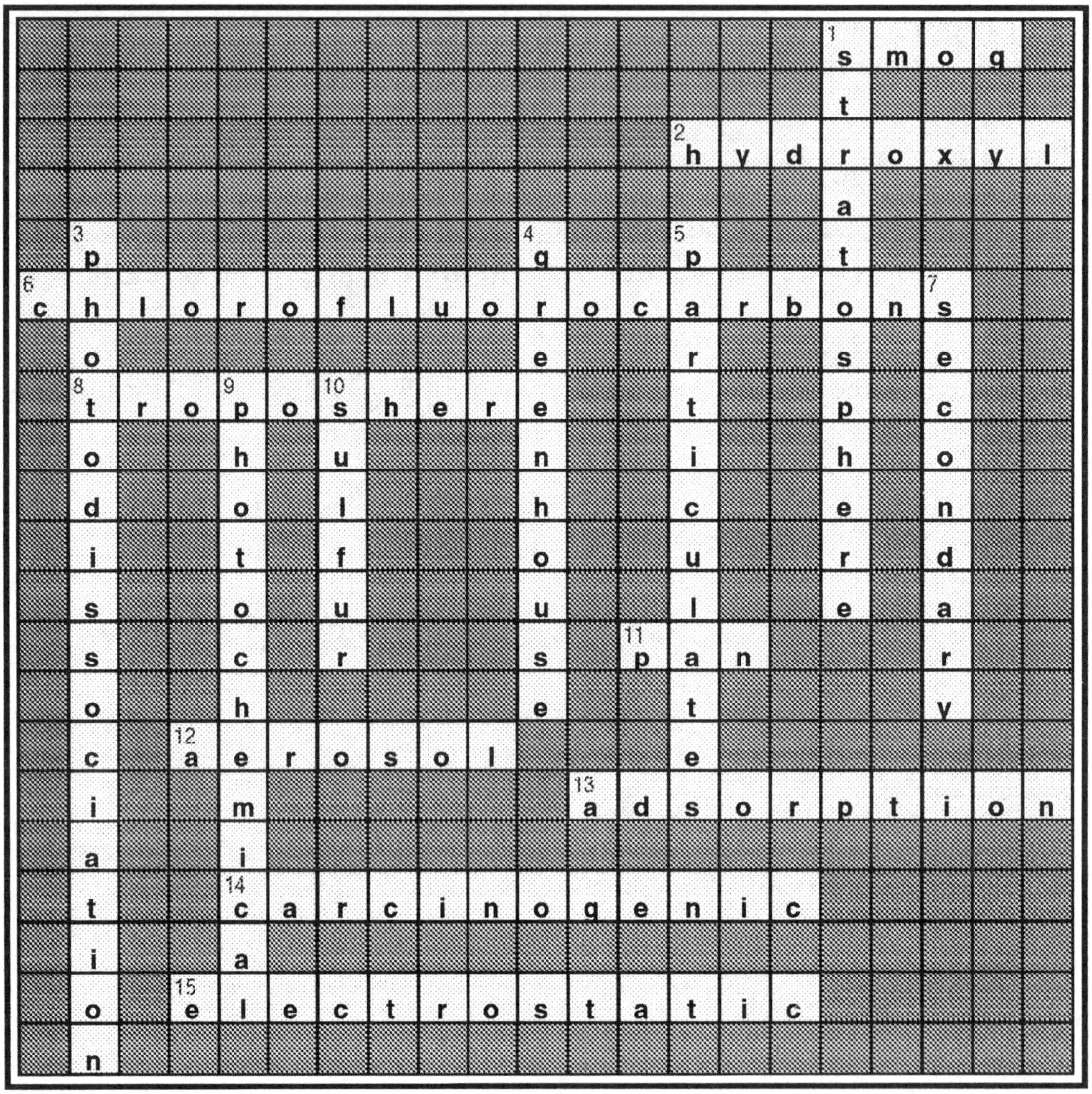

Some Supplementary Readings Relating to This Chapter:

1. Global warming trend interrupted in 1992, Chemical & Engineering News , January 25, 1993, page 7.

2. CFC Users Seek Exemptions from Ban on Ozone-Depleting Substances, Chemical & Engineering News , August 16, 1993, page 15.

3. Perfluorocarbons Use, Emissions May Face Restriction, Chemical & Engineering News , August 9, 1993, page 16.

4. Zapping Acid Rain With Microwaves, Science , Vol. 257, 11 September 1992, page 1479.

5. Ozone Depletions's Recurring Surprises Challenge Atmospheric Scientists, Chemical & Engineering News , May 24, 1993, page 8.

6. Ocean Carbon Cycle, Chemical & Engineering News , May 31, 1993, page 30.

7. Your Personal Greenhouse, Chem Matters , December 1990, page 8.

8. The Ozone Backlash, Science , Vol. 260, 11 June 1993, page 1580.

9. Greenhouse Science Survives Skeptics, Science , Vol. 256, 22 May, 1992, page 1138.

10. Massive Ozone and Smog Defile South Atlantic Sky, The New York Times , October 12, 1992.

11. As Halon Ban Nears, Researchers Seek a New Miracle Firefighter, The New York Times , December 15, 1992.

12. Help for Cleaner Air From a Mystery Coal, The New York Times , August 10, 1992.

13. A Promising Way to Cut Nitrogen Oxides at Power Plants, The New York Times , June 28, 1992.

14. Early Zap for Auto Pollution, The New York Times , March 22, 1992.

15. Looming Ban on Production of CFCs, Halons Spurs Switch to Substitutes, Chemical & Engineering News , November 15, 1993, page 12.

16. The Ozone Hole Reaches a New Low, Science , Vol. 262, 22 October 1993.

17. Growing Exchange of Information Spurs Pollution Prevention Efforts, Chemical & Engineering News , July 26, 1993, page 8.

Chapter 21 Chemistry of Life

This chapter develops some of the chemistry involved with life processes and in living cells. The common biochemical classes studied include: proteins, lipids, carbohydrates, nucleic acids, and energy compounds. Discussion includes how several of these chemicals relate to each other and are used in living systems.

Focus Your Study Efforts

Amino acids and proteins:

Objectives:

1. You should be able to recognize amino acids by their structure and functional groups.
2. You should be able to illustrate how condensation reactions lead to polypeptides and proteins.
3. You should be able to identify the various roles that proteins function in.
4. You should be able to explain the various structural formats proteins are found in.

Amino acids are the building blocks for protein polymers. Twenty natural amino acids are used, however, only the L-enantiomer of each.

Most amino acids have the amine group, NH_2, bonded on the carbon adjacent to the carboxylic acid group. This carbon is known as the alpha carbon. From this we get the term alpha amino acid.

Amino acids are classified into several types: nonpolar (hydrocarbon R groups), polar (alcohols or ketones on R group), acidic (2 acid groups present), or basic (2 amine groups present). Table 21.1 in the text identifies the class each fits into. For convenience the amino acids are usually identified with an abbreviation.

Draw the structure for: serylalaylcysteine.

The first thing to do is identify which amino acids are involved in the structure and write their structures out. They are joined with a bond between the amine group and acid group with elimination of a water molecule.

The amino acids are serine and alanine and cysteine.

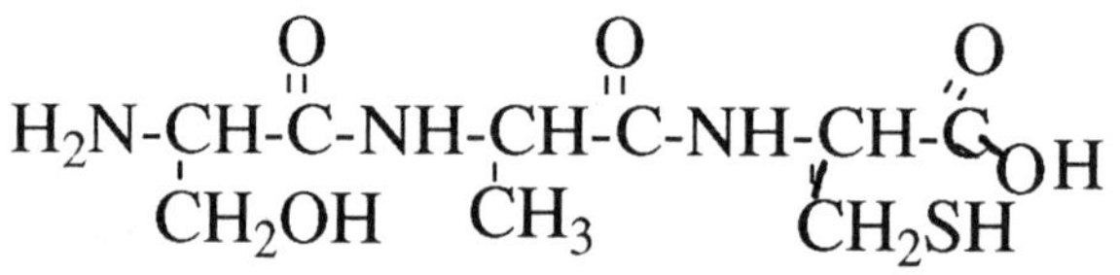

When placed in water the proton from the acid functional group of an alpha amino acid can be donated to the amine group. When this happens a double salt known as a zwitterion if formed. This form results in the amino acid having ionic properties.

Draw the zwitterion form of the peptide serylalaylcysteine.

The zwitterion is formed when the proton from the acid group is transferred to the amine group. This leaves a double ion charge on the molecule.

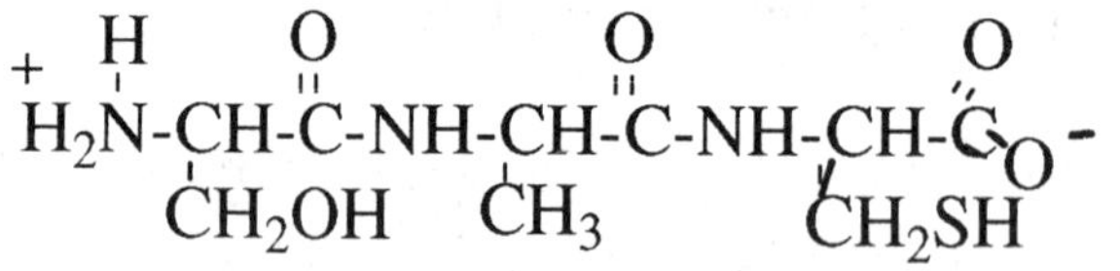

Amino acids undergo a condensation reaction to form dipeptides. The acid group of one amino acid reacts with the amine group of a second amino acid forming the peptide bond. An amino acid sequence is the pattern that is produced as a result of the series of these condensation reactions.If condensation reactions continue a polypeptide (up to 50 amino acid residues) or a protein (more than 50 residues) will form.

What is the name of the peptide drawn below?

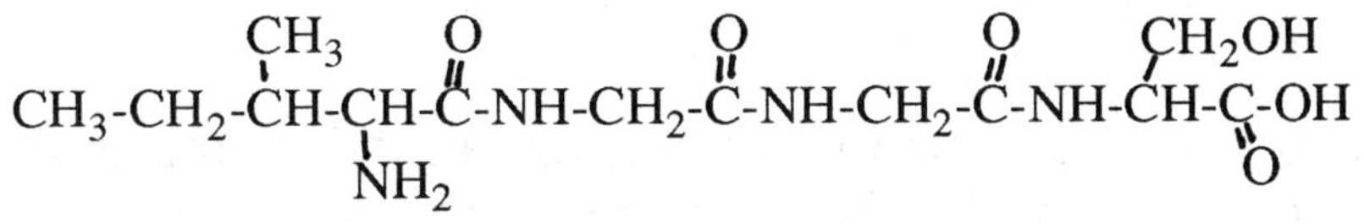

First identify the amino acids present. Look for the C=O group since that identifies where a new amino acid will start.

The first group is isoleucine, Ile, the second is glycine, Gly, the third is also glycine, and the forth is serine, Ser.

IleylGlyclGlyclSerine

Proteins are capable of functioning in several roles:

- Enzymes - catalysts for biological systems.
- Hormones - regulators of body reactions.
- Antibodies - protect us from disease.
- Transport medium - hemoglobin carries oxygen in blood.
- Membrane materials - such as hair or fingernails.

Of the 20 amino acids used, 8 are not produced in the body. These are known as the essential amino acids.

Proteins may have several structural formats:

- Primary - what sequence the amino acids are joined in.

Proteins may have several structural formats:

- Primary - what sequence the amino acids are joined in.
- Secondary - shape the molecule has, such as a helix.
- Tertiary - the way the protein is folded.
- Quaternary - when several chains are present they have a shape.

When a protein structure is changed so it can not function, the process of denaturation has occurred. This may be done by heat or chemical reactions.

When proteins function as enzymes they affect reaction rates by lowering the activation energy. The enzyme has an active site that is affected by charge, size, and shape. The reactant molecule, known as the substrate, must have a region that compliments the enzyme active site. Because of these criteria, enzymes are very selective catalysts.

It is relatively easy to modify the shape or charge in an enzyme. Changes in temperature, pH, or solution ionic strength, can affect the enzymes activity.

Monosaccharides and carbohydrates:

Objectives:

1. You should be able to identify various kinds of mono- saccharides, aldoses, ketoses, hexoses, and pentoses.
2. You should be able to recognize and explain the various structures of glucose.
3. You should be able to illustrate the common reactions of carbohydrates; condensation and hydrolysis.
4. You should be able to describe how starch and cellulose vary in structure and the significance of this variation.

Carbohydrates, like proteins, are condensation polymers. The monomer units for carbohydrates are sugars. These sugars usually are hexoses, six carbon sugars, or pentoses, five carbon atom sugar molecules.

All monosaccharides contain the alcohol functional group. Some will also contain the aldehyde group and consequently are called aldose sugars. The ketose sugars have a ketone functional group rather than the aldehyde. Because all three functional groups have polarity, sugars are water soluble.

Note in Figures 21.10 and 21.11 of the text, that several formats are used to represent sugars. You should become familiar with how the carbon atom numbering occurs, how to determine if the sugar is an aldose or ketose, and what the symbols α and ß mean.

Would the glycoside shown below have an alpha or beta linkage?

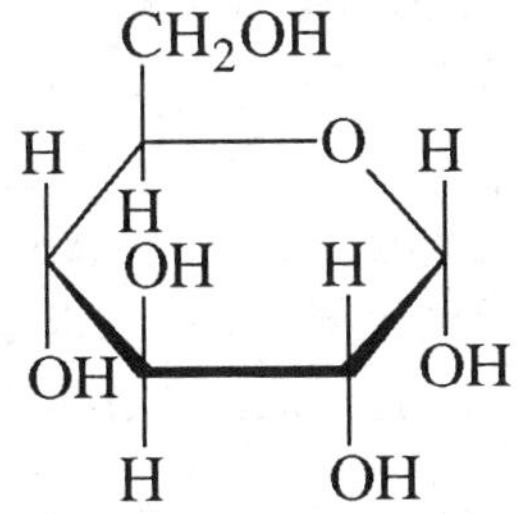

The linkage would be from the oxygen on the bottom at the right. Since it is on the bottom the linkage would be alpha.

Plants are the natural source of carbohydrates. Plants, via photosynthesis, convert CO_2 and H_2O to carbohydrates and oxygen.

Plants produce large polymers known as polysaccharides. These polymers store large amounts of energy. Some of the polymers, such as cellulose, also provide structural material to the plant.

The number of monosaccharides is not large. Only 20 are natural, and D-glucose is extremely important. Two monosaccharides can combine yielding a disaccharide. Only three disaccharides are of importance in this chapter - sucrose, maltose, and lactose.

The reverse of a condensation reaction is the hydrolysis reaction. This is outlined in the text in Figure 21.12. Hydrolysis reactions are very important to our ability to use carbohydrates.

Glucose is commonly called blood sugar. As such, it is an immediate energy source for our metabolism. It is, however, very important for us to have a regulated supply of glucose in our bloodstream. If the body can not effectively regulate glucose concentration a disease known as diabetes results.

Artificial sweeteners were developed for several reasons. As you look at the structures of four common products on pages 923 and 924 of the text, the similarity between alitame and aspartame can be seen. Sucralose is most similar to the disaccharides.

Of the polysaccharides; glycogen, starch, and cellulose, the first two are readily usable by animals but cellulose is not. Starch has two structural forms - amylose and amylopectin. These two materials show different properties and have different uses.

Glycogen has much chain branching. This molecule serves a quick energy source for animals.

Cellulose and starch differ in small way. Figures 21.13 and 21.15 in the text illustrate this well. Look at the bonding on the oxygen joining adjacent glucose units. This illustrates how significant even slight structural changes can be. Most animals can not hydrolyze cellulose.

Lipids - fats and oils and others:

Objectives:

1. You should be able to describe the difference between fats and oils.
2. You should be able to write equations for the formation of triglycerides from glycerol and fatty acids, and the reaction of fats with a base.
3. You should be able to illustrate the differences between saturated and unsaturated fats and oils.
4. You should be able to identify the structures of common steroids.

Lipids differ from protein and carbohydrate materials in several ways. Lipids are not polymers. Several lipids are esters. Lipids are not water soluble but dissolve in many nonpolar solvents.

Fats and oils are triesters of the alcohol glycerol. One difference between fats and oils is the physical state. Oils usually are liquid and fats are a semi-solid.

The acids making fats and oils are known as fatty acids. These acids may be either saturated or unsaturated. It is the type of acid present that determines if a fat is considered to be saturated, unsaturated, or poly unsaturated.

A fat made from three palmitic acids is compared to a fat made from three linoleic acids. What differences would be expected?

The main difference between the acids is in the amount of unsaturation. Linoleic has two sites of unsaturation and palmitic is completely saturated.

Unsaturation leads to oils rather than semi-solids. The linoleic would be a liquid and the palmitic would be solid.

A saponification reaction is the result of reacting a strong base, such as NaOH, with a fat or oil. This produces glycerol and three soap molecules. Thus the name saponification.

When unsaturated fats or oils are reacted with hydrogen under catalytic conditions the reaction is hydrogenation. The advantage of removing multiple bonds is that when cooking with multiple bonds present, they can undergo reactions at high temperatures and pick up food tastes.

Steroids are another main group of compounds classified as lipids. Steroid structures are much different than the triglyceride structure. One identifying characteristic of the steroids is the presence of four rings. See text page 933 for the skeletal structure.

Energy for biochemical reactions:

Objectives:

1. You should be able to discuss the process of photosynthesis and what the terms light and dark reactions refer to.
2. You should be able to describe the structure of ATP and its role in energy exchange.

The ultimate source of energy for biochemical systems is the sun. Photosynthesis converts CO_2 and H_2O to glucose and oxygen using visible radiation energies. The oxygen produced by this reaction is the source of oxygen present in the atmosphere.

Figure 21.17 in your text shows an important structure. This molecule, ATP, along with ADP, and AMP, are biological energy transfer compounds. Energy exchange occurs as phosphate groups are added to or removed from this series of molecules.

Photosynthesis is a sequence of reactions, some of which require sunlight, and some of which do not. The coordination compounds of manganese, the chlorophylls, capture light energy in a process known as the light reactions. This energy is then used in a series of dark reactions to complete the synthesis of glucose. The word dark here does not mean the absence of light, just that light energy is not required for them to occur.

Nucleic acids and molecule replication:

Objectives:

1. You should be able to explain the terms nucleic acids, DNA, RNA, and nucleotides.
2. You should be able to illustrate the differences between the structures of DNA and RNA.
3. You should be able to explain the differences between replication and transcription.
4. You should be able to explain the function of the different types of RNA.

Information needed to allow replication of molecules is found in the nucleic acids. Nucleic acids, like proteins and carbohydrates are condensation polymers.

Major important nucleic acids include DNA and a series of RNA molecules. They all contain a sugar, organic bases, and a phosphate unit. A major difference between DNA and RNA is the sugar base present. These molecules are illustrated in Figure 21.20 of the text. Again note the importance of slight differences in structure.

Another striking feature of these molecules is the use of only five organic bases. The base units, as shown in the text figure 21.21, are of two general formats, purine and pyrimidine.

A nucleotide unit contains one of two sugars, the nitrogenous base and a phosphoric acid component. When these units polymerize, a polynucleotide is formed.

A very important component of polynucleotides is the presence of hydrogen bonding. This bonding occurs between base units of the two strands, which in turn, holds the strands together.

A significant point to remember is that only certain bases will pair with certain other bases. These are known as complimentary base pairs.

Almost all DNA is found in the cell nucleus. This also is where the DNA strands unwind in a process known as replication. This allows them to function as templates.

After replication, the single DNA strand allows a strand of messenger RNA, mRNA, to be transcribed on its surface. This is where only complimentary base pairs will fit together and form hydrogen bonds.

A DNA molecule has a base sequence of: ATGCTAGGAT. What is the base sequence that compliments this in mRNA?

Only certain base pairs can join together. This is regulated to some extent by the ability to form hydrogen bonding. A table of pair arrangements is given below:

DNA		RNA	
Adenine	A	Uracil	U
Thymine	T	Adenine	A
Cytosine	C	Guanine	G
Guanine	G	Cytosine	C

The complementary base pairs can now be determined.

UACGAUCCUA

Replication and transcription will occur only if energy and enzymes are present.The newly formed mRNA carries the template to the cell ribosome where it becomes a template for transfer RNA, tRNA, to synthesize a protein.

Each individual amino acid has a three base code, known as a codon, that identifies it. See table 21.5 in the text for details. The transfer RNA molecule uses this code to bring in the proper amino acid to the protein molecule under construction.

Use of a host material to synthesize a needed substance is becoming more common in the field of biochemistry. The process involves placing specific genes in the host material. The common term for this process is biogenetic engineering. This technique has had mixed results and is currently carefully regulated.

Symbols found in Chapter 21

Page	Symbol	
906	D-	One of the enantiomer structures.
906	L-	A second common enantiomer structure.
906	R & R'	Symbols for hydrocarbon groups.
907	Pro etc.	Symbols for amino acids (see table 21.1).
920	α-D β-L	Glucose enantiomers.
921	-ose	Names of saccharides end in this.
926	⸝O⸜ & ˋOˊ	Linkages found in starch and cellulose.
931	C	Food Calorie, unit of energy equivalent to a kcal.
934	LDL	Low density lipoprotein.

934	HDL	High density lipoprotein.
935	ATP	Adenosine triphosphate.
935	ADP	Adenosine diphosphate.
938	DNA	Deoxyribonucleic acid.
938	RNA	Ribonucleic acid - there are various kinds.
939	A G T C U	Symbols for organic bases in DNA or RNA.
945	ATZ	Azidothymidine, drug used in treatment of AIDS.

Terms in Chapter 21

The following terms are listed for your convenience. They should not become a list for you to memorize. You should have a working definition for each. If you have doubt about any particular term check its definition in the glossary and/or review it in the chapter. Some of these terms are included in the crossword puzzle later in the chapter. This chapter also makes use of several terms introduced in chapters 10 and 11 and 13.

Biochemistry	sec. amine group	amino acid
alpha carbon	zwitterion	peptide bond
dipeptide	n-terminal end	c-terminal end
polypeptide	protein	amino acid sequence
enkephalins	ß-endorphin	enzymes
hormones	antibodies	hemoglobin
primary structure	secondary structure	tertiary structure
quaternary structure	α-helix	ß-pleated sheet
denaturation	activation energy	hydrolysis reactions
substrate	active site	carbohydrate
monosaccharide	disaccharide	polysaccharide
aldose	ketose	saccharin
aspartame	sucralose	alitame
hypoglycemia	hyperglycemia	diabetes mellitus
starch	glycogen	cellulose
amylose	amylopectin	lipid

triglyceride	linoleic acid	prostaglandins
saponification	atherosclerosis	hydrogenated
biogenetic engineering	steroids	Simplesse
Olestra	lipoproteins	progesterone
testosterone	estrogens	light reactions
dark reactions	chlorophyll a & b	nucleotide
genome	replication	codon
anticodon	retrovirus	

complementary hydrogen bonding

Concept Maps for Chemistry of Life

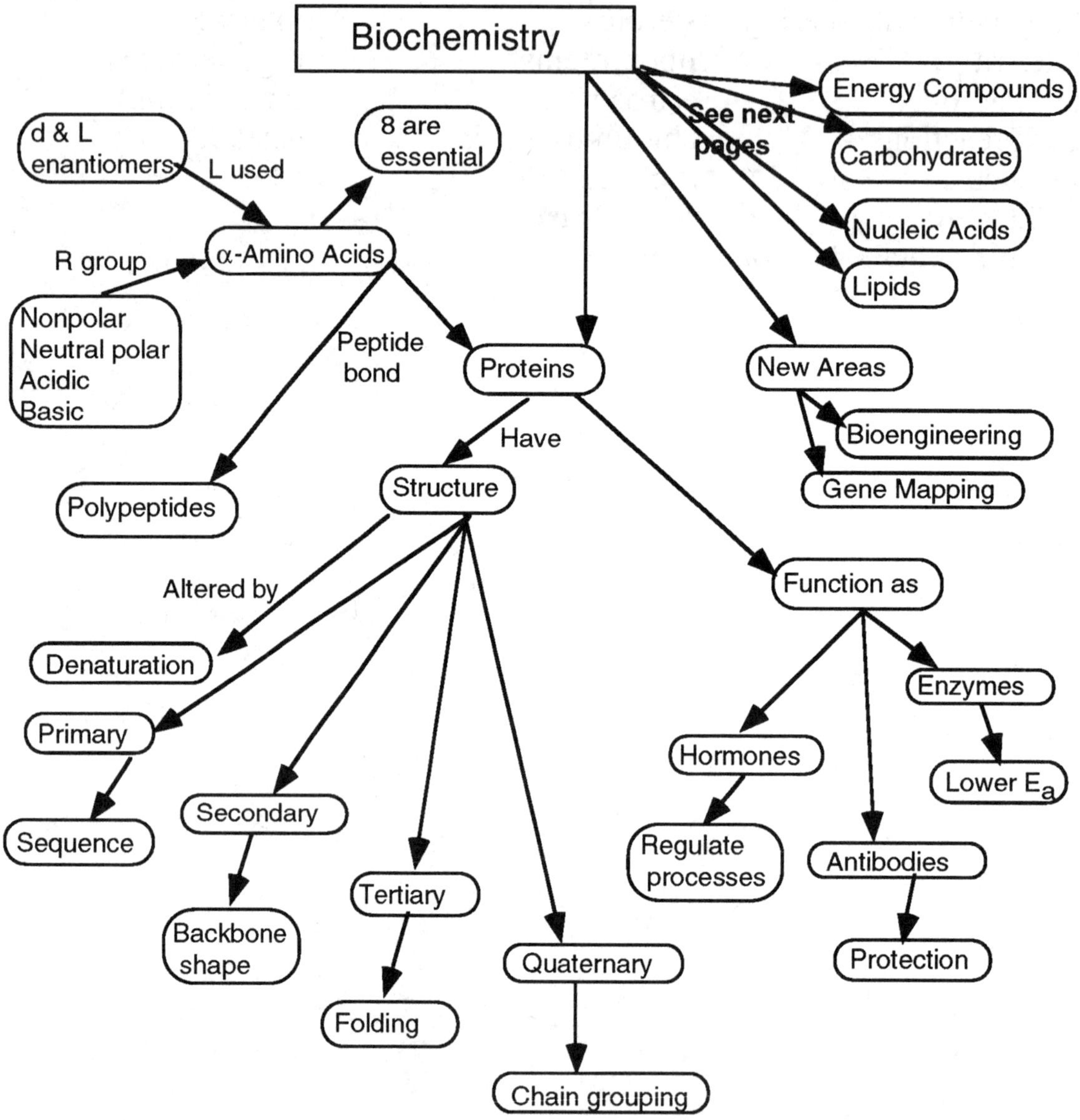

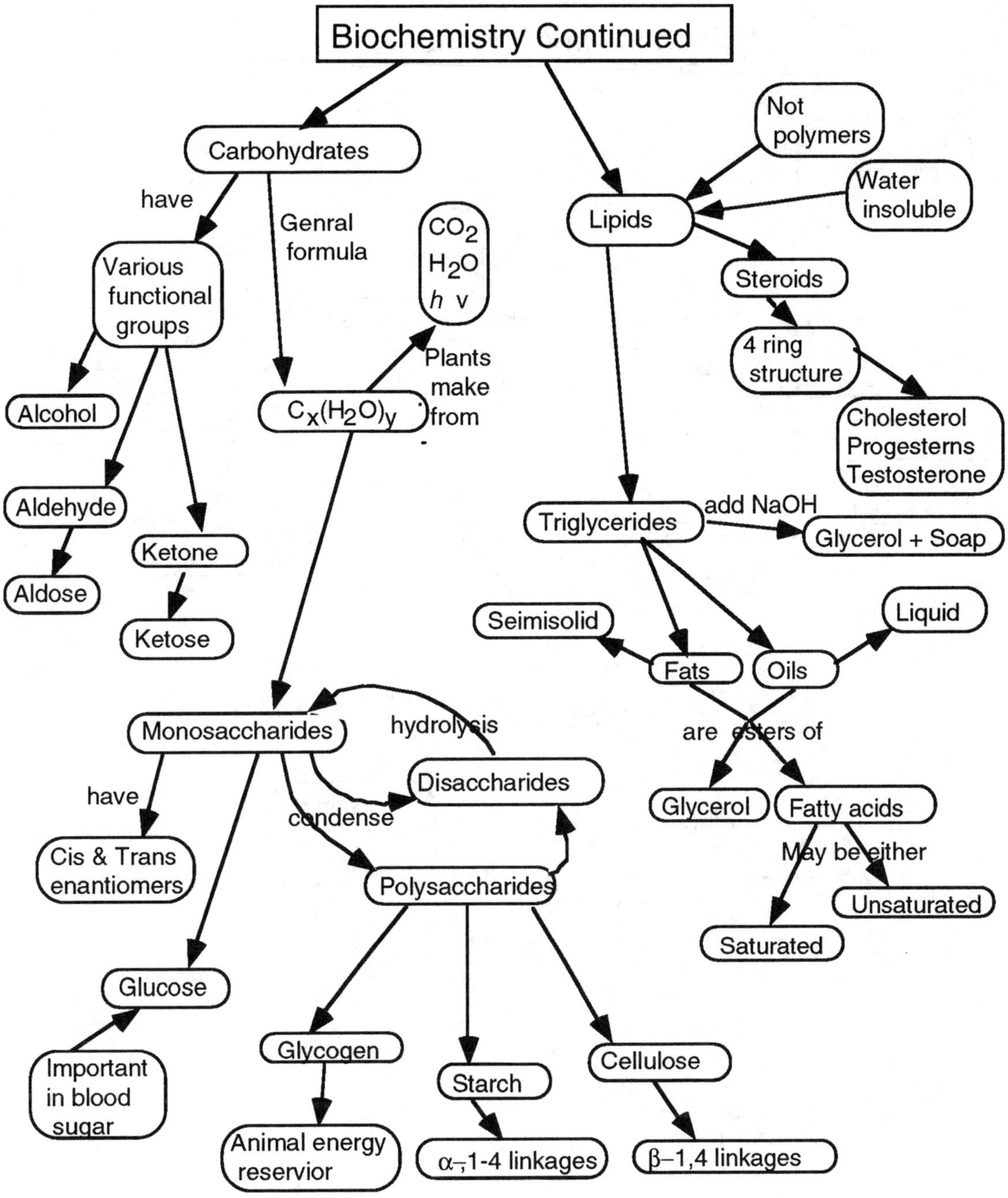

Biochemistry Continued
Carbohydrates
have
Various functional groups
Alcohol
Aldehyde
Aldose
Ketone
Ketose
Genral formula
$C_x(H_2O)_y$
CO_2
H_2O
h v
Plants make from
Monosaccharides
hydrolysis
condense
Disaccharides
Polysaccharides
have
Cis & Trans enantiomers
Glucose
Important in blood sugar
Glycogen
Animal energy reservior
Starch
α-1-4 linkages
Cellulose
β-1,4 linkages
Lipids
Not polymers
Water insoluble
Steroids
4 ring structure
Cholesterol
Progesterns
Testosterone
Triglycerides
add NaOH
Glycerol + Soap
Seimisolid
Fats
Oils
Liquid
are esters of
Glycerol
Fatty acids
May be either
Saturated
Unsaturated

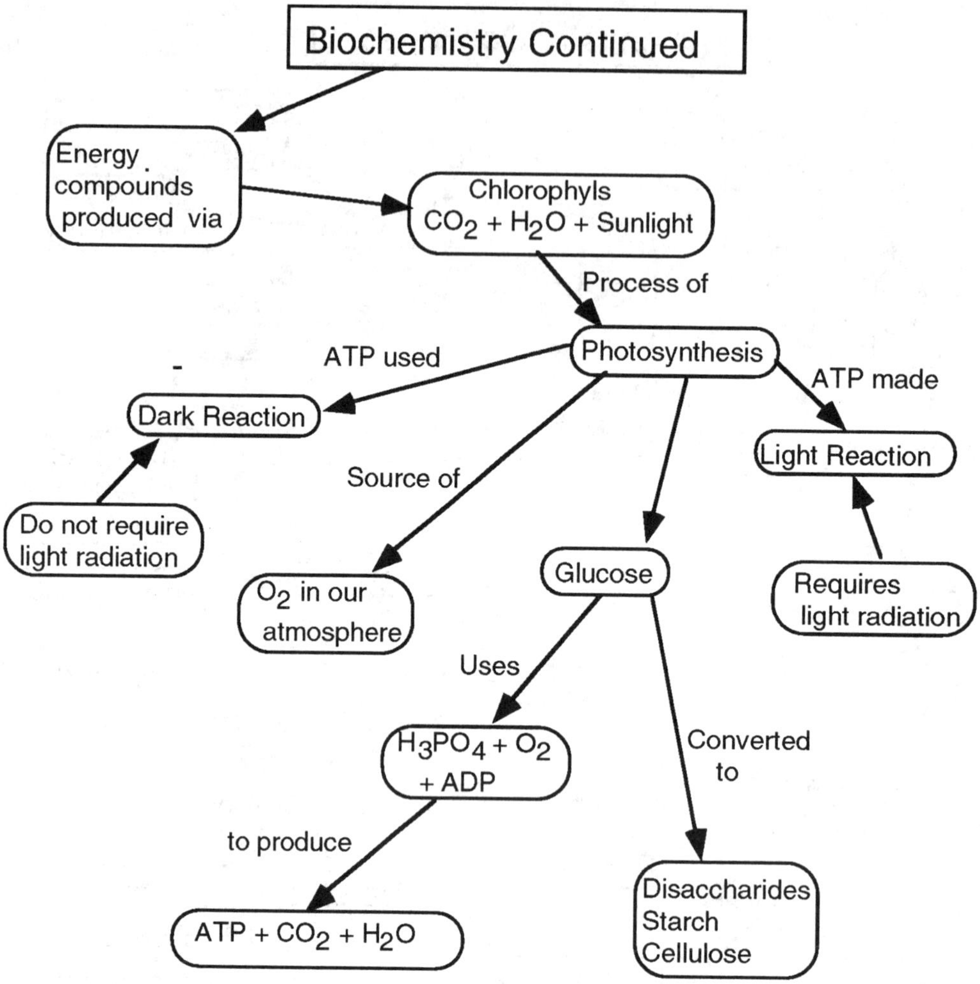
Biochemistry Continued
Energy compounds produced via
Chlorophyls
CO_2 + H_2O + Sunlight
Process of
Photosynthesis
ATP used
ATP made
Dark Reaction
Light Reaction
Do not require light radiation
Requires light radiation
Source of
O_2 in our atmosphere
Glucose
Uses
H_3PO_4 + O_2 + ADP
to produce
ATP + CO_2 + H_2O
Converted to
Disaccharides
Starch
Cellulose

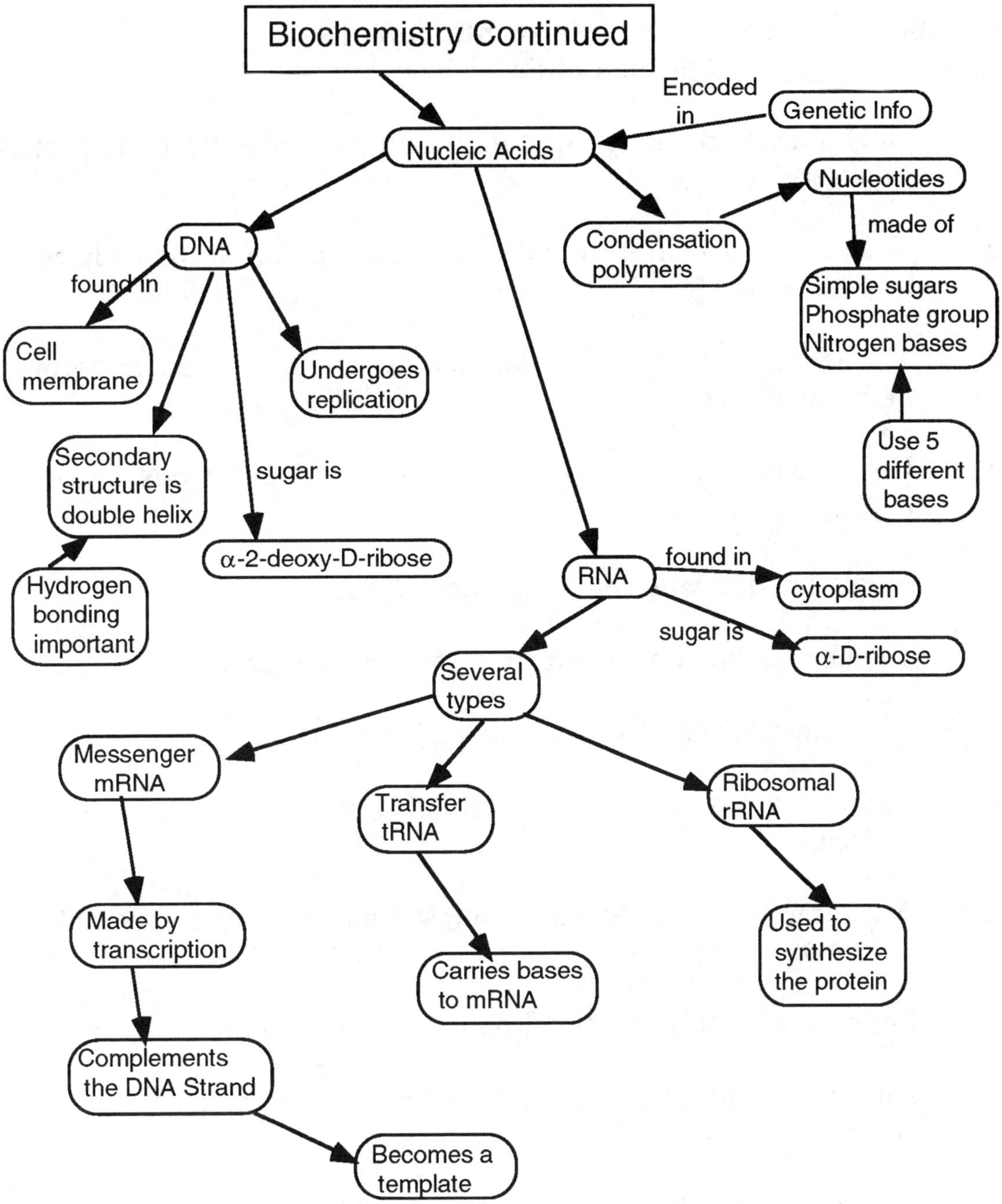
Biochemistry Continued
Encoded in
Genetic Info
Nucleic Acids
Nucleotides
made of
Condensation polymers
Simple sugars
Phosphate group
Nitrogen bases
Use 5 different bases
DNA
found in
Cell membrane
Undergoes replication
Secondary structure is double helix
sugar is
α-2-deoxy-D-ribose
Hydrogen bonding important
RNA
found in
cytoplasm
sugar is
α-D-ribose
Several types
Messenger mRNA
Transfer tRNA
Ribosomal rRNA
Made by transcription
Carries bases to mRNA
Used to synthesize the protein
Complements the DNA Strand
Becomes a template

Chemistry of Life True/False Statements

Answers are at the end of the chapter.

1. Only a racemic mixture of amino acid isomers are used in protein synthesis.

2. α- amino acids will have the amine group on the carbon adjacent to the carboxylic acid group.

3. The amino acid serine is polar, neutral, and has an alcohol group in the molecule.

4. Histidine is a basic amino acid because it contains two NH_2 groups.

5. Formation of zwitterions is pH dependent.

6. The peptide bond results from a condensation reaction.

7. The name for Ser-Gly-Ala is serylglycylalanine.

8. Amino acid residues are the materials left over when a peptide molecule is formed.

9. A natural pain reliever may come from a group of compounds known as endorphins.

10. Proteins have only one function in the human body.

11. Any amino acid used to make proteins is said to be an essential amino acid.

12. Disulfide bridge bonds are used to develop a tertiary structure in proteins.

13. Hydrogen bonding determines a proteins primary structure.

14. Enzymes, when functioning as catalysts, are very specific.

15. The primary reaction in digestion of proteins is saponification.

16. Regions of chemical reactivity in a molecule are known as active sites.

17. Maltose is made from two ketose monosaccharides.

18. Carbohydrates are produced by the photosynthesis reactions.

19. Only the D- enantiomer of glucose is used in the human body.

20. Carbohydrates may undergo a hydrolysis reaction during digestion.

21. Artificial sweeteners are structurally very similar to the carbohydrates.

22. Glycogen and starch are the same material.

23. Soluble starch is made of amylopectin.

24. One function of cellulose could be to act as structural material.

25. Lipids are not condensation polymers.

26. Two fatty acids with the same number of carbon atoms may be liquid or solid depending on the degree of saturation in the acid chain.

27. Fats and oils are esters.

28. Prostaglandins are triglycerides.

29. Fats are high energy foods compared to carbohydrates.

30. All artificial fats are not digestible.

31. Lipoproteins carry cholesterol in the bloodstream.

32. ADP, adenosine diphosphate, is the high energy compound in biochemical reactions.

33. Photosynthetic dark reactions occur only at night.

34. Chlorophyll is a coordination compound of magnesium ion with a coordination number of 4.

35. RNA is usually found in the cell nucleus.

36. Not all base types used in nucleic acids will be found in the nucleic acid RNA.

37. Replication and transcription are the same thing.

38. Codons identify amino acid ordering in the synthesis of a protein.

39. AZT is a retrovirus.

Testing Your Understanding of Concepts in Chapter 21.
Answers are at the end of the chapter.

1. Which classes of materials discussed in this chapter are polymers?

2. The human body uses only specific enantiomers of amino acids and carbohydrates. Compare them.

3. What effect do different R groups have on amino acid properties?

4. Why might the formation of a zwitterion be pH sensitive?

5. What is meant by the term “ amino acid sequence”?

6. Identify several functions proteins are involved in.

7. What does the term “protein structure” mean?

8. How can a protein be denatured?

9. What are carbohydrates, and what are some of their properties?

10. Which artificial sweetener is similar in structure, and in what way, to a carbohydrate sweetener such as sucrose?

11. Why can’t humans digest cellulose?

12. What makes lipids more soluble in nonpolar solvents than in polar solvents?

13. What makes cysteine so important in tertiary protein structure?

14. When double bonds are present in fats how do they affect the properties of the fat?

15. Why are triglycerides so important to humans?

16. The steroids, cholesterol, estrogens and sex hormones have much in common. They are also different in some fundamental structure aspects. What are these major differences?

17. What is the major difference between ATP, ADP, and AMP?

18. If photosynthesis "dark reactions" can occur in light why are they called "dark reactions"?

19. How are nucleotide bases related where complementary hydrogen bonding occurs?

20. What is the function of a codon?

Testing You on Some Analytical Concepts.
Answers are at the end of the chapter.

1. Write the zwitterion of cysteine.

2. Identify the amino acids present in the hexapeptide shown below. What amino acid is at the C-terminal end?

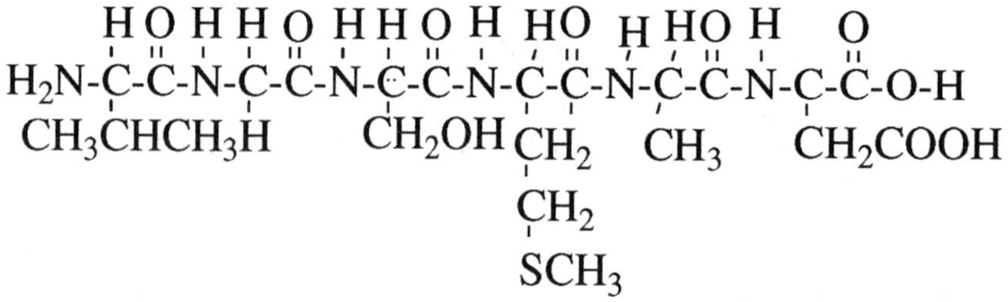

3. The reaction below occurs in the cell.:

$$\text{Glucose + phosphate} \Longrightarrow \text{Glucose 6-phosphate} + H_2O$$
$$\text{The } \Delta G \text{ value is } 13.3 \text{ kJ / mol}$$

Is the reaction endothermic or exothermic? Is the reaction product favored as written?

4. Would glucose 6-phosphate be an energy source for humans?

5. Two DNA strands are complimentary. The base sequence on one strand is TATGCAG. What is the base sequence in the complimentary strand?

6. A certain food contains 9.0 grams of fat and 72.0 grams of carbohydrate per serving. Determine the number of Calories present in a serving. What percent is fat?

Crossword - Chemistry of Life

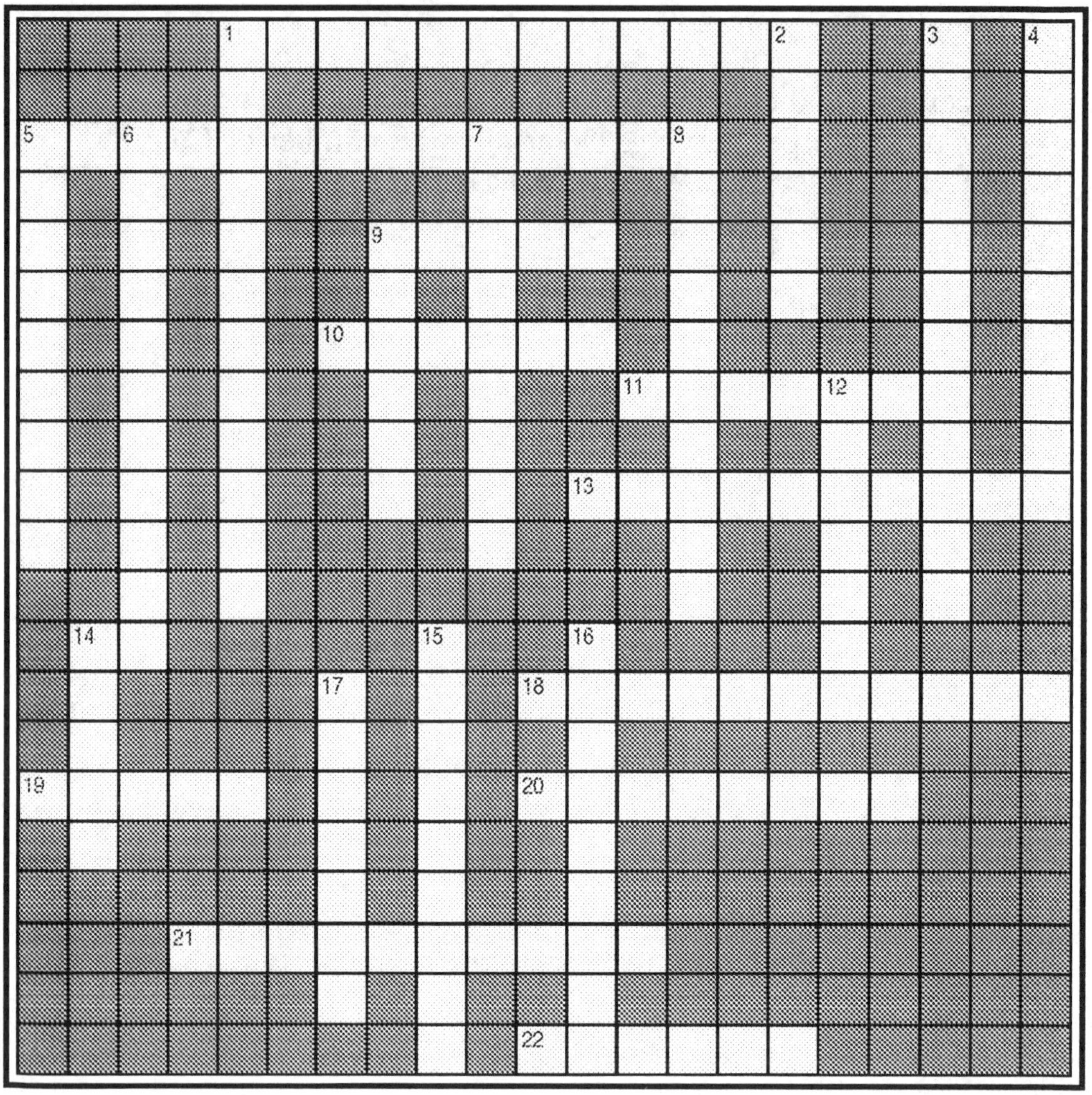

Crossword - Chemistry of Life Clues

ACROSS

1 When two monomer units, such as glucose, condense a unit known as a ________ forms.
5 The process of reacting a strong base with fats or oils is given the name ___________.
9 This carbon is next to the carboxylic acid group in amino acids.
10 A carbohydrate polymer that humans can digest is ______.
11 Progesterone is an example of a compound classified as a _______.
13 When the carboxylic acid proton is donated to the amine group in an amino acid we form a ________.
14 Symbol for the first noble gas.
18 This process happens to DNA in the cell nucleus.
19 An organic substance found in living systems that is not water soluble but is soluble in non-polar organic solvents.
20 This carbohydrate is stored in the liver and is an "instant" energy source.
21 Enzymes lower this energy barrier to reactions.
22 Fructose is this type of monosaccharide.

DOWN

1 May result when hydrogen bonding causing secondary structures is broken.
2 Name given to catalysts in living systems.
3 Fats or oils made with the alcohol glycerol are called this.
4 Oxygen transfer material found in the blood.
5 The ß-pleated sheet structure is known as a protein _____structure.
6 This is a large chain of amino acid residues.
7 The trade name for this material is "NutraSweet".
8 When these polymerize they make DNA and RNA.

9 This part (site) of the molecule is where the important reaction occurs.
12 This material is a fake fat made from emulsified starch and protein.
14 A form of secondary structure in proteins is the ______.
15 The name given to the material that interacts with the enzyme.
16 A carbohydrate polymer that humans can not digest is _______.
17 These fats have a four ring structure.

Ties to Previous Chapters

Chapters 11 and 13 provide a very significant background necessary for the topics in this chapter. The discussion of isomers in chapter 11 will be a key many ideas in this chapter and it will be worth your time to go back and review that section. You also may want to review the naming and identification of functional groups.

Answers to True/False Statements:

1. False Only the L enantiomer of amino acids are utilized.

2. True The terminology used with amino acids and proteins defines the alpha carbon as the one next to the carboxylic acid. All but one of the amino acids used is an alpha type.

3. True All three characteristics are true. The alcohol gives the molecule polarity. It is neutral since no extra acid or base groups are present.

4. False There are three nitrogens in the molecule but only one NH_2 group. The other two nitrogens are in a ring.

5. True The zwitterion will form only when the proton from the acid functional group is transferred to the amine group. This transfer is pH dependent.

6. True The bond is formed when the amine group of one molecule and the acid group of another react.

7. True This naming procedure is outlined on page 912 of the text.

8. False Amino acid residues are the amino acids that have combined to form the peptide.

9. True See the text, page 910, for details.

10. False Proteins may function in several ways. Some of these are as enzymes, hormones, antibodies, or as membrane materials.

11. False The essential amino acids are those which the body is not capable of synthesizing and must be part of our diet.

12. True See page 915 of the text for a discussion of the four protein structures and characteristics of each.

13. False The primary structure is the sequence of the bonded amino acids.

14. True Enzymes are very specific catalysts. They are dependent on an exact fit of the substrate to the enzyme at the active site.

15. False Hydrolysis is the process of breaking the large molecules into smaller ones. Saponification refers to a reaction with lipids.

16. True Only a small part of some molecules are involved in the chemistry of the reaction. This region is called the active site.

17. False Maltose is dimer of two glucose units.

18. True This may involve several reaction steps.

19. True Compare this to the amino acids where only the L enantiomer is used.

20. True This is the primary reaction to convert the large molecules to smaller units.

21. False Of the four common artificial sweeteners only sucralose has a structure similar to the disaccharides.

22. False Glycogen is a quick energy source and is a more highly branched molecule than starch.

23. False Soluble starch is amylose.

24. True The woody part of plants, for example, is cellulose.

25. True Lipids are esters. The triglycerides are triesters of the alcohol glycerol.

26. True Two factors are important in the physical state of a fat. One is chain length, longer chains would be solids. The other is the number of multiple bonds, with more unsaturation giving the fat liquid properties.

27. True They are esters of the alcohol glycerol and any number of long chain acids. These acids are commonly referred to as fatty acids.

28. False These are several related compounds that require a fatty acid in their synthesis but they are not triglycerides. See page 930 in the text for their structure.

29. True Fats furnish about 9000 calories per gram and sugars furnish only about 3800 calories per gram.

30. False Several are not digestible. Simplesse for example is made from egg white or milk protein.

31. True When cholesterol combines with proteins it forms lipoproteins.

32. False ATP is the high energy molecule.

33. False The dark reactions are named this because they do not need light energy to proceed. They use energy made in the light reactions but readily occur when light is present.

34. True See Figure 21.19 in the text for the structure.

35. False Most RNA is found in the cytoplasm.

36. True The base thymine is in DNA only.

37. False Replication occurs before transcription. These two terms are differentiated on page 942 of the text.

38. True These are groups of three bases. Each amino acid has its own code.

39. False ATZ is a drug used to treat the retrovirus AIDS.

Answers to Understanding of Concepts Questions:

1. Proteins, carbohydrates, and nucleic acids all have polymer structures. Fats are not polymeric.

2. The body uses the L- enantiomer of amino acids and the D-enantiomer of sugars.

3. The type of R group on lipids can change several aspects including: polarity, acid/base properties, charge density, or size.

4. Zwitterion forms occur when a proton shifts from the carboxylic acid functional group to the amine functional group. The solution pH can affect how easily protons can be added or removed from the acid or amine groups. At low pH values, the -COO- will accept a proton and at high pH, the $-NH_2^+$ group will give up the proton.

5. When proteins are formed the amino acids link in a specific sequence determined by the codon code. This is known as the amino acid sequence.

6. Proteins can function as enzymes, antibodies, hormones, transport mediums, and as structural materials.

7. Protein structure refers to several situations. There actually are four structures associated with proteins. The primary structure is the amino acid sequence. Secondary structure is a repetitive backbone pattern and is affected by hydrogen bonding. Tertiary structure is the result of many interactions and weak bonds and determines the folding in the protein. If several chains are present they assume a shape which is the quaternary structure. Not all four structures are always involved.

8. Denaturation occurs when a protein structure changes to the extent that normal functioning is altered. Denaturation could be affected by heat, pH change, or precipitation. What commonly happens, is that secondary structure forces are disrupted.

9. Carbohydrates are molecules of carbon, hydrogen, and oxygen. A unique aspect is that the ratio of hydrogen and oxygen is such that it appears like specific numbers of water molecules could be present. The general formula is $C_x(H_2O)_y$. Carbohydrates contain the alcohol functional group and either an aldehyde or ketone group. When the monomer units polymerize the result can be glycogen, starch, or cellulose depending on the type of linkage between monomers and how much side chain formation is present.

10. Sucralose has a 6 member ring and a 5 member ring. This is similar to some disaccharide molecules.

11. The ability of our body to use a material in metabolism depends on having an enzyme available that matches the substrate structure. Cellulose has alternating linkages between glucose units. The enzymes present in our system only accept a linkage that is aligned in one direction only.

12. Lipids have large hydrocarbon parts in their structure. These are hydrophobic so water does not interact well with them. The hydrocarbon chains, do however, interact with nonpolar solvents very well.

13. The amino acid cysteine has a sulfur atom present. The sulfur atom is capable of forming bridge bonds between strands giving a tertiary structure.

14. The double bonds fix shapes of the hydrocarbon chains in a much more rigid pattern. This prevents strong interactions from forming between chains so there are few intermolecular forces. As a result the material remains a liquid rather than being semisolid.

15. The common fats & oils are triglycerides. They are very prevalent in foods and also in animal fat storage.

16. The sex hormones have a ketone group in place of an alcohol. The estrogens have an aromatic ring in the molecule that the others do not have.

17. They vary in the number of phosphate groups present. ATP has 3, ADP has 2, and AMP just one. The more phosphate groups the more energy content in the molecule.

18. The term "dark reaction" just means that the energy from light is not required for that reaction. The reaction energy is furnished from other sources, such as ATP.

19. Complementary bases involve pairing a purine base with a pyrimidine type base. These combinations allow for more effective hydrogen bonding.

20. The codon is a 3 base code that identifies the amino acid to be added at that site in the protein as it is being synthesized.

Answers to Testing the Analytical Concepts

1. The zwitterion would have the proton of the acid shifted to the amine group.

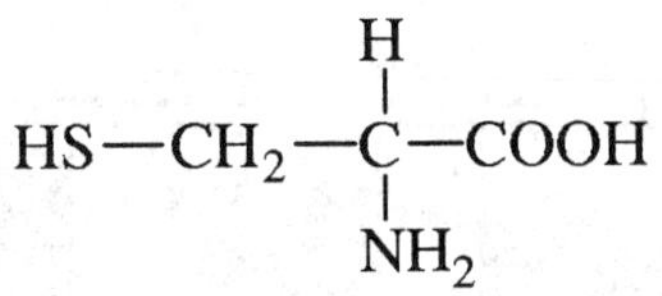

becomes

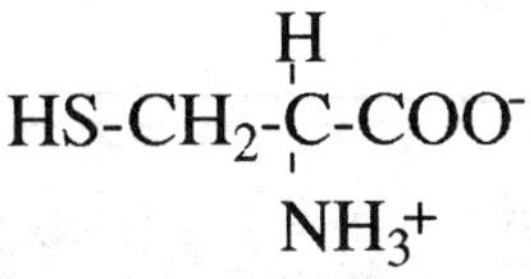

2. The amino acids in sequence starting at the N-terminal end (on the left) are valine, glycine, serine, methionine, alanine, and aspartic acid. The C-terminal amino acid is aspartic acid.

3. The value of ΔG is positive and that means the energy change is not spontaneous. The reaction is endothermic and is reactant favored. This reaction will occur only if 13.3kJ/mol of energy is put into the system.

4. Since energy is needed to make glucose 6-phosphate this same amount of energy will be available. The material is a possible energy source.

5. The pairing with DNA strands is A and C with T and C.

   The result will be ATACGTA

6. The energy released from fats is about 9.0 Calories per gram and that from carbohydrates is 3.8 Calories per gram. (see page 931)

   Calories = 72.0 g x 3.8 Cal/g + 9.0 g x 9.0 cal/g
   = 274 Calories + 81 Calories = 354 Calories

   $$\text{Percent fat} = \frac{\text{81 Calories from fat}}{\text{354 Calories total}} \times 100\% = 22.9\%$$

Answer to Chapter 21 Crossword

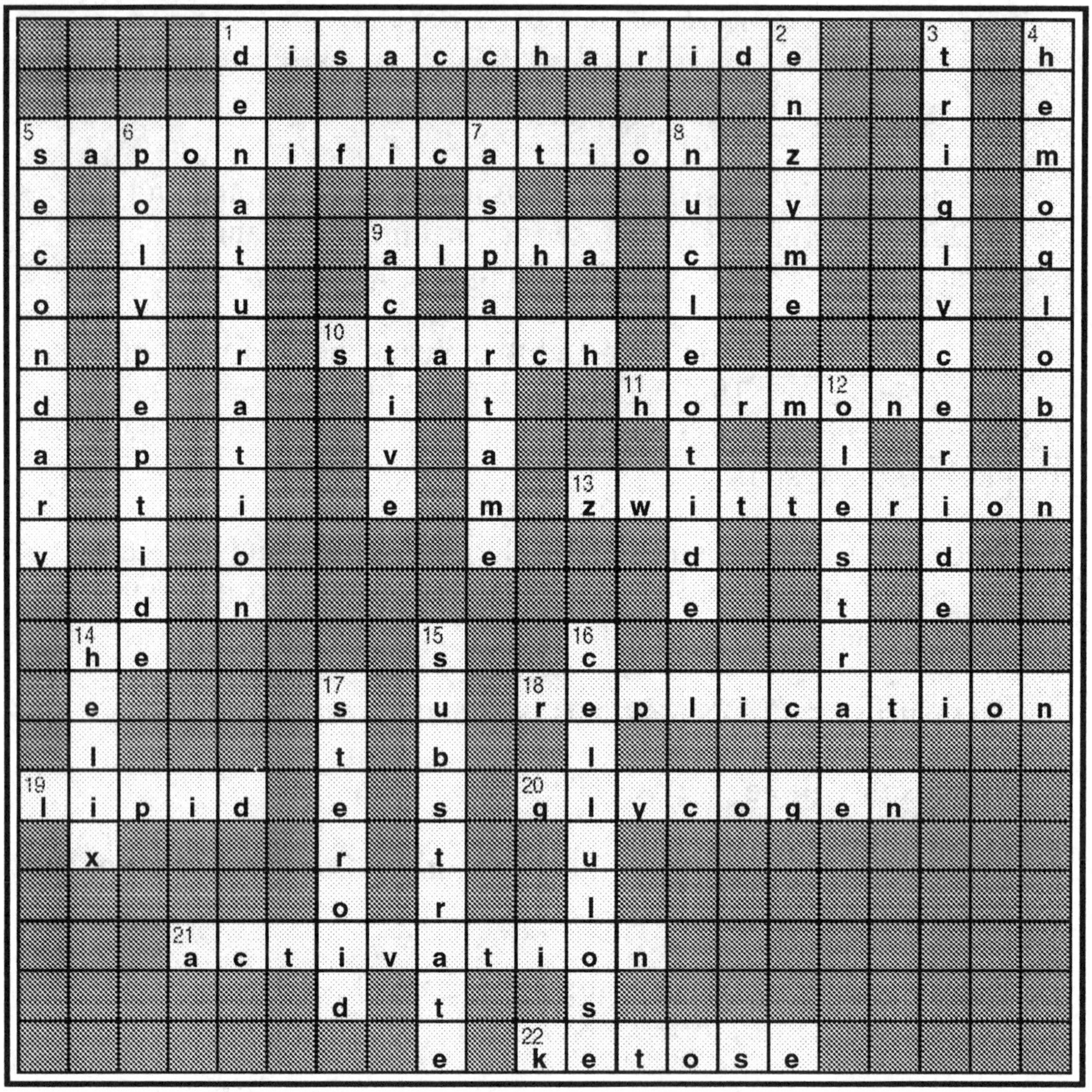

Some Supplementary Readings Relating to This Chapter:

1. The Gene That Knows Left from Right, Discover , August 1993, page 20.

2. Attack of the Superbugs, Time , August 31, 1992, page 62.

3. Chiral Drugs, Chemical & Engineering News , September 27, 1993, page 38.

4. Secrets of human life yielding to genome project detectives, USA Today , December 28, 1993, page 5D.

5. The surprising life of NITRIC OXIDE, Chemical & Engineering News , December 20, 1993, page 26.

6. Abnormal Insulin, Chem Matters , February 1988, page 16.

7. Oil Changes, Chem Matters , December 1989, page 7.

8. Artificial Sweeteners, Chem Matters , February 1988, page 4.

9. Permanent Waves, Chem Matters , April 1993, page 8.

10. The Safety of Foods Developed by Biotechnology, Science , Vol. 256, 26 June 1992, page 1747.

11. Separating 'Mirror' Molecules for Better Drugs, The New York Times , February 12, 1992.

12. Carbon Monoxide Gas Is Used by Brain Cells As a Neurotransmitter, The New York Times , January 26, 1993.